Plant Adaptation to Environmental Stress

Edited by

Leslie Fowden, Terry Mansfield
and John Stoddart

CHAPMAN & HALL

London · Glasgow · New York · Tokyo · Melbourne · Madras

Co-published by James & James (Science Publishers) Ltd, 5 Castle Road,
London NW1 8PR, UK and Chapman & Hall, 2–6 Boundary Row, London
SE1 8HN, UK

Chapman & Hall, 2–6 Boundary Row, London SE1 8HN, UK

Blackie Academic & Professional, Wester Cleddens Road, Bishopbriggs,
Glasgow G64 2NZ, UK

Chapman & Hall Inc., One Penn Plaza, 41st Floor, New York, NY 10119

Chapman & Hall Japan, Thomson Publishing Japan, Hirakawacho Nemoto
Building, 6F, 1-7-11 Hirakawa-cho, Chiyoda-ku, Tokyo 102, Japan

Chapman & Hall Australia, Thomas Nelson Australia, 102 Dodds Street,
South Melbourne, Victoria 3205, Australia

Chapman & Hall India, R. Seshadri, 32 Second Main Road, CIT East,
Madras 600 035, India

First edition 1993
© 1993 James & James (Science Publishers) Ltd

Typeset by Colset Pte Ltd, Singapore
Printed in Great Britain by St Edmundsbury Press, Bury St Edmunds, Suffolk

ISBN 0 412 49000 5

A catalogue record for this book is available from the British Library

Library of Congress Cataloging-in-Publication data available

∞ Printed on permanent acid-free text paper, manufactured in accordance with ANSI/NISO
Z39.48–1992 and ANSI/NISO Z39.48–1984 (Permanence of Paper)

Preface

The chapters of this book represent the Proceedings of an International Symposium on 'Plant Adaptation to Environmental Stress' sponsored by the Rank Prize Funds. The Symposium attempted a broad coverage of knowledge concerning the ways in which plants respond and adjust to environmental variables, and sought unifying concepts spanning different levels of organization, from the subcellular to the whole natural plant community or field crop.

Stress has become a multipurpose word in English usage. It has a strictly defined physical context describing the force per unit area acting upon a material, inducing strain which leads to dimensional change but, beyond that, it is applied to speech (syllable emphasis) and to psychological disorder. More generally, it is used to describe the impact of adverse forces or influences, and it is in this sense that it is applied to biological systems, with particular emphasis on environmental circumstances producing a physiological response in individual organisms. The Symposium considered how departures from optimal conditions are reflected across all orders of organization from the genome to the population, and attempted to assess whether there is commonality or linkage in the mechanisms exposed.

It is particularly timely to consider such issues because of the range of related public concerns. Organisms have evolved progressively in response to environmental changes during the development of life-forms on Earth, and we know that adaptations have sometimes occurred quickly in terms of geological time. Human activities are, however, producing changes on a global scale at a rate that is quite exceptional, and there are agents (e.g. man-made chemicals) that are exerting unprecedented effects of which the long-term consequences are difficult to predict. For example, predicted global environmental change over the next century will drive shifts in climatic zone boundaries, widening the range of growth conditions and, at the margins, reaching the extremes of adaptation and flexibility for present communities and crop genetic structures. Our ability to feed human populations in the most affected areas could well hinge on our understanding of, and ability to manipulate, plant and animal stress responses. In the opposite vein, the opportunities for adopting new crops in higher latitudes will, in many circumstances, depend on modifying low temperature tolerance and growth performance.

There is increasing exposure to anthropogenic pollutant outputs emitted to the gaseous and aqueous environments and transferred to the land. Presence in any phase has effects on the biosphere and the consequences can ramify into food and environmental quality, and contribute to atmospheric forcing of the greenhouse effect. Our ability to design cropping systems to tolerate such effects, assessments of the feasibility of using biological systems to sequester or degrade pollutants (including the reversal of legacies of previous industrial activities), and the alleviation of impacts on natural flora and fauna again require a better understanding of mechanisms and limitations in the appropriate organisms.

Beyond the mechanistic considerations at the level of the individual there is growing interest in the ways in which environmental factors interact with biological systems to shape populations and communities. The available genetic diversity for stress response characters is an important component of this, as are the means whereby individuals deploy their genomes to obtain competitive advantage under adverse conditions. Within this area we can also consider the further political dimension of the public will to preserve and enhance species diversity and the stresses imposed by governmental devices aimed at controlling the costs of agricultural production (e.g. reduced inputs, lower animal populations, reduced cultivation).

In considering the form, frequency and development of stress tolerance and avoidance strategies we are considering some of the important basic processes of evolution, now more accessible than ever because of the powerful new molecular technologies available to biologists, but we are also addressing fundamental questions relating to the world food supplies in the 21st century.

One of the aims of the Rank Prize Funds is to promote knowledge of, and interest in, nutrition and crop husbandry, in accordance with the wishes of the late Lord Rank, who in his lifetime had been an entrepreneur in the food and agricultural industries. Although the book presents much fundamental knowledge, authors were encouraged also to focus on the agricultural implications of research findings in line with the Funds' objectives. We hope our readers appreciate this approach, and enjoy the material now collected together.

Leslie Fowden
Terry Mansfield
John Stoddart

Contents

CONTENTS

List of Participants

Dr. T.W. Ashenden,
Institute of Terrestrial Ecology,
Bangor Research Unit,
Deiniol Road, U.C.N.W.
Bangor,
Gwynedd, LL57 2UP

Dr. C.J. Atkinson,
Horticulture Research International,
East Malling,
West Malling,
Kent,
ME19 6BJ

Professor P.G. Ayres,
Institute of Environmental and Biological
Sciences,
Division of Biological Sciences,
The University,
Lancaster, LA1 4YQ

Professor K. Bachmann,
Hugo de Vries Laboratorium,
Universiteit van Amsterdam,
Kruislaan 318,
NL-1098 SM Amsterdam,
The Netherlands

Professor D.A. Baker,
Wye College,
University of London,
Ashford,
Kent,
TN25 5AH

Professor N.R. Baker,
Department of Biology,
University of Essex,
Colchester,
Essex,
CO4 3SQ

Dr. J.D. Barnes,
Department of Agriculture and Environmental
Science,
Ridley Building,
The University,
Newcastle-upon-Tyne, NE1 7RU

Dr. P.V. Biscoe,
A.F.R.C. Silsoe Research Institute,
Wrest Park,
Silsoe,
Beds.,
MK45 4HS

Dr. C.R. Black,
Department of Physiology and Environmental
Science,
Nottingham University,
Sutton Bonington,
Loughborough, Leics., LE12 5RD

Professor D. Boulter,
Department of Biological Sciences,
Science Laboratories,
University of Durham,
South Road,
Durham, DH1 3LE

LIST OF PARTICIPANTS

Dr. C. Brownlee,
Marine Biological Association,
The Laboratory,
Citadel Hill,
Plymouth,
PL1 2PB

Dr. S.J.M. Caporn,
Department of Environmental Biology,
University of Manchester,
Oxford Road,
Manchester,
M13 9PL

Dr. M.M. Chaves,
Dept. de Botanicea e Engenharia Biologica,
Instituto Superior de Agronomia, Universidade
Tecnica de Lisboa, Tapada da Ajuda,
1399 Lisboa Codex, Portugal

Professor D.T. Clarkson,
AFRC/IACR Long Ashton Research Station,
University of Bristol,
Long Ashton,
Bristol, BS18 9AF

Professor E.C. Cocking,
Department of Life Science,
University of Nottingham,
University Park,
Nottingham,
NG7 2RD

Dr. C.F. Eagles,
AFRC Institute of Grassland and
Environmental Research,
Plas Gogerddan,
Aberystwyth,
Dyfed, SY23 3EB

Dr. J.F. Farrar,
School of Biological Sciences,
University College of North Wales,
Bangor,
Gwynedd,
LL57 2UW

Sir Leslie Fowden,
Rothamsted Experimental Station,
Harpenden,
Herts.,
AL5 2JQ

Dr. P.H. Freer-Smith,
Site Studies South,
Forest Research Station,
Alice Holt Lodge,
Farnham,
Surrey, GU10 4LH

Professor J.P. Grime,
Unit of Comparative Plant Ecology,
Deparment of Animal and Plant Sciences,
University of Sheffield,
Sheffield, S10 2TN

Dr. C.J. Howarth,
AFRC Institute of Grassland and
Environmental Research,
Plas Gogerddan,
Aberystwyth,
Dyfed, SY23 3EB

Professor W.J. Davies,
Institute of Environmental and Biological
Sciences,
Division of Biological Sciences,
The University,
Lancaster, LA1 4YQ

Professor W.H.O. Ernst,
Dept. of Ecology & Ecotoxicology,
Faculty of Biology,
Vrije Universiteit,
De Boelelaan 1087,
1081 HV Amsterdam, The Netherlands

Professor T.J. Flowers,
School of Biology,
University of Sussex,
Falmer,
Brighton,
Sussex, BN1 9QG

Dr. R.S.S. Fraser,
Horticulture Research International,
Worthing Road,
Littlehampton,
BN17 6LP

Dr. H. Griffiths,
Department of Agricultural and Environmental
Science,
Ridley Building,
The University,
Newcastle-upon-Tyne, NE1 7RU

Dr. W. Hartung,
Julius von Sachs Institüt für Biowissenschaften
der Universität,
Lehrstuhl Botanik 1,
Mittlerer Dallenbergweg 64,
D 8700 Würzburg, Germany

Dr. M.A. Hughes,
Department of Biochemistry and Genetics,
The Medical School,
The University,
Newcastle upon Tyne, NE2 4HH

Dr. M.B. Jackson,
Department of Agricultural Sciences,
University of Bristol,
AFRC Inst. of Arable Crops
Research,
Long Ashton Research Station,
Bristol, BS18 9AF

Professor H.G. Jones,
Horticulture Research International,
Wellesbourne,
Warwick,
CV35 9EF

Professor A.J. Karamanos,
Athens Agricultural University,
75, Iera Odos,
118 55 Athens,
Greece

Dr. D.T. Krizek,
Climate Stress Laboratory,
ARS, USDA,
Room 206, B-001, BARC-W, Beltsville,
MD 20705–2350,
U.S.A.

Dr. D.W. Lawlor,
Biochemistry & Physiology Department,
Rothamsted Experimental Station,
Harpenden,
Herts.,
AL5 2JQ

Professor R.A. Leigh,
Biochemistry & Physiology Dept.,
Rothamsted Experimental Station,
Harpenden,
Herts.,
AL5 2JQ

Professor E.A.C. MacRobbie,
Department of Plant Sciences,
Downing Street,
Cambridge,
CB2 3EA

Professor P.G. Jarvis,
Institute of Ecology and Resource
Management,
University of Edinburgh,
Darwin Building, Mayfield Road,
Edinburgh, EH9 3JU

Dr. M.B. Jones,
Department of Botany,
University of Dublin,
Trinity College,
Dublin 2,
Ireland

Dr. A.J. Keys,
AFRC Institute of Arable Crop Research,
Rothamsted Experimental Station,
Harpenden,
Herts., AL5 2JQ

Professor J.V. Lake,
European Environmental Research
Organisation,
P.O. Box 191,
6700 AD Wageningen,
The Netherlands

Professor J.H. Lawton,
Centre for Population Biology,
Imperial College,
Silwood Park,
Ascot,
Berks., SL5 7PY

Professor T. Lewis,
AFRC Institute of Arable Crops Research,
Rothamsted Experimental Station,
Harpenden,
Herts., AL5 2JQ

Dr. M. Malone,
Horticulture Research International,
Wellesbourne,
Warwick,
CV35 9EF

Professor T.A. Mansfield,
Institute of Environmental and Biological
Sciences,
Division of Biological Sciences,
University of Lancaster,
Lancaster, LA1 4YQ

Professor J.L. Monteith,
Institute of Terrestrial Ecology,
Bush Estate,
Penicuik,
Midlothian,
EH26 0QB

Dr. J.I.L. Morison,
Department of Meteorology,
University of Reading,
2, Earley Gate,
Whiteknights,
Reading, RG6 2AU

Dr. L.K. Oksanen,
Department of Plant Ecology,
University of Umea,
S-90187 Umea,
Sweden

LIST OF PARTICIPANTS

Professor J.S. Pate,
Department of Botany,
The University of Western Australia,
Nedlands, Perth,
Western Australia 6009,
Australia

Professor P.J. Peterson,
GEMS-Monitoring and Assessment Research Centre,
The Old Coach House,
Campden Hill,
London, W8 7AD

Dr. C.J. Pollock,
AFRC-IGER Welsh Plant Breeding Station,
Aberystwyth,
Dyfed,
SY23 3EB

Dr. J. McDonald,
Swedish University of Agricultural Sciences,
Dept. of Ecology & Environmental Research,
P.O. Box 7072,
S-750 07 Uppsala, Sweden

Professor J. Moorby,
Dept. of Agriculture, Horticulture and the Environment,
Wye College,
Wye, Ashford,
Kent, TN25 5AH

Dr. S.J. Neill,
Department of Biological Sciences,
Faculty of Applied Sciences,
Bristol Polytechnic,
Coldharbour Lane,
Bristol, BS16 1QY

Dr. A.D. Parry,
Department of Biological Sciences,
Science Laboratories,
University of Durham,
South Road,
Durham, DH1 3LE

Dr. J.M. Peacock,
ICARDA,
P.O. Box 5466,
Aleppo,
Syria

Dr. A. Polle,
Fraunhofer-Institüt für Umweltforschung,
Kreuzeckbahnstrasse 19,
D-8100 Garmisch-Partenkirchen,
Germany

Dr. M.C. Press,
Department of Environmental Biology,
The University,
Manchester,
M13 9PL

Dr. J. Pritchard,
S.B.S. UCNW, Memorial Building,
Deiniol Road,
Bangor,
Gwynedd,
LL57 2UW

Dr. W.P. Quick,
Department of Animal and Plant Sciences,
University of Sheffield,
P.O. Box 601,
Sheffield, S10 2UQ

Dr. D. Robinson,
Department of Physiology,
Scottish Crop Research Institute,
Dundee,
DD2 5DA

Professor R.K. Scott,
Department of Agriculture and Horticulture,
University of Nottingham,
Sutton Bonington, Loughborough,
Leics., LE12 5RD

Dr. N. Smirnoff,
Department of Biological Sciences,
University of Exeter,
Hatherly Laboratories,
Exeter,
EX4 4PS

Dr. F. Tardieu,
INRA Agronomie,
Thiveral Grignon,
F78850,
France

Dr. Howard Thomas,
Cell Biology Department,
IGER-Welsh Plant Breeding Station,
Plas Gogerddan,
Aberystwyth,
Dyfed, SY23 3EB

Dr. S.A. Quarrie,
Cambridge Laboratory,
JI Centre,
Colney Lane,
Norwich,
NR4 7UJ

Professor H. Rennenberg
Institüt für Forstbotanik und Baumphysiologie,
Albert-Ludwigs Universität Freiburg,
Werderring 8, D-7800 Freiburg,
Germany

Professor F. Schöffl,
Lehrstuhl für Allgemeine Genetik,
Universität Tübingen,
Auf der Morgenstelle 28,
W-7400 Tübingen 1,
Germany

Dr. J. Shalhevet,
Institute of Soils and Water,
Agricultural Research Organisation,
Bet Dagan,
P.O. Box 6,
50250, Israel

Professor J.L. Stoddart,
AFRC Institute of Grassland and
Environmental Research,
Plas Gogerddan,
Aberystwyth,
Dyfed, SY23 3EB

Dr. Henry Thomas
Acclimatory Physiology,
IGER-WPBS,
Plas Gogerddan,
Aberystwyth,
Dyfed, SY23 3EB

Dr. P.B.H. Tinker,
N.E.R.C.,
Polaris House,
North Star Avenue,
Swindon,
SN2 1EU

Dr. D. Tomos,
Ysgol Gwyddorau Bioleg,
Coleg Y Brifysgol,
Bangor,
Gwynedd,
LL57 2UW

Dr. P.R. van Gardingen,
Institute of Ecology and Resource
Management,
University of Edinburgh,
School of Agriculture Building,
Edinburgh, EH9 3JG

Dr. H.M. West,
Institute of Environmental and Biological
Sciences,
Division of Biological Sciences,
The University,
Lancaster, LA1 4YQ

Professor B.A. Wharton,
Old Rectory,
Belbroughton,
Worcs.,
DY9 9TF

Dr. S.J. Woodin,
Department of Plant and Soil Science,
University of Aberdeen,
Aberdeen,
AB9 2UD

Dr. A.R. Yeo,
School of Biological Sciences,
University of Sussex,
Brighton,
Sussex,
BN1 9QG

Dr. Lesley Turner,
AFRC Institute of Grassland and
Environmental Research,
Plas Gogerddan,
Aberystwyth,
Dyfed, SY23 3EB

Professor J.C. Waterlow,
15, Hillgate Street,
London,
W8 7SP

Dr. J.D.B. Weyers,
Department of Biological Sciences,
University of Dundee,
Dundee,
DD1 4HN

Dr. J.N. Wingfield,
AFRC Central Office,
Polaris House,
North Star Avenue,
Swindon,
Wiltshire, SN2 1EU

Professor R.G. Wyn-Jones,
Countryside Council For Wales,
Plas Penrhos,
Bangor,
Gwynedd,
LL57 2LQ

Community and ecosystem level

CHAPTER 1

Climatic constraints on crop production

J. L. MONTEITH and J. ELSTON

1.1. Crops and 'meteorological phenomena'

It is 30 years since the Commonwealth Scientific and Industrial Research Organization sponsored a symposium on the Environmental Control of Plant Growth to mark the opening of the Canberra Phytotron. D. J. Watson's contribution on 'Climate, weather and yield' made this distinction: 'Climate determines what crops a farmer can grow; weather influences the annual yield, and hence the farmer's profit, and more important, especially in underdeveloped and overpopulated countries, how much food there is to eat' (Watson, 1963).

Agriculturally, Watson's contrast is appropriate but, in meteorology, the climate of a region encompasses all the states of weather it experiences. This includes not only long-term *mean* values of variables such as temperature or rainfall but also a set of *extremes*. It is the mean values ('climate' in Watson's sense), coupled with the soil environment, that mainly determine what crops a farmer can grow, accepting the risk that there will be years when harvests are reduced or even destroyed by extremes in the form of frosts, droughts, typhoons or epidemics of pests and diseases linked to weather in complex ways. Farmers, like plants, may be able to adapt to the stresses of bad years; but in the tropics and subtropics, where stresses are particularly frequent and severe, the resources needed to combat them are often meagre.

In his paper, Watson also quoted Lawes and Gilbert who wrote, in 1880, 'As yet the connection between meteorological phenomena and the progress of vegetation is not so clearly comprehended as to enable us to estimate with any accuracy the yield of a crop by studying the statistics of weather during the period of its growth'.

Comprehension advanced substantially between 1880 and 1962, but progress *since* 1962 has been as great, partly as a consequence of observations in controlled environments like the Canberra phytotron; partly because of major developments of instrumentation that is reliable in the field as well as in the laboratory; and partly because computers, accessible to few agricultural scientists in 1962, are now ubiquitous. Today, Lawes and Gilbert might be impressed by our ability to relate the growth of vegetation to 'meteorological phenomena', but they would be shrewd enough to identify weaknesses underlying much of our analysis.

It is true that we can describe ways in which many crop species respond to temperature, radiation and even water supply when we examine one variable at a time. It is encouraging that through bottom-up models we can attempt to integrate the impact of several variables acting and interacting together. However, most of these models are little more than ingenious schemes for summarizing what we have already observed. When we attempt to *predict* how yield might depend on several variables changing simultaneously, we are troubled by many gaps in understanding and tend to shelter behind the convenient fact that such forecasts are usually untestable. Moreover, models cannot readily simulate the impact of extremes such as high temperature – often coupled with drought – let alone the complexities of adaptation to stress.

The theme of this introductory paper is complementary to the topics that follow because it is primarily concerned with the *potential* for growth and production established by the mean climate of different regions and with the impact of the relatively small deviations from the mean responsible for differences in production from year to year. This top-down treatment is designed to provide a baseline for assessing the impact of other environmental stresses and the ways in which plants respond to them.

Four elements dominate the response of plants to climate. Two of these are resources: energy in the form of radiation and mass in the form of rainfall. Two are states: temperature and saturation vapour pressure deficit, which can also be described as 'rate modifiers' because they interact with physiological processes to determine how fast the resources can be used. All plants, whether in managed or in undistrubed systems, respond to these elements but examples will be drawn exclusively from simple agricultural systems in which differences between climates and between species are more readily compared.

Within the compass of a short review, it is impossible to deal with the indirect impact of climate on production through the water and nutrient status of soil, for example, or the prevalence of pests and diseases.

1.2. Climatic and physiological interactions

The interaction of climatic elements with each other has a strong influence on plant growth and many other biological processes. Major interactions can be summarized as follows.

1. *Solar radiation and rainfall.* Clouds absorb radiation and so, in all climates, months that are particularly wet tend also to be relatively sunless.
2. *Rainfall, temperature and saturation deficit.* During spells of drought, evaporation from the land surface (including transpiration from vegetation) decreases, temperature increases and saturation deficit increases. One

consequence is that potential evaporation increases when actual evaporation decreases, a source of frequent confusion in agroclimatology.

Responding both to climate and to weather, plants have evolved complex sets of physiological interactions, some of which are considered in more detail later. Here, it is sufficient to draw attention to the central role of stomata, whose behaviour is tightly coupled to the processes of carbon assimilation and transpiration, which proceed at rates determined both by weather and by soil conditions. This coupling appears to be strongly interactive through feedback mechanisms that stabilize the carbon : water ratio and conserve water. The assimilation of carbon and of nutrients provides material for the accumulation of biomass, allocated within plants according to relative rates of development.

In most temperate maritime climates, growth from early spring to late autumn is usually limited by radiation, but long spells of drought occur in some years. Winter temperatures are so low that little photosynthesis or development is possible. Temperate continental climates have some features more favourable for growth (sunshine and warmth), but others are less favourable (less rain and drier air).

In the tropics, there is a clear distinction between 'humid' areas within a few degrees of the equator receiving rain throughout the year and the semi-arid (perhaps better described as 'seasonally humid') tropics with unreliable periods of intense rainfall. In the fully humid tropics, radiation is usually the only climatic element significantly limiting growth. The semi-humid tropics are more complex. In the wet season, radiation is often the main constraint because cloud is so prevalent, but water takes over during dry spells, which are common and erratic. During the rainless season, crops can be produced with irrigation but are prone to damage by exposure to high temperature or very dry air.

Between temperate and tropical lie Mediterranean climates with a wet winter season in which growth is limited by radiation and, in some areas, by low temperature. Irrigation is widely practised during summers, in which rainfall is scanty or even absent.

A simple quantitative treatment of growth in different climates is presented later.

1.3. Crop distribution

The relation between climate and crop distribution was dealt with in detail by Bunting et al. (1982) in a previous Rank Prize Funds Symposium. One figure in their paper illustrated how the distribution of major cereals is related to rainfall and temperature. The original figure contains several hundred individual points grouped, as in Fig. 1.1, within four boxes (which exclude a small number of outliers).

Wheat and rice occupy contrasting climatic niches corresponding to cool relatively dry environments (wheat) and warm wet environments (rice). The other two cereals, maize and bulrush millet, occupy the warm dry corner of

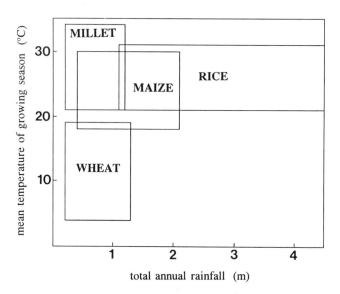

FIGURE 1.1 *Climatic niches for four major cereals. (Adapted from Bunting et al., 1982.)*

the diagram, and sorghum straddles these two boxes. Limiting temperatures for these three C4 species are similar, but time from germination to maturity ranges from about 60 days for millet to about 160 days for maize. The exact length of this period – a genotypic characteristic responding mainly to temperature – determines how much water is needed. Mean yields are largest (but not necessarily most stable) when the length of the biologically determined *growth period* closely matches the length of the physically determined *growing season* set, for example, by the end of a rainy season or the onset of a cool one.

In regions where temperature is not a limiting factor, as in much of the lowland tropics, the distribution of crops depends strongly on rainfall and ranges from plantation crops where rain is abundant to native grassland species used for grazing where rainfall is marginal. In regions where the rainfall gradient is steep, the spectrum of cropping is very closely matched to the supply of water, as demonstrated by a survey in south-east Kenya (Downing *et al.*, 1988). Fig. 1.2 shows the relation between annual rainfall and the corresponding estimate of growing season length for major crops and for grazing. This is a good demonstration of how farmers have adapted systems of production to environmental stresses through skills that are based on centuries of experience.

1.4. Impact of individual constraints

Where one climatic constraint dominates, it is sometimes possible to establish a consistent relation between daily or monthly mean values of the element

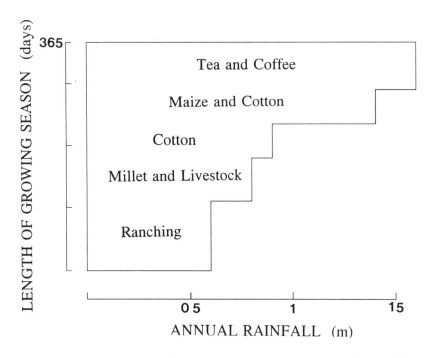

FIGURE 1.2 *Climatic niches for agricultural systems in south-east Kenya. (Adapted from Downing* et al., *1988.)*

and growth rate or yield. Examples follow for the four major constraints already listed.

1.4.1. Radiation

Plant growth and development depend on several features of radiant energy. Irradiance in the visible spectrum (universally but ambiguously referred to as photosynthetically active radiation or PAR) is the principal determinant of photosythesis rates and therefore of biomass accumulation. Over a whole growing season, however, the receipt of PAR does not usually change by more than a few per cent from year to year (Monteith, 1981), implying that *received* (as distinct from *absorbed*) PAR is not usually an important discriminant of dry matter production.

Somewhat exceptionally, Seshu *et al.* (1989) were able to demonstrate that the yield of one rice variety grown at 32 sites in south-east Asia increased with the income of solar radiation during ripening, presumably because irradiance was often much reduced by cloud.

Many field trials have shown that biomass accumulation over periods ranging from days to months is strongly correlated with *absorbed* PAR. Biomass produced per unit of absorbed PAR is often referred to as a 'radiation

7

use efficiency', but it is not an efficiency in the proper sense of the term. 'Biomass radiation coefficient' is better, but the symbol e will be retained for this quantity because it is so widely used. It appears that, for healthy crops with a good supply of nutrients and water, e remains almost constant from soon after emergence until the onset of senescence. Even when nutrients (Green, 1987) or water (Singh and Sri Rama, 1989) are in short supply, e is effectively constant for long periods at less than its potential value. The spectral quality of daylight changes somewhat with climate because of changes in the absorbing and scattering properties of the atmosphere, but much larger changes occur within vegetation because the selective absorption of PAR by leaf pigments reduces energy in the blue and red bands of the spectrum much more than in the near infrared. This change of quality appears to have little impact on biomass production in monocultures but has implications for competition in natural mixed communities, which deserve more attention in studies of intercropping and agroforestry.

The duration of daylight is not usually regarded as a feature of climate but has a major influence on the length of developmental phases in many cereals and legumes and is referred to later under temperature, with which it often interacts.

1.4.2. Rainfall

The primary function of rain in most agricultural systems is to recharge the soil profile so that water and nutrients become available to roots. Many attempts have therefore been made to correlate yield with *amounts* of rain. In the dry tropics, however, the *timing* of seasonal rains has great significance. If the first rains are 'good', sowing and establishment can proceed unchecked; but, when they are scanty, germination may be so poor that one or more further sowings are needed. In some climates, timing of the first rain is also a major constraint because it is correlated with the length of the subsequent growing season and therefore the ability of grain-bearing crops to reach maturity (Sivakumar, 1990).

Physiologically, it is the amount of water that plants can transpire that determines how fast they can grow when other elements of climate are not limiting, but in very dry climates there is a strong correlation between transpiration and rainfall and therefore between rainfall and biomass or production. Fig. 1.3 demonstrates this correlation for maize and also for forage grasses in south-east Kenya (Parry and Carter, 1988) with a difference in scale that reveals, for this environment, the advantage of systems based on perennial species that can exploit water at any time of year.

Another example of yields dominated by seasonal rainfall comes from semi-arid districts in India as reported by Sivakumar et al. (1984). Correlations between seasonal rainfall and yield exist not only on research stations where other constraints are minimized, but also on farmers' fields where yields are much less because the impact of drought is exacerbated by nutrient deficiencies, by poor soil structure, and by pests and diseases. More detailed discussion of the relation between biomass and transpired water is postponed

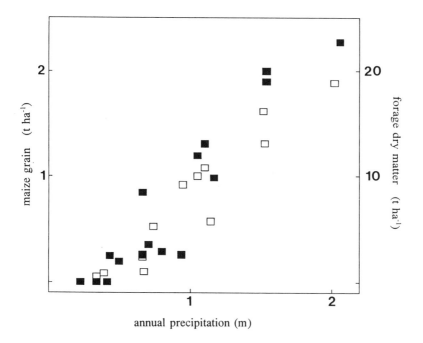

FIGURE 1.3 *Relation between maize grain yield (■), forage production (□) and annual precipitation in south-east Kenya. (Adapted from Parry and Carter, 1988.)*

to the section on saturation deficit because of the central role played by this variable.

1.4.3. Temperature

Temperature was categorized as a rate modifier because so many physiological processes depend on biochemical reactions that respond to the temperature of cells in which they occur. In laboratory and phytotron studies of these responses, it is usually necessary to assume that all the cells within a leaf, or sometimes within a whole plant, are at the same temperature but this can rarely be true. In field trials and in a temperate climate, mean cellular temperature is usually within a degree or two of mean screen temperature but much larger differences can occur in the tropics, particularly during drought.

Because of such problems, one of the simplest and most consistent systems to investigate is germination. For non-dormant seeds, it appears to be a general rule that germination rate is zero below a base temperature and increases linearly with temperature up to an optimum temperature at which the rate is maximal and beyond which it declines, again almost linearly, to zero (Squire, 1990). Other processes of development such as leaf appearance and floral initiation respond to temperature in the same way, except when the rate is controlled by day length as well as or in place of temperature (Summerfield et al., 1991). The existence of three 'cardinal' temperatures was well known to de Candolle and other 19th century physiologists, but despite their central importance in ecology and agricultural climatology their biochemical basis remains obscure.

Simple empiricism has allowed some progress, however. Below the optimum temperature for any rate of development, the reciprocal of the increase in rate per unit increase in temperature has the dimensions of temperature × time, and units of degree-days have been used for many years by agroclimatologists – not always correctly. It is pointless to quote the number of degree-days in a particular month or season: this is simply an inconvenient way of specifying a difference between mean and base temperatures. Instead, it is important to know how many degree-days a plant needs to complete a discrete stage of development or an entire life-cycle. Dividing this quantity by the mean temperature during the phase gives an equivalent number of calendar days. At Nottingham we coined the term 'thermal time' to emphasize that degree-days should be regarded not as an integral of temperature over a fixed time but rather as an integral of time as measured by plants whose biological clocks go slow in cold weather and faster as temperature rises.

Field measurements of germination on a range of species by Angus et al. (1981) yielded values which are grouped into three discrete ranges of base temperature and reveal a negative correlation between base temperature and thermal time for germination (Table 1.1). Tropically adapted species growing in soil with a mean temperature of, say, 21 °C (10 °C above a base of 11 °C) germinated twice as fast as temperate species growing in soil with a mean temperature of 10 °C above a base of 2 °C. This rapid response (which is not confined to the C4 group) may be an adaptation to an environment in which rainfall is erratic and soil dries rapidly after wetting.

Because development rate increases with temperature (below an optimum), the life-cycle of *determinate* species, measured by the calendar, is shorter in a warm season than in a cool one and, in the same season, is shorter at a relatively warm site. This appears to be the main reason why regional mean wheat yields in the UK decrease with mean temperature during the growing season by about 5% per °C (Monteith, 1982). In contrast, the yield of *indeterminate* species increases with seasonal mean temperature because of faster leaf expansion, photosynthesis, etc. In Iceland, for example,

TABLE 1.1. *Base temperature and thermal time for germination of 44 species in the field* (data from Angus et al., 1981).

Range of base temperature (°C)	Mean base temperature (°C)	Mean thermal time (°Cd)
1–3	2.1 (0.2)	88 (4)
6–10	9.0 (0.4)	63 (3)
10–14	11.4 (0.3)	44 (2)

Values in parentheses are standard deviations.

hay production between September and June increases by about $0.6\,t\,ha^{-1}$ per °C increase of mean temperature in the range 0–5 °C (Bjornsson and Helgadottir, 1988).

1.4.4. Saturation deficit

The saturation vapour pressure deficit of the atmosphere (D) plays two distinct but closely related roles as a discriminant of crop production. First, the amount of dry matter produced per unit of water transpired is inversely related to D. So, in an environment where water is the limiting resource, the production of biomass is also inversely related to D. Secondly, the stomata of many species tend to close when the atmosphere increases its demand for water (by increasing D) and to re-open when demand decreases. In consequence, rates of photosynthesis are faster in humid than in dry air even when radiation and not water is the limiting resource. The processes will be examined separately.

(a) Biomass : water ratio

The amount of biomass accumulated by a crop per unit of water transpired by plants or evaporated by the plant/soil system is often referred to as a 'water use efficiency', but this is yet another misnomer and biomass : water ratio (BWR) is the term used here. The BWR depends primarily on the ratio of carbon dioxide molecules to water molecules. Because stomata provide a common pathway for the two gases, the BWR is determined mainly by the pressure gradients that drive carbon dioxide into leaves and water vapour out of them. Both gradients change during the course of a day and through the growing season but changes in the carbon dioxide gradient are much smaller than changes in the water vapour gradient, which depend mainly on changes in D (Monteith, 1990a). This is the physical and physiological basis for the fact that BWR, in most environments, is almost inversely proportional to D.

For analysis, it is convenient either to treat the product of BWR and D as a constant or to work with 'normalized' water, i.e. the amount of water received as rain and subsequently transpired, *divided by* the mean daytime value of D. The potential for biomass production in terms of rainfall is then proportional to normalized rainfall, and the actual biomass production is proportional to normalized transpiration (Monteith, 1990a; Squire, 1990).

(b) Stomatal closure

Many laboratory experiments and field studies using leaf cuvettes (Schulze and Hall, 1982) have demonstrated an inverse and non-linear relation between stomatal conductance and D, which in many cases conceals a *linear* increase of resistance with increasing D and a *linear* decrease of conductance with increasing transpiration rate. Increasing D increases the transpiration rate to a point where it becomes constant or even decreases as air becomes drier. In contrast, increasing D invariably decreases the rate of photosynthesis in response to closing stomata.

The full significance of D for crop production in dry regions has still to be explored. It is clear, however, that although a temporary or long-term shortage of rain (the resource) may be a major constraint to production in all climates, this will always be associated with a reduction of evaporation from land surfaces which, in turn, is responsible for an increase of temperature and of D near the ground. In terms of crop production, the increase of D exacerbates stresses caused by the decrease of rain, but in terms of adaptation it is clear that plants can conserve water by partly closing their stomata.

1.5. Synthesis

The previous discussion of plant responses to individual climatic constraints is only a first step towards understanding the relation between climate and the productivity of systems. Integration of responses through bottom-up models has helped to focus attention on aspects of physiology, climatology and soil science where measurements are scanty and understanding weak. Less attention has been paid to what might be called 'balanced' models which, like bottom-up models, are based on physiological understanding but which share the advantage of top-down models in combining a minimal amount of environmental information with a small number of physiological parameters.

Constructing a balanced model from the four climatic variables examined above is a relatively straightforward operation, summarized as follows:

1. Calculate the dry matter production for each month from the corresponding receipt of total solar radiation multiplied by a coefficient derived from field measurements at a comparable site, e.g. $1.6 \, \mathrm{g \, MJ^{-1}}$ for a C4 species and $1.2 \, \mathrm{g \, MJ^{-1}}$ for a C3.
2. Also calculate the dry matter production for each month from the rainfall divided by the mean saturation deficit multiplied by a BWR derived from measurements at a comparable site and normalized for saturation deficit, e.g. $5 \, \mathrm{g \, kPa \, kg^{-1}}$ for C3 species and $10 \, \mathrm{g \, kPa \, kg^{-1}}$ for C4 (Monteith, 1990b).

   The actual rate of dry matter production is assumed to be the smaller of the estimates based on radiation and on water.
3. Calculate the thermal time for each month from the mean temperature and a base temperature, e.g. $0 \, °\mathrm{C}$ for temperate C3 species and $10 \, °\mathrm{C}$ for C4 and tropical C3 species.

Examples follow for five sites representative of the climates already distinguished. For each, the same thermal time-scale is used for the horizontal axis, on which calendar months appear long in warm weather and short or virtually absent in cold weather. A constant vertical scale gives the biomass accumulation in each month.

1. *Temperate maritime* (Sutton Bonington, East Midlands, UK) The annual thermal time for C3 species ($3530 \, °\mathrm{Cd}$) is ample for crops grown in temperate maritime (i.e. cool) climates but development of overwintering

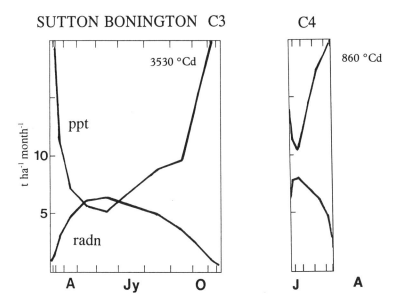

FIGURE 1.4 *Potential monthly biomass production for eastern England, estimated as a function of radiation or precipitation in terms of thermal time for C3 and C4 species.*

cereals is very slow (little thermal time passes in winter months). Growth is limited by radiation rather than by water through most of the year, but the two constraints almost balance from May to July (Fig. 1.4).

The thermal time for C4 species, shown for contrast, is only 860 °Cd on average, too little for maize except possibly in unusually warm, sunny summers when water would be likely to limit production.

2. *Temperate continental* (Amarillo, north-west Texas) The growing season for C4 species is limited by low temperature (i.e. below 10°C) for 6 months per year. Thermal time of 2090 °Cd between April and October is adequate for the growth of maize and sorghum but, in the absence of irrigation or on light soils, yield is severely limited by water (Fig. 1.5).

3. *Mediterranean* (Bet Dagan, Israel, Central Coastal Plain) Irrigation is essential for arable crops except in the three or four winter months, when growth is limited by radiation. Thermal time is abundant (Fig. 1.6).

4. *Semi-humid tropics* (Hyderabad, Central India) There is a sharp distinction between short monsoon and long dry seasons, with radiation levels lower in the former because of cloud. Thermal time is adequate for the production of C4 cereals and tropical legumes between June and October. Crop production is possible without irrigation during the monsoon and in the post-monsoon season, given a good reserve of water in the soil profile (Fig. 1.7).

5. *Humid tropics* (Kuala Lumpur, Malaysia) Rainfall is abundant throughout the year and growth is limited by radiant energy with little annual

AMARILLO C4

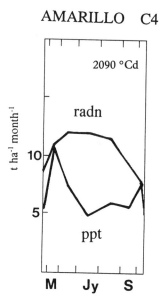

FIGURE 1.5 *Potential monthly biomass production for north-west Texas, estimated as a function of radiation or precipitation in terms of thermal time for C4 species.*

BET DAGAN C3

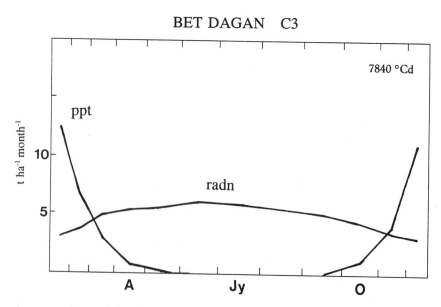

FIGURE 1.6 *Potential monthly biomass production for central Israel, estimated as a function of radiation or precipitation in terms of thermal time for C3 species.*

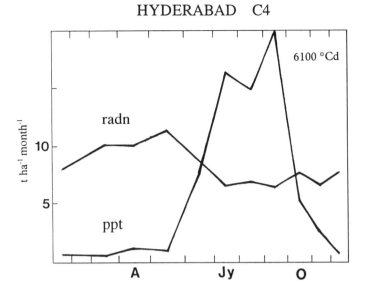

FIGURE 1.7 *Potential monthly biomass production for central India, estimated as a function of radiation or precipitation in terms of thermal time for C4 species.*

variation. Temperature and therefore thermal time per month are also very uniform throughout the year (Fig. 1.8).

1.6. Potential and actual yields

At the end of this type of review, it is salutary to return from estimates of growth based on hypotheses and assumptions to figures obtained in the real world. Comparisons cannot be precise, however, because of the difficulty of specifying regionally representative figures for climate on the one hand and yield on the other.

Table 1.2 contains figures for four major cereals in regions where they are dominant. Yields were estimated on the basis of climatic data already used to obtain biomass, assuming a common harvest index of 0.4. Average national yields for 1985–90 were obtained from Food and Agriculture Organization Production Yearbooks. The ratio of estimated to reported yield, a 'figure of merit', is deliberately quoted to one digit only.

Record yields are substantially larger than those quoted in the table. For UK wheat and barley, the *Guinness Book of Records* reports 14 and 12 t ha^{-1} for wheat and barley, respectively, both grown on farms in southern Scotland, where relatively low temperature favours a long growing season. There are verbal reports of similar wheat yields in parts of China where low temperature is associated with high radiation. At the International Crops Research Institute

15

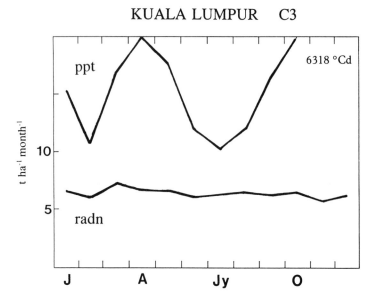

KUALA LUMPUR   C3

FIGURE 1.8 *Potential monthly biomass production for Malaysia, estimated as a function of radiation or precipitation in terms of thermal time for C3 species (but with a base temperature of 10 °C).*

for the Semi-Arid Tropics, sorghum yields have reached 7–8 t ha^{-1}; this is less than that of temperate cereals in Europe because the disadvantage of a shorter growing season more than outweighs the advantages of C4 metabolism.

These figures suggest that the best cereal yields achieved in Western Europe, in the USA and in tropical research stations may be moving close to the climatic limit for the genetic material currently available. In Asia, where average yields are still far below the climatic limit because of many

TABLE 1.2. *Regional yields of major cereals in relation to climate.*

Region	Crop	Limit to growing season	Main growth period	Estimated total biomass (t ha^{-1})	Reported average yield (t ha^{-1})	Figure of merit
US Great Plains	Maize	Temperature	May–September	34	7.1	0.5
Eastern England	Wheat	Temperature	April–August	26	6.5	0.6
Philippines	Rice	–	January–December	76	2.7	0.1
Central India	Sorghum	Water	July–October	26	0.7	0.1

other physical, biological and economic constraints, there is a clear distinction between rice (which can be harvested twice or even three times a year when water is available) and sorghum (which is restricted by rainfall to a 3–4 month growing season).

REFERENCES

Angus, J.F., Cunningham, R.B., Moncur, M.W. and Mackenzie, D.H. (1981). Phasic development in field crops. I. Thermal response in the seedling phase. *Field Crops Res.* **3**, 365–378.

Bjornsson, H. and Helgadottir, A. (1988). The effect on grass yield and their implications for dairy farming. In: *The Impact of Climatic Variations on Agriculture*, Vol. 1. pp. 445–474, eds. M.L. Parry, T.R. Carter and N.T. Konijn. Dordrecht: Kluwer.

Bunting, A.H., Dennett, M.D., Elston, J. and Speed, C.B. (1982). Climate and crop distribution. In: *Food, Nutrition and Climate*, pp. 43–74, eds. K.L. Blaxter and L. Fowden. London: Applied Science Publishers.

Downing, T.E., Akong'a, J., Mungai, D.N., Muturi, H.R. and Potter, H.L. (1988). Drought climatology and the development of the climatic scenarios. In: *The Impact of Climatic Variations on Agriculture*, Vol. 2, pp. 149–173, eds. M.L. Parry, T.R. Carter and N.T. Konijn. Dordrecht: Kluwer.

Green, C.F. (1987). Nitrogen nutrition and wheat growth in relation to absorbed radiation. *Agric. Meteorol.* **41**, 207–248.

Monteith, J.L. (1981). Does light limit crop production? In: *Physiological Processes Limiting Plant Productivity*, pp. 23–38, ed. C.B. Johnson. London: Butterworths.

Monteith, J.L. (1982). Climatic variation and the growth of crops. *Q. J. R. Meteorol. Soc.* **107**, 749–774.

Monteith, J.L. (1990a). Conservative behaviour in the response of crops to water and light. In: *Theoretical Production Ecology*, pp. 3–16, eds. R. Rabbinge, J. Goudriaan, H. van Keulen and F.W.T. Penning de Vries. Wageningen: PUDOC.

Monteith, J.L. (1990b). Steps in crop climatology. In: *Proceedings of International Conference on Dryland Farming*, Texas Agricultural Experiment Station, College Station, Texas, eds. P.W. Unger, W.R. Jordan, T.V. Sneed and R.W. Jensen.

Parry, M.L. and Carter, T.R. (1988). The assessment of climatic variations on agriculture: a summary of results for semi-arid regions. In: *The Impact of Climatic Variations on Agriculture*, Vol. 2, pp. 9–60, eds. M.L. Parry, T.R. Carter and N.T. Konijn. Dordrecht: Kluwer.

Schulze, E.-D. and Hall, A.E. (1982). Stomatal responses, water loss and CO_2 assimilation. In: *Encyclopedia of Plant Physiology*, Vol. 12B, *Physiological Ecology*, Part II, pp. 181–230. eds. O.L. Lange, P.S. Nobel, C.B. Osmond and H. Ziegler. New York: Springer Verlag.

Seshu, D.V., Woodhead, T., Garrity, D.P. and Oldeman, L.R. (1989). Effect of weather and climate on production and vulnerability in rice. In: *Climate and Food Security*, pp. 93–113. Los Banos, Philippines: International Rice Research Institute/American Association for the Advancement of Science.

Singh, P. and Sri Rama, Y.V. (1989). Influence of water deficit on transpiration and radiation use efficiency of chickpea. *Agric. Forest Meteorol.* **48**, 317–330.

Sivakumar, M.V.K. (1990). Exploiting rainy season potential from the onset of rains in the Sahelian zone. *Agric. Forest Meteorol.* **51**, 321–332.

Sivakumar, M.V.K., Huda, A.K.S. and Virmani, S.M. (1984). Physical environment of sorghum and millet growing areas in South Asia. In: *Agrometeorology of Sorghum and Millet*, pp. 63–83, eds. S.M. Virmani and M.V.K. Sivakumar. Hyderabad, India: International Crops Research Institute for the Semi-Arid Tropics.

Squire, G. (1990). *The Physiology of Tropical Crop Production.* Wallingford, UK: CAB International.

Summerfield, R.J., Roberts, E.H., Ellis, R.H. and Lawn, R.J. (1991). Towards the reliable prediction of time to flowering in six annual crops. *Exp. Agric.* **27**, 11–31.

Watson, D.J. (1963). Climate, weather and plant yield. In: *Environmental Control of Plant Growth*, pp. 337–350, ed. L.T. Evans. New York: Academic Press.

CHAPTER 2

Population dynamics, evolution and environment: adaptation to environmental stress

W. H. O. ERNST

2.1. Introduction

Plant populations react to environmental changes consisting of short-term fluctuations and more permanent changes within various time-scales. In the northern hemisphere, the great glacial–interglacial cycles have not only diminished and enlarged the space that species can occupy, but have caused changes in the atmospheric chemistry (Neftel *et al.*, 1982; Stauffer *et al.*, 1985) and in the annual temperature and precipitation cycles (Jouzel *et al.*, 1987). Weathering of parent rock and pedogenesis have continuously affected the metabolism of the individual, and its intraspecific and interspecific interaction at the autotrophic and heterotrophic level of communities and ecosystems. During the recolonization of great parts of Europe after the last glaciation, the great differences in parent rock chemistry demanded genetically based metabolic differentiation because they could not be balanced by acclimation. Such evolutionary changes are well documented alongside the chemistry of the very different environments (Kinzel, 1982).

The shift in the ecological amplitude of species and populations as a consequence of environmental changes within a geological time-scale has been accentuated during this century by anthropogenic changes of the atmosphere (gaseous emissions), the pedosphere and hydrosphere (eutrofication, acidification, desertification, pesticide application) and the biosphere (removal and introduction of species, destruction and creation of new ecosystems). Part of these anthropogenic changes may be present in the evolutionary memory of a species (local air pollution from volcanoes, increased carbon dioxide levels in the tertiary period, heavy metals on ore outcrops), but others are completely new, such as the application of pesticides.

Dynamics of populations will be the result of the physiological potential of all genotypes of a population to react to these temporal and spatial patterns of changes. The coincidence of multiple changes, even if the intensity of each individual change may be low, may be as selective as a strong change of one environmental factor. As soon as the physiological potential of a genotype, the repair system, cannot properly react to the change, the plant will be stressed.

Let me first elaborate on some principles in population dynamics in relation to stress, to facilitate judgement on the prediction of the evolution of stress resistance.

2.2. Plant population dynamics

For most ecologists a population is defined as the total number of individuals of a single species in a certain area. Variation of population size is therefore measured in a quantitative way, i.e. change of number of individuals per unit time upon a fixed surface area. To understand selection, i.e. the impact of stress factors on the various genotypes building up natural populations, it is necessary to follow the changes in genotype frequency, i.e. qualitative and quantitative changes.

2.2.1. Quantitative plant population dynamics

Changes in population size can be described by the logistic (Verhulst–Pearl) growth equation:

$$\frac{\mathrm{d}X}{\mathrm{d}t} = rX\left(\frac{K - X}{K}\right) \tag{2.1}$$

where r is the intrinsic rate of increase, X the number of individuals, and K the carrying capacity supporting the number of individuals in a particular environment.

If that plant structure remained fixed under changing environmental conditions, a stress or stresses resulting from a shortage of an environmental source, e.g. water or nutrients, may diminish the carrying capacity, i.e. $K_S < K$, so that the population size will decrease. If an environmental resource increases, e.g. fertilization, the carrying capacity will increase, $K_f > K$, so that the population will expand, often at the expense of the population of other species (Davy and Bishop, 1984; Bobbink et al., 1988; Bakker, 1989).

Examples of increasing carrying capacity of the environment for certain species and populations and reducing capacity for others, i.e. red list species, are the eutrophication of the environment by emission of nitrogen and sulphur dioxide. Sulphur dioxide emissions have not only changed atmospheric chemistry, injured plant populations (Guderian, 1985) and caused evolution of resistance to sulphur dioxide in affected populations (Taylor and Murdy, 1975; Horsman et al., 1979; Ernst et al., 1985), but sulphur-enriched precipitation and dry deposition has also increased the sulphur status of the soil (Wookey and Ineson, 1991). The highest concentrations of sulphate in rain water were found in the Netherlands between 1960 and 1975 (Stuyfzand, 1991). After that year there was a steady decrease due to the change from coke gas to natural gas and desulphurization of fossil coal and oil. Principally, the 1960s and 1970s can be characterized as a time of sulphur fertilization. Additional inputs of sulphur may benefit plant species with a high sulphur demand, such as members of the Brassicaceae and the Allioideae. In contrast to other plant species, which have decreased by one-third in their occurrence during the last 40 years, the Brassicaceae have enlarged their area in the Netherlands and the Western part of Germany tremendously (Table 2.1; Ernst 1993), followed by a population increase of the white butterfly (*Pieris rapae*) which displays a high preference for Brassicaceae as hosts for its larvae (Tax, 1989). This tremendous change in population size due to sulphur dioxide fumigation provides a new opportunity to study the interface between qualitative and quantitative population dynamics in relation to so-called stress.

DeAngelis (1992) has tried to relate the carrying capacity to the nutrient amount in the environment. He has introduced a nutrient factor into the logistic growth equation (2.1):

$$\frac{dX}{dt} = r(N)X - (d + e)X \qquad (2.2)$$

where $r(N)$ is the intrinsic rate of growth of the number of X individuals, which depends on the amount of available nutrient N; d is the rate coefficient for the loss of biomass (by senescence, herbivory, plant pathogens, frost, etc.) but where the nutrient mass stays in the system (biotic system turnover);

TABLE 2.1. *Increase or decrease in the occurrence of Brassicaceae in the inventory quadrats (each 25 km²) in the flora of the Netherlands in the period before and after 1950.*

Ecosystem	Species no.	Decrease (−) or increase (+) of inventory quadrats	Change of inventory quadrats per species
(Semi)-natural ecosystems			
Coastal floodline	2	+45	+23
Heavy metal vegetation	1	−3	−3
Semi-natural grassland	6	−84	−14
Wells	2	+467	+234
Wetlands	9	+1140	+127
Man-made ecosystems			
Walls	1	−18	−18
Arable fields	16	+2309	+144
Ruderal sites	26	+3158	+121

The dataset is based on the maps, as published by Mennema *et al.* (1980, 1985) and van der Meijden *et al.* (1989). The species are divided between (semi)-natural ecosystems and man-made ecosystems.

and e is the rate coefficient for the loss of biomass, where the biomass and thus also the nutrients are lost from the system, e.g. by harvest, erosion or washout.

The nutrient aspect has widely been investigated, using the concept of nutrient limitation (Tilman, 1988), based on Von Liebig's law of the minimum (Von Liebig, 1840). This concept is a typically one-sided approach, because it ignores the favourable effects of nutrient limitation on the relative stability of natural ecosystems (Rosenzweig, 1971) and on species diversity (Grime, 1979). It re-emphasizes some well known evidence from agriculture, that only monocultures guarantee maximization of harvest (De Wit, 1960). Most statements made on nutrient limitations say that nitrogen is the most limiting factor in ecosystems (Tamm, 1968). However, the human race has thoroughly changed the nitrogen supply to the environment. Since the late 1960s, in some places atmospheric nitrogen deposition has increased up to 50 kg ha^{-1} year^{-1} because of volatization of ammonia from farm manure and NO$_x$ in traffic exhausts (Van Breemen et al., 1982; Heij and Schneider, 1991), although it remained lower than the amount of nitrogen applied in fertilizer experiments, which were set up to abolish nitrogen limitation (Tamm, 1968; Kachi and Hirose, 1983; Dougherty et al., 1990; Kozlowski et al., 1991; Olff, 1992). Nevertheless, and ironically, removal of nitrogen limitation results in tree injury, because moderate nitrogen availability increases the frost sensitivity of plants (Dueck et al., 1991).

In addition, an increased nutrient supply may increase the vegetative biomass, but it may result in the plant's not flowering at the normal time, as in the annual dune grass *Phleum arenarium* (Ernst, 1983), or at all, as in the perennial herb *Arnica montana* (van der Eerden, 1992). In the latter case, enhanced fumigation with ammonia stimulated biomass production, but diminished plant survival and reproduction (Fig. 2.1).

For the implementation of this nutrient-related logistic growth equation,

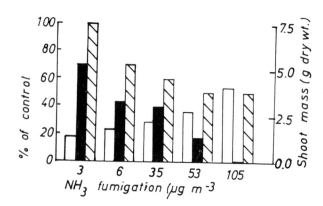

FIGURE 2.1 *Effects of 16 months' exposure to increasing fumigation with ammonia of shoot mass (□: 1 s.d.), flowering (■) and survival (▧) of* Arnica montana. *(Data from van der Eerden, 1992.)*

not only is a life-history-based approach necessary, but another aspect of plants – the plasticity of their performance – also has to be considered. Because of the high plasticity of plants, carrying capacity must also be related to plant density and plant biomass. The mean plant dry weight (W) is related to the density of surviving plants per unit area (X) by the equation (Watkinson, 1984):

$$W = W_m(1 + aX)^{-b} \qquad (2.3)$$

where W is the plant dry weight of each population member, W_m the mean yield of an isolated plant, a the area required to achieve W_m, b the efficiency of resource utilization, and X the number of individuals per study area.

If $b = 1$, the total yield per unit area becomes independent of plant density. This situation is known as 'the law of constant yield' (Shinozaki and Kira, 1956).

With regard to plasticity and genetic diversity among and between populations, in wild plants it will be difficult to find a real mean yield of an isolated plant (Fig. 2.2). In research into the density-independent reaction of perennial

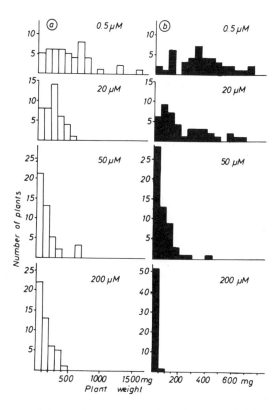

FIGURE 2.2 *Impact of increasing concentration of copper (µM copper l^{-1} nutrient solution) on biomass production (mg dry weight) of (a)* Phleum pratense *and (b)* Poa pratensis *after 7 weeks of growth. (After Lolkema, 1985.)*

grasses to an increase of micronutrient concentrations, Lolkema (1985) has shown that copper affects the growth of these grasses above the repair level up to sublethal concentrations by strongly causing dwarfism. The response remained species specific and, among species, genotype specific even under severe stress.

2.2.2. Qualitative plant population dynamics

Wild plant populations are characterized by a high degree of genetic diversity, as demonstrated by analysis of enzyme polymorphism (Loveless and Hamrick, 1984). Populations of the same species, having adapted to 'extreme' environmental conditions such as a surplus of a combination of various heavy metals or sulphur dioxide, can have the same polymorphic index (Table 2.2) as metal-sensitive populations (Verkleij et al., 1985). Therefore the logistic growth equation (2.1), taking the i genotypes of a population into account, has to be modified to:

$$\frac{dX}{dt} = \sum r_i X_i \left(\frac{K - X_i}{K} \right) \tag{2.4}$$

In a large random-mating population with no selection, mutation or migration, the gene frequencies and the genotype frequencies are constant from generation to generation, i.e. the population is in Hardy–Weinberg equilibrium. The contribution of offspring to the next generation is defined by the fitness of the individual. Individuals in a population differ in viability and fertility; recently published examples are those of the annual plant species *Arabidopsis thaliana* (Aarssen and Clauss, 1992), *Capsella bursa-pastoris* (Neuffer and Hurka, 1986), *Erophila verna* (Van Andel et al., 1986) and *Raphanus raphanistrum* (Stanton, 1984).

TABLE 2.2. *Genetic variation based on 16 loci including five monomorphic ones in 12 populations of* Silene vulgaris (data from Verkleij et al., 1985).

Adaptation	Number of populations	Average number of alleles per locus	Polymorphic index
No metal tolerance	2	2.28 (0.03)	0.264 (0.056)
Copper tolerance	2	2.22 (0.09)	0.272 (0.033)
Zinc/cadmium tolerance	2	2.22 (0.09)	0.253 (0.005)
Zinc/copper tolerance	3	2.37 (0.09)	0.342 (0.006)
Zinc/cadmium/copper tolerance	2	2.38 (0.07)	0.347 (0.006)
Sulphur dioxide tolerance	1	2.07	0.260

Values in parentheses are standard deviations.

(a) Impact of multiple stresses on genotype dynamics

Changes in gene and genotype frequencies may be excellent evidence of the impact of stress on population dynamics, and may help to identify stress components. An example is the dynamics of *Senecio sylvaticus*, a potentially wind-dispersed herb with winter-annual and summer-annual phenotypes co-occurring in open pine forests on the Dutch coastal dunes (Ernst, 1989). The winter-annual phenotype, germinating without cold treatment in autumn, is sensitive to the period of snow cover and frost intensity, to winter drought and to herbivory by the larvae of Sciaridae; the summer-annual phenotype, germinating after cold treatment in spring, is affected by two pest insects (aphids in the vegetative stage, tephrids as fruit consumers) and summer drought (Fig. 2.3). Co-occurrence of more than one stress factor has dramatically affected the phenotype frequency, especially that of the winter-annual phenotype, and also the total population size. From an estimated size of a million plants per hectare in 1975, the long-lasting droughts in the winter of 1986/87 and in the summer of 1987 have allowed only 189 individuals of the summer-annual phenotype to survive and just two from the winter-annual phenotype. Obviously, singular events are very decisive for the equilibrium of both life forms in the populations.

(b) Impact of one stress on the autotrophic and heterotrophic population cycle

Trends in the genetic composition of plant populations over time have to be ranked within the life-history of a species, including its often long-lived

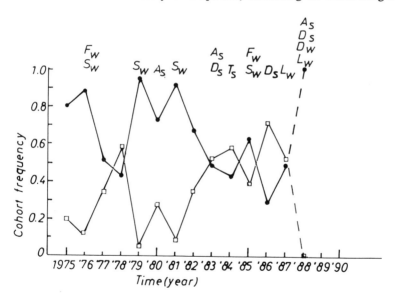

FIGURE 2.3 *Population dynamics of the winter-annual* ($_W$, □) *and summer-annual* ($_S$, ●) *phenotype of* Senecio sylvaticus *in thinned pine forest on the Dutch coastal dunes. Selecting factors: A, aphids; D, drought; F, frost; L, larvae of Sciaridae; S, snow; T, Tephritidae.*

seed bank. Unfortunately, ecologists often restrict their perception to the autotrophic phase of a population, and misinterpret the absence of this phase as extinction (van der Meijden *et al.*, 1992) despite the persistence of a viable seed bank. One example of a plant species with a naturally short autotrophic phase and a long-lasting heterotrophic phase is *Digitalis purpurea*. Harper (1977, p. 543) gave the first misinterpretation of the population dynamics of this plant species by writing: 'It is a curious feature of the dynamics of these populations that when they are conspicuous enough to be chosen for study they are usually already in a condition of decline!' Wild types of *D. purpurea* can be recognized by the red colour of their flowers, which is absolutely dominant, when a population establishes its autotrophic phase after clear-felling, windfalls or fire in forests, from a long-living persistent seed bank (Van Baalen, 1982; Van Baalen and Prins, 1983). After two autotrophic generations of this biennial plant, as a consequence of forest succession, the vegetation cover becomes closed so that only leaf-filtered daylight penetrates to the seeds on the soil surface, enforcing seed dormancy. Therefore a further autotrophic generation does not come into existence. Mowing of the vegetation on forest clearings, thus preventing vegetation succession and allowing penetration of non-filtered daylight to the seeds on the soil surface, provides a situation where *D. purpurea* can maintain its autotrophic population for a long time. However, this artificial prolongation of the autotrophic phase results in a strong change of flower colour phenotypes. After 10 years the wild type was nearly completely substituted by a pink phenotype (Fig. 2.4), often combined with selection against the hairy, i.e. recessive

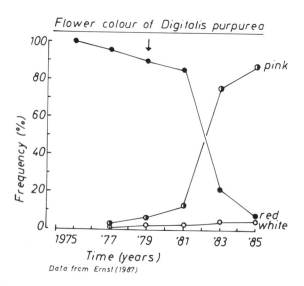

FIGURE 2.4 *Frequency changes of flower colour of* Digitalis purpurea *in woodland clearing with a mowing regimen:* ●, *red;* ◑, *pink;* ○, *white. The arrow indicates the change from the autotrophic to the heterotrophic (seed bank) phase of the population without a mowing regimen. (Data from Ernst, 1987.)*

homozygote, phenotypes (Ernst, 1985a, 1987). Survival of seeds of non-wild-type parents is inferior to that of wild-type ones; thus removal of the environmental restraint stimulates the expression of hidden genetic diversity, but at the same time favours the mortality of non-wild-type seeds in the seed bank. Selection on a high survival time of the seed of *D. purpurea* in the seed bank, and on a rapid germination after exposure to full daylight, seems to be an excellent adaptation to natural cycles of forest succession. By analysis of pollen and macrofossils in a forested area with small fens in the Sauerland (Germany), Pott (1985) has shown that such short autotrophic periods of *D. purpurea* have occurred in Central Europe for more than 4000 years (Fig. 2.5). Thus, from a phytocentric view, short autotrophic and long heterotrophic phases in the population dynamics of *D. purpurea* seem to be such a perfect result of selection that rapid succession from clearing to forest may not be an environmental constraint. Enlarging the autotrophic phase causes stress.

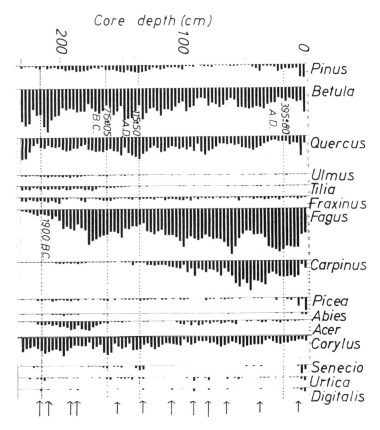

FIGURE 2.5 *Natural cycles of* Digitalis purpurea *in a forested (without clear-felling) area in the Central Mountains of Germany. (After Pott, 1985.)*

(c) Gene flow between adjacent but differently adapted populations

Gene flow between adjacent populations can be realized by exchange of pollen and/or exchange of seeds. I will restrict my considerations to heavy metal soils. Pollen flow between adjacent populations occurring on chemically distinct soils has been documented for wind pollinated grasses, e.g. *Agrostis capillaris* on a copper mine and an area without copper contamination (McNeilly, 1968), and for the insect pollinated herb *Silene vulgaris* (Ernst, 1969). Exchange of seeds of these autochorous species occurs very seldomly, except through transport by humans. Gene flow may increase metal tolerance in metal-sensitive populations and decrease it in metal-resistant ones. The success of alien genes will depend on the fitness of the phenotypes, which is determined by the direct metabolic cost for maintenance of the resistance and by the indirect costs by linkage of resistance genes to other genes influencing plant performance, such as relative growth rate, seed mass and drought tolerance (Table 2.3). The direct cost may be related to the production of storage capacity for toxic ions (vacuoles), to the synthesis of carriers to transport the ions across the tonoplast into the vacuole, to the synthesis of complexing agents detoxifying ions in the cytosol, to the establishment of cellular repair mechanisms or to the biochemical changes within biomembranes. All or part of these processes will be associated with a different allocation of resources in resistant plants compared with non-tolerant ones.

It is very obvious that resistant plants have less fitness in environments where the selecting factor is absent, being outcompeted by sensitive genotypes (Cook *et al.*, 1972; Hickey and McNeilly, 1975). Their biomass is often at least 20% less than that of non-resistant genotypes (Ernst, 1983); this reduction is independent of the metal involved and is not restricted to metals alone (Table 2.4). Part of this diminished productivity can be explained by the additional demand for essential heavy metals, due to the resistance mechanisms (Ernst *et al.*, 1992), and another part may be caused by associated performance characters (Ernst *et al.*, 1990b). In every case, non-resistant phenotypes have zero fitness on metal-contaminated soils. As a general prin-

TABLE 2.3. *Multiple stresses associated with copper resistance of* Silene vulgaris *on copper mines in Germany.*

Environmental factor	Plant response	Reference
Copper enrichment	Biomembrane protection	De Vos *et al.* (1991)
Low status of available major nutrients	Low growth rate	Lolkema *et al.* (1984)
?	Small seed size	Ernst (1985b)
Low water supply (drought)	Small leaf size	Schwanitz and Hahn (1954)
High seed predation by *Hadena* sp.	Small capsule size Delayed (4 week) flowering	Ernst (1985b) Ernst *et al.* (1990b)

TABLE 2.4. *Diminished biomass production in resistant genotypes compared with non-resistant ones* (after Ernst, 1983; Lolkema *et al.*, 1984; Verkleij and Prast, 1989).

Resistance	Plant species	Diminished biomass (% of sensitive genotype)	Reference
Zinc	*Silene cucubalus*	20	Baumeister (1954)
Cadmium	*Silene cucubalus*	44	Verkleij and Prast (1989)
Copper	*Silene cucubalus*	45	Lolkema *et al.* (1984)
Nickel	*Festuca cinerea*	25	Sasse (1976)
Sodium chloride	*Festuca rubra*	40	Rozema *et al.* (1978)
Atrazin	*Amaranthus reflexus*	40	Conard and Radosevich (1979)
	Senecio vulgaris	25	Conard and Radosevich (1979)

ciple, costs of resistance are evident only in environments without stress. Adaptation to the stressed environments ultimately results in greater fitness than in plants that do not exhibit resistance mechanisms.

2.2.3. *Conclusion: population dynamics*

Changes of environmental factors will affect the various genotypes of a population to a different extent. Understanding the dynamics of populations under stress will only be promoted by elaborating on quantitative *and* qualitative aspects.

2.3. Stress

2.3.1. *Definition of stress*

The term 'stress' has to be judged from the perception of an organism, and not from that of a scientist. Two such anthropocentric views of stress can be illustrated. Cushman *et al.* (1990, p. 197), in a review on gene expression during adaptation to salt stress, state: 'Yet halophytes exist and thrive under conditions of high salinity, which suggests that life under salt stress is not necessarily life on the edge.' These authors have overlooked, for instance, that *Halobacterium* has evolved such a high adaptation to a saline environment that its ribosomes are functional only in the presence of 3–4 M KCl, in contrast to 1 mM Mg in non-halophytic *Escherichia coli* (Conte, 1973). A similar anthropocentric understanding of stress is given by Roose (1991,

p. 114): 'For CO_2, the opportunity for studies of the effects of pollution on fitness components of natural populations from truly unpolluted environments is lost since all populations may well have been influenced by increasing CO_2 levels.' Later, I will demonstrate that a lot of plant populations are exposed to somewhat higher concentrations of carbon dioxide than that of the present ambient air.

Stress should be defined according to the physiological (Chapin, 1991) and ecological requirements of an organism throughout its life-cycle. If an environmental factor surpasses the lower and upper limit of these requirements, it will induce 'a potentially injurious strain in living organisms' (Levitt, 1972, p. 3). As a consequence of this definition we have to admit that 'extreme environments are not stressful to the species occupying these environments' (Chapin, 1991, p. 69). Plant species growing in so-called extreme environments, which may be defined as the tails of the normal distribution of all environments on a regional, continental or global scale, exhibit specific physiological adaptations that enable them to survive and contribute to the next generation. These adaptations to an extreme environment may be achieved by avoidance and/or tolerance mechanisms *sensu* Levitt (1972, 1980). The physiological requirements may change during the life-history of a plant; for example, there may be a low-temperature requirement by seeds to break dormancy and a high-temperature requirement for growth and reproduction of a summer annual plant. The study of plant responses to stress is complicated by the fact that a lot of abiotic and biotic factors interact, often simultaneously.

Changes in the environmental factors that cause stress in plants may occur within different time-scales, varying in duration and intensity during the life-span of an organism. With the increase of longevity of an organism, the chance of exposure to stress will increase, formerly only because of natural changes of climatic and edaphic factors, but recently because of anthropogenic changes. As long as the plant can repair or compensate for the stress, and the energy used for repair and maintenance of the metabolic pools is small, the plant will restore its metabolic homeostasis. If metabolism is impaired, selection will favour genotypes with the least metabolic destabilization. As a result, the genetic composition of the population will change. The predictability of the occurrence of stress will influence the reaction patterns of the individual and thus the selection process in a population, leading to a new adjustment of the population to the new circumstances.

Based on time-scale and predictability, we can distinguish the following stress-inducing factors, which may operate as environmental forces (DeAngelis, 1992) affecting the population dynamics:

1. Predictable gradual changes of environmental factors within the geological time-scale: climatic changes and pedogenesis.
2. Unpredictable sudden, but temporally repeating, changes within a time-scale of decennia and centennia: severe frost, storms, shortage and surplus of water, fire, pest epidemics, anthropogenic river pollution.
3. Unpredictable gradual changes of environmental factors within a time-

scale of years and decennia: eutrophication, acidification, pesticide application, air pollution.
4. Predictable seasonal changes of environmental factors: annual cycles of temperature, precipitation and irradiation. Whether these seasonal changes will really cause stress to the plants will depend on the degree of predictability.

Stress symptoms can be grouped into the following categories:

1. Invisible, but measurable injury (e.g. changes in biomembranes, enzyme activities)
2. Visible injury of plant parts (e.g. necrosis, chlorosis, discoloration)
3. Retarded growth, development and reproduction
4. Changes in the interference caused by heterotrophs
5. Changes in the genotypes of which populations are composed
6. Change in species composition of an ecosystem

One of the problems is a clear-cut analysis of stress. Generally, a plant only suffers from stress if an experiment allows us to demonstrate how the supposed stress factors affect its metabolism, in conditions as near as possible to the realistic settings of its habitat.

2.3.2. Reactivity of plant populations to stress

Abundant evidence exists for the occurrence of genetic variation in plant populations of a broad range of species (Loveless and Hamrick, 1984). Obviously, most populations can resist environmental conditions where low-level variation around a mean prevents high mortality rates that would otherwise eliminate large numbers of individuals within a population. Many studies on the resistance potential of plant populations have established that only a few species, often with a generally broad amplitude of variation enabling them to persist in various environments, are able to evolve resistance to various kinds of new environmental and man-made stress. Environments with exceptionally low or high concentrations of nutrients, non-nutritional elements, disturbance by abiotic and biotic factors are monopolized by a few species belonging to a few taxa and families. In Europe the families Brassicaceae, Caryophyllaceae, Chenopodiaceae and Poaceae are well known for their evolutionary potential, although within each family only a few genera possess this ability. The genera *Agrostis* and *Festuca* have evolved resistance to drought, flooding, salt and heavy metals, but are unable to cope with herbicides and high concentrations of air pollutants (Ellenberg *et al.*, 1991). In the genus *Silene* there is sound evidence of resistance to salt, heavy metals and air pollutants, but a high sensitivity to flooding and attack by white flies (Aleurodina). However, not every population of these genera contains sufficient genetic variation to provide a basis of selection on exposure to stress, as demonstrated for *Agrostis capillaris* populations surrounding areas recently polluted by metals in England (Al-Hiyaly *et al.*,

1988) and in the Netherlands (Lolkema, 1985). Unfortunately, the physio-logical principles of most resistance mechanisms are not clear, so that changes in the genome due to stress will be difficult to predict.

2.3.3. Establishment of stress

(a) Predictable gradual changes within the geological time-scale: carbon dioxide increase – an environmental stress?

There is growing concern about the increase in atmospheric carbon dioxide levels resulting from an accelerated turnover of fixed carbon dioxide from fossil energy sources to the atmosphere. In the discussion of greenhouse gases, carbon dioxide is supposed to be the dominant environmental force that could change the species composition of ecosystems (Bazzaz, 1990). This assumption is based on experimental results showing a growth stimula-tion at exposure to elevated carbon dioxide levels in phytotrons, open-top chambers and greenhouses (Enoch and Hurd, 1979; Rozema *et al.*, 1991; Strain, 1991). The history of carbon dioxide, in the global atmosphere revealed by ice cores from the Arctic and Antarctic (Neftel *et al.*, 1982; Barnola *et al.*, 1987), shows a strong variation of the atmospheric carbon dioxide concentration during the past 160 000 years. Several times and often for a long period it has approached 190–200 ppmv, the global minimum based on the ocean's gypsum equilibrium (Holland, 1965), but only twice has it surpassed 300 ppmv (Fig. 2.6), thus being far from the upper ocean carbonate equilibrium at 1300 ppmv (Holland, 1965). Considering the life-cycle of plants, annual plant species have experienced more than 50 000 generations at low (200 ppmv) carbon dioxide levels, and long-lived trees were exposed for several hundred to a thousand generations to such conditions. However, in all research published to date (for a review, see Bazzaz, 1990), there was and still is no *a priori* test of the physiological requirements for carbon dioxide by plants, based on the real ambient carbon dioxide concen-tration in their habitat. It is well known that plants having leaves near the soil surface receive 30–70% of their carbon dioxide from soil respiration, at concentrations up to 1000 ppmv (Amundson and Davidson, 1990). Thus, experiments in a greenhouse or phytotron create a carbon dioxide-deficient environment (at so-called ambient air with 350 ppmv carbon dioxide), depriv-ing the plants of their real ambient concentrations. Therefore it is not sur-prising that plants respond positively to carbon dioxide enrichment up to 600 ppmv (Fig. 2.7) and may already show a tendency to a negative response at higher, thus surplus (800 ppmv) carbon dioxide levels (Forstreuter, 1991).

The outcome of isotopic fractionation studies gives further support to the significance of the supply of soil-derived carbon dioxide. Organic material in C3 plants has an isotopic carbon composition ($\delta^{13}C$ values) of around $-28‰$, because ribulose bisphosphate carboxylase discriminates against car-bon dioxide molecules containing ^{13}C to a considerable degree. In contrast, C4 plants have values of about $-14‰$ (Farquhar and Richards, 1984), because phosphoenol pyruvate carboxylase does not discriminate against $^{13}CO_2$. Soil-derived carbon dioxide is poor in $^{13}CO_2$, because of either strong discrimina-

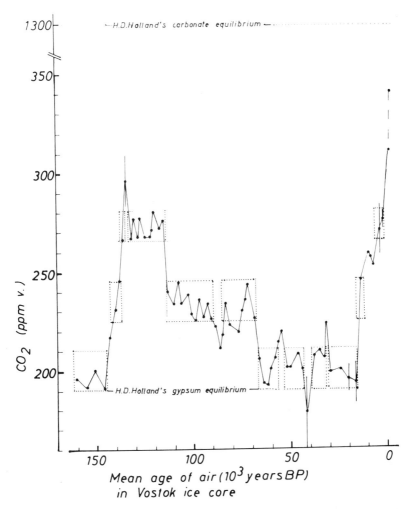

FIGURE 2.6 *Atmospheric carbon dioxide concentration as reconstructed from the ice core at Vostok. The lower and upper carbon dioxide equilibria as calculated by Holland (1965) are indicated. (Modified from Barnola et al., 1987.)*

tion by the respiratory enzymes of fungi, bacteria and soil fauna or a lower diffusion rate of the heavier ^{13}C from the soil atmosphere to the soil surface. As a result, ^{13}C values of plants using soil respiratory carbon dioxide are more negative than those of plants and plant parts at a higher elevation above the soil surface (Fig. 2.8), as demonstrated for a leaf gradient in beech trees (Schleser, 1990, 1991). Interpretation of carbon isotope discrimination by plants in latitudinal and altitudinal settings (Körner *et al.*, 1991) and in

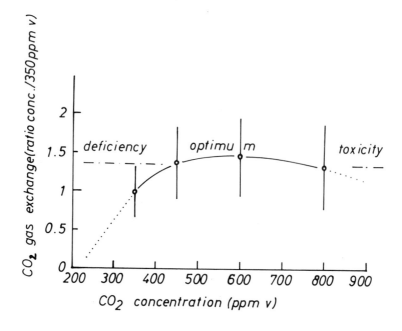

FIGURE 2.7 *Response of a* Trifolium pratense–Festuca pratensis *grass plot to four concentrations of carbon dioxide. (After Forstreuter, 1991.)*

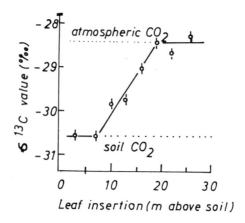

FIGURE 2.8 $\delta^{13}C$ *values of leaves of a beech (*Fagus sylvatica*) in a height gradient. (After Schleser, 1991.)*

relation to the growing season (Smedley *et al.*, 1991) lacks a critical evaluation of the impact of soil-borne carbon dioxide.

It is of considerable interest to understand the genetic potential of plants to adapt to different carbon dioxide concentrations and to discriminate between ^{13}C and ^{12}C at the intraspecific and interspecific level. With the exception of the interspecific and intraspecific differences in carbon dioxide fixation by C3 and C4 species (Fladung and Hesselbach, 1989; Arp *et al.*, 1993), genetic differences in the response to elevated carbon dioxide levels at the intraspecific level have been demonstrated only for the perennial herb *Plantago lanceolata* (Wulff and Alexander, 1985); changes in leaf area, however, were strongly related to the daily temperature regime (Fig. 2.9), so that evidence for relevant genotypic differentiation may be camouflaged by other environmental factors. Also in *Plantago lanceolata*, with its leaf rosette near the soil surface, the carbon dioxide concentration in the control (350 ppmv) may be quite different from the ambient concentration in its natural habitat, so that this experiment has only demonstrated that the plant can cope with an environment deficient in carbon dioxide.

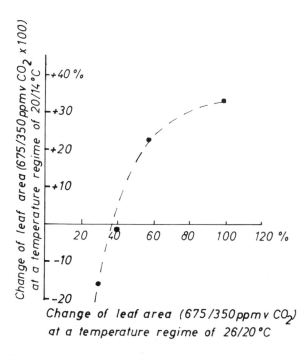

FIGURE 2.9 *Change of leaf area (−, negative; +, positive) of genotypes of* Plantago lanceolata *grown at two carbon dioxide concentrations (350 and 675 ppm) and at two daily temperature regimens (20/14 °C and 26/20 °C). (Based on Wulff and Alexander, 1985.) The low concentration may be a deficient one for a plant with a typical rosette structure.*

Increase of ambient carbon dioxide concentration may affect only plant species occurring in pioneer ecosystems poor in soil organic matter (e.g. dunes and deserts) and tall plant species (i.e. trees). It has been suggested that a carbon dioxide increase during the last 200 years has resulted in a 40% decrease in the density of stomata of trees (Woodward, 1987). Extrapolation of these data to the carbon dioxide concentration from the last glacial period onwards will double the stomatal density. If this is the case, a tremendous variation in genetic responses may be expected. Up to now, however, there is no evidence available from macrofossils. If we take Woodward's extrapolation to the ambient carbon dioxide concentration of 1000–5000 ppmv some 10^8 years ago (Strain, 1991), leaves of prepleistocene trees will have had an epidermis nearly devoid of stomata.

A second aspect of evolutionary potential is the carbon-isotope discrimination. In the short-lived desert perennial *Gutierrezia microcephala*, the heritability of carbon-isotope discrimination was above 80% (Schuster *et al.*, 1992). Although the authors discuss a positive correlation between growth and discrimination potential, the results presented are not convincing ($r = 0.20$).

Therefore the research on carbon dioxide has to be thoroughly reorientated, first measuring the real carbon dioxide environment of the plant and then analysing the carbon dioxide response curve, from its lower oceanic equilibrium level (190 ppmv; Holland, 1965) to its upper equilibrium level (1300 ppmv). As long as increasing levels do not affect the diffusion rate of carbon dioxide emitted from the soil, the herbs, grasses and seedlings of shrubs and trees in ecosystems with a well developed litter layer and humus horizon will not undergo stress caused by the level of carbon dioxide.

(b) Unpredictable sudden but repeating changes: reparation of frost stress during the life-history

Not every form of stress, even if it results in leaf loss, will influence population dynamics negatively. The best examples of complete defoliation just after the start of the growing season are those of *Quercus robur*, *Q. petraea* and *Q. ilex* by the Green oak leafroller (*Tortrix viridiana*) (Schütte, 1957; DuMerle, 1988) or that of *Euonymus europaea* by *Yponomeuta plumbellus*. By mid-summer, however, all initially dormant buds have lost their dormancy and developed into new leaves, which survive nearly without herbivorous injury up to autumnal senescence. There is not even a measurable effect on seed production, so that the repair system, despite severe losses of photosynthetic capacity and nutrients lasting for a month, remains in balance.

Sudden loss of leaves can occur within a day in Southern and Central Africa, when black frost kills all green leaves and dormant buds in the upper crown of trees and shrubs (Hattle, 1968; Ernst, 1971). Reparation of the crown takes a few years, primarily by the development of new twigs from stem-borne dormant buds. During this period trees do not flower and reproduce, so that potential population increase $dX/dt = 0$. The high seed loss through predation by insects (Ernst *et al.*, 1989), the low probability that seeds of savanna shrubs and trees will germinate and establish (Tolsma, 1989) and the strong recurrent shoot die-back of miombo seedlings (Ernst, 1988;

Chidumayo, 1991) will obviously not affect population dynamics in the miombo woodlands and savanna of Southern and Central Africa as long as these frosts occur only two or three times in a centennium.

Frost resistance of plant species does not fully coincide with the lowest temperature that can occur at a particular site. The metabolic costs of complete frost resistance, i.e. the alterations in the lipid composition of the plasma membrane (Steponkus, 1990) are too high, and so perennial plants take the risk of being exposed to severe losses of their photosynthetic capacity during their life-cycle. Restoration of a population can occur quite rapidly in the temperate zone, as described for the salt marsh herb *Halimione portulacoides* (Beeftink *et al.*, 1978). Although in the years immediately after heavy frost damage other species on a Dutch salt marsh increased in percentage cover, after 3 years *H. portulacoides* recovered fully.

As long as the survival of the individual is ensured, the level of stress does not go beyond the resistance limit, even if parts of the individual plant are lost.

(c) Unpredictable gradual changes during decennia

Industrial and agricultural activities have changed all the important 'spheres' of plants:

1. The atmosphere, as a result of emissions of gaseous pollutants
2. The pedosphere, by deposition of pollutants as acids and particulates, by fertilization and by pollution
3. The biosphere, by removal and addition of species

Interspecific and intraspecific variation of species and populations are known to occur as a result of nearly all changes of the kinds mentioned above. Sulphur dioxide emissions, the deposition of sulphate and the deposition of metal fallout near point sources are examples which show the response not only of special populations, but of whole ecosystems to environmental stress. The best analysed example can be taken from areas polluted by heavy metals. In the vicinity of a zinc/cadmium smelter near Budel in the Netherlands, there is proven metal resistance for the heterotrophic bacterium *Alcaligenes eutrophus*, a vesicular–arbuscular mycorrhizal fungus, the grass *Agrostis capillaris*, and two soil invertebrates (the isopod *Porcellio scaber* and the collembolan *Orchesella cincta*) (Table 2.5). In this special case a completely functional ecosystem has evolved after a centennium of exposure to a point-emission source of heavy metals.

2.4. Conclusions

As soon as stress factors surpass the repair potential of a plant, its physiological and ecological performance will diminish. Because of the high genetic variability among and between populations and species of the same or bordering ecosystems, selection will favour the best stress-adapted genotypes and thus will have an impact on the qualitative and quantitative aspects of

TABLE 2.5. *Organisms with a proven resistance to zinc and cadmium in the vicinity of a zinc/cadmium smelter, composing a functional resistant ecosystem.*

Species	Resistance mechanisms	Reference
Bacteria	Plasmid encoded	Diels *et al.* (1989)
Alcaligenes eutrophus	Metal sequestration	
Fungi		
Vesicular–arbuscular mycorrhiza by *Scutellospora dipurpurascens*	Unknown	Ietswaart *et al.* (1992)
Angiosperms		
Agrostis capillaris	Diminished uptake Accumulation in the vacuole	Dueck *et al.* (1984)
Soil fauna		
Porcellio scaber (isopod)	Enhanced excretion	Van Capelleveen (1985)
Orchesella cincta	Metallothionein	Donker and Bogert (1991)
(Collembola)	Enhanced excretion Immobilization by metal crystal formation	Posthuma *et al.* (1992)

population dynamics. To predict changes, it is necessary to know the physiological and ecological range of a population and its genotypes in the non-stressed situation. Otherwise, stress-oriented research will misinterpret stress, as argued above in relation to exposure to increased carbon dioxide levels.

On the other hand, long-term exposure to stress allows resistant genotypes to evolve which can build up a completely resistant ecosystem, not only in the case of high exposure to a naturally occurring damaging agent, but also within a centennium of anthropogenic pollution.

REFERENCES

Aarssen, C.W. and Clauss, W.J. (1992). Genotypic variation in fecundity allocation in *Arabidopsis thaliana*. *J. Ecol.* **80**, 109–114.

Al-Hiyaly, S.A., McNeilly, T. and Bradshaw, A.D. (1988). The effects of zinc contamination from electricity pylons: evolution in a replicated situation. *New Phytol.* **110**, 571–580.

Amundson, R.G. and Davidson, E.A. (1990). Carbon dioxide and nitrogenous gases in the soil atmosphere. *J. Geochem. Explor.* **38**, 13–41.

Arp, W.J., Drake, B.G., Pockman, W.T., Curtis, P.S. and Wigham, D.F. (1993). Interactions between C_3 and C_4 salt marsh plant species during four years of exposure to elevated atmospheric CO_2. *Vegetatio* **104/105**, 133–143.

Bakker, J.P. (1989). *Nature Management by Cutting and Grazing*. Dordrecht: Junk.

Barnola, J.M., Raynaud, D., Korotkevich, Y.S. and Lorius, C. (1987). Vostok ice core provides 160 000-year record of atmospheric CO_2. *Nature* **329**, 408–414.

Baumeister, W. (1954). Über den Einfluss des Zinks bei *Silene inflata* Sm. *Ber. Dtsch. Bot. Ges.* **67**, 205–213.

Bazzaz, F.A. (1990). Response of natural ecosystems to the rising global CO_2 level. *Annu. Rev. Ecol. Syst.* **21**, 167–196.

Beeftink, W.G., Deane, M.C., De Munck, W. and Nieuwenhuize, J. (1978). Aspects of population dynamics in *Halimione portulacoides* communities. *Vegetatio* **36**, 31–43.

Bobbink, R., Bik, L. and Willems, J.H. (1988). Effect of nitrogen fertilization on vegetation structure and dominance of *Brachypodium pinnatum* (L.) Beauv. in chalk grassland. *Acta Bot. Neerl.* **37**, 231–242.

Chapin III, F.S. (1991). Effects of multiple environmental stresses on nutrient availability and use. In: *Response of Plants to Multiple Stresses*, pp. 67–88, eds. H.A. Mooney, W.E. Winner, E.J. Pell and E. Chu. San Diego, Calif.: Academic Press.

Chidumayo, E.N. (1991). Seeding development of the miombo woodland tree *Julbernardia globiflora*. *J. Vegetation Sci.* **2**, 21–26.

Conard, S.G. and Radosevich, S.R. (1979). Ecological fitness of *Senecio vulgaris* and *Amaranthus retrofluxus* biotypes susceptible or resistant to atrazine. *J. Appl. Ecol.* **16**, 171–177.

Conte, F. (1973). *Biochemical Adaptation*. Chicago: University of Chicago Press.

Cook, S.C.A., Lefèbvre, C. and McNeilly, T. (1972). Competition between metal tolerant and normal plant populations on normal soil. *Evolution* **26**, 366–372.

Cushman, J.C., De Rocher, E.J. and Bohmert, H.J. (1990). Gene expression during adaptation to salt stress. In: *Environmental Injury to Plants*, pp. 173–203, ed. F. Katterman. San Diego, Calif.: Academic Press.

Davy, A.J. and Bishop, G.F. (1984). Response of *Hieracium pilosella* in Breckland grass-heath to inorganic nutrients. *J. Ecol.* **72**, 319–330.

De Vos, C.H.R., Schat, H., De Waal, M.A.M., Vooijs, R. and Ernst, W.H.O. (1991). Increased resistance to copper-induced damage of the root cell plasmalemma in copper tolerant *Silene cucubalus*. *Physiol. Plant.* **82**, 523–528.

De Wit, C.T. (1960). *On competition*. Versl. Landbouwk. Onderzoek. 66.8. Wageningen.

DeAngelis, D.L. (1992). *Dynamics of Nutrient Cycling and Food Webs*. London: Chapman and Hall.

Diels, L., Sadunk, A. and Mergeay, M. (1989). Large plasmids governing multiple resistances to heavy metals: a genetic approach. *Toxicol. Environ. Chem.* **23**, 79–89.

Donker, M.H. and Bogert, C.H. (1991). Adaptation to cadmium in three populations of the isopod *Porcellio scaber*. *Comp. Biochem. Physiol.* **100C**, 143–146.

Dougherty, K.M., Mendelsohn, J.A. and Monteferrante, F.J. (1990). Effect of nitrogen, phosphorus and potassium additions on plant biomass and soil nutrient content in a swale barrier strand community in Louisiana. *Ann. Bot.* **66**, 265–271.

Dueck, T.A., Ernst, W.H.O., Faber, J. and Pasman, F. (1984). Heavy metal immission and genetic constitution of plant populations in the vicinity of two metal emission sources. *Angew. Bot.* **58**, 47–59.

Dueck, T.A., Dorel, F., Ter Horst, R. and Van der Eerden, L.J. (1991). Effects of ammonia and sulphur dioxide on the frost sensitivity of *Pinus sylvestris. Water, Air, Soil Pollut.* **54**, 35–49.

DuMerle, P. (1988). Phenological resistance of oaks to the Green oak leafroller, *Tortrix viridiana* (Lepidoptera: Tortricidae). In: *Mechanisms of Woody Plant Defenses Against Insects. Search for Pattern*, pp. 215–226, eds. Mattson, W.J., Levieux, J. and Bernard-Dagan, C. New York: Springer Verlag.

Ellenberg, H., Weber, H.E., Düll, R., Wirth, V., Werner, W. and Paulissen, D. (1991). Indicator values of plants in Central Europe. *Scripta Geobotanica* 18. Göttingen: Verlag Goltze.

Enoch, H.Z. and Hurd, R.G. (1979). The effect of elevated CO_2 concentrations in the atmosphere on plant transpiration and water use efficiency: a study with potted carnation plants. *Int. J. Biometeorol.* **23**, 343–351.

Ernst, W. (1969). *Die Schwermetallvegetation Europas.* Habilitationsschrift Math. Nat. Fak. Westf. Wilhelms-Univ. Münster, Münster.

Ernst, W. (1971). Zur Ökologie der Miombo-Wälder. *Flora* **160**, 317–331.

Ernst, W.H.O. (1983). Element nutrition of two contrasted dune annuals. *J. Ecol.* **71**, 197–209.

Ernst, W.H.O. (1985a). The effects of forest management on the genetic variability of plant species in the herb layer. In: *Population Genetics in Forestry*, Lecture Notes in Biomathematics 50, pp. 200–211, ed. H.R. Gregorius.

Ernst, W.H.O. (1985b). Schwermetallimmissionen-ökophysiologische und populationsgenetische Aspekte. *Düsseldorfer Geobot. Kolloqu.* **2**, 43–57.

Ernst, W.H.O. (1987). Scarcity of flower colour polymorphism in field populations of *Digitalis purpurea. Flora* **179**, 231–239.

Ernst, W.H.O. (1988). Seed and seedling ecology of *Brachystegia spiciformis*, a predominant tree component in miombo woodlands in South Central Africa. *For. Ecol. Manage.* **25**, 195–210.

Ernst, W.H.O. (1989). Selection of winter and summer annual life forms in populations of *Senecio sylvaticus. Flora* **182**, 221–231.

Ernst, W.H.O. (1993). Ecological aspects of sulfur metabolism: the impact of SO_2 and the evolution of the biosynthesis of sulfur compounds on populations and ecosystems. In: *Proc. 2nd Sulfur Workshop*, Garmisch Partenkirchen, 1992.

Ernst, W.H.O., Tonneijck, A.E.C. and Pasman, F.J.M. (1985). Ecotypic response of *Silene cucubalus* to air pollutants (SO_2, O_3). *J. Plant Physiol.* **118**, 439–450.

Ernst W.H.O., Tolsma, D.J. and Decelle, J.E. (1989). Predation of seeds of *Acacia tortilis* by insects. *Oikos* **54**, 294–301.

Ernst, W.H.O., Decelle, J.E. and Tolsma, D.J. (1990a). Dispersal seed predation in native leguminous shrubs and trees in southeastern Botswana. *Afr. J. Ecol.* **28**, 45–54.

Ernst, W.H.O., Schat, H. and Verkleij, J.A.C. (1990b). Evolutionary biology of metal resistance in *Silene vulgaris*. *Evol. Trends Plants* **4**, 45–51.

Ernst, W.H.O., Verkleij, J.A.C. and Schat, H. (1992). Metal tolerance in plants. *Acta Bot. Neerl.* **41**, 229–248.

Farquhar, G.D. and Richards, R.A. (1984). Isotopic composition of plant carbon correlates with water-use efficiency of wheat genotypes. *Aust. J. Plant Physiol.* **11**, 539–552.

Fladung, M. and Hesselbach, J. (1989). Effects of varying environment on photosynthetic parameters of C_3, C_3-C_4 and C_4 species in *Panicum*. *Oecologia* **79**, 168–173.

Forstreuter, M. (1991). Langzeitwirkungen der atmosphärischen CO_2-Anreicherung auf den Kohlenstoff- und Wasserhaushalt von Rotklee-Wiesenschwingel-Gemeinschaften. *Verh. Ges. Ökol.* (Osnabrück) **19**, 265–279.

Grime, J.P. (1979). *Plant Strategies and Vegetation Processes.* Chichester: Wiley.

Guderian, R. (1985). *Air Pollution by Phytochemical Oxidants.* Berlin: Springer Verlag.

Harper, J.L. (1977). *Population Biology of Plants.* London: Academic Press.

Hattle, J.B. (1968). The big freeze. *Rhod. Farm.* **29**, 22–23.

Heij, G.T. and Schneider, T., eds. (1991). *Acidification Research in The Netherlands.* Amsterdam: Elsevier.

Hickey, D.A. and McNeilly, T. (1975). Competition between metal tolerant and normal plant populations: a field experiment on normal soil. *Evolution* **29**, 458–464.

Holland, H.D. (1965). The history of ocean water and its effect on the chemistry of the atmosphere. *Proc. Natl. Acad. Sci. USA* **53**, 1173–1183.

Horsman, D.C., Roberts, T.M. and Bradshaw, A.D. (1979). Studies on the effect of sulphur dioxide on perennial ryegrass (*Lolium perenne* L.). *J. Exp. Bot.* **30**, 495–501.

Ietswaart, J.H., Griffioen, W.A.J. and Ernst, W.H.O. (1992). Seasonality of VAM infection in three populations of *Agrostis capillaris* (Gramineae) on soil with and without heavy metal enrichments. *Plant Soil* **139**, 67–73.

Jouzel, J., Lorius, C., Petit, J.R., Geuthon, C., Barkov, N.I., Kotlyakov, V.M. and Petrov, V.M. (1987). Vostok ice core: a continuous isotope temperature record over the last climatic cycle (160 000 years). *Nature* **329**, 403–408.

Kachi, N. and Hirose, T. (1983). Limiting nutrients for plant growth in coastal sand dune soil. *J. Ecol.* **71**, 937–944.

Kinzel, H. (1982). *Pflanzenökologie und Mineralstoffwechsel.* Stuttgart: E. Ulmer.

Körner, C., Farquhar, G.D. and Wong, S.C. (1991). Carbon isotope discrimination by plants follows latitudinal and altitudinal trends. *Oecologia* **88**, 30–40.

Kozlowski, T.T., Kramer, P.J. and Pallardy, S.G. (1991). *The Physiological Ecology of Woody Plants.* San Diego, Calif.: Academic Press.

Levitt, J. (1972). *Responses of Plants to Environmental Stresses.* New York: Academic Press. 2nd edn, 1980.

Levitt, J. (1980). *Responses of Plants to Environmental Stresses,* Vol. I, *Chilling, Freezing, and High Temperature Stresses.* New York: Academic Press.

Lolkema, P.C. (1985). *Copper Resistance in Higher Plants.* Thesis, Vrije Universiteit, Amsterdam.

Lolkema, P.C., Donker, M.H., Schouten, A.J. and Ernst, W.H.O. (1984). The possible role of metallothioneins in copper tolerance of *Silene cucubalus. Planta* **162**, 174–179.

Loveless, M.D. and Hamrick, J.L. (1984). Ecological determinants of genetic structure in plant populations. *Annu. Rev. Ecol. Syst.* **15**, 65–95.

McNeilly, T. (1968). Evolution in closely adjacent plant populations. III. *Agrostis tenuis* on a small copper mine. *Heredity* **23**, 99–108.

Mennema, J., Quené-Boterenbrood, A.J. and Plate, C.L. (1980). *Atlas van de Nederlandse Flora. I. Uitgestorven en Zeer Zeldzame Planten.* Amsterdam: Kosmos.

Mennema, J., Quené-Boterenbrood, A.J. and Plate, C.L. (1985). *Atlas van de Nederlandse Flora. 2. Zeldzame en Vrij Zeldzame Planten.* Utrecht: Bohn, Scheltema and Holkema.

Neftel, A., Oeschger, H., Schwander, J., Stauffer, B. and Zumbrunn, R. (1982). Ice core sample measurements give atmospheric CO_2 content during the past 40 000 yr. *Nature* **295**, 220–223.

Neuffer, B. and Hurka, H. (1986). Variation of development time until flowering in natural populations of *Capsella bursa-pastoris* (Cruciferae). *Plant Syst. Evol.* **152**, 277–296.

Olff, H. (1992). On the mechanisms of vegetation succession. Thesis, State University, Groningen.

Posthuma, L., Hogervorst, R.F. and van Straalen, N.M. (1992). Adaptation to soil pollution by cadmium excretion in natural populations of *Orchesella cincta* (L.) (Collembola). *Arch. Environ. Contam. Toxicol.* **22**, 146–156.

Pott, R. (1985). Vegetationsgeschichtliche und pflanzensoziologische Untersuchungen zur Niederwaldwirtschaft in Westfalen. *Abhandl. Westfälisches Museum Naturkde* **47**, 1–75.

Roose, M.L. (1991). Genetics of response to atmospheric pollutants. In: *Ecological Genetics and Air Pollution,* pp. 111–126, eds. G.E. Taylor, L.F. Pitelka and M.T. Clegg. New York: Springer Verlag.

Rosenzweig, M.L. (1971). Paradox of enrichment: destabilization of exploitation ecosystems in ecological time. *Science* **171**, 385–387.

Rozema, J., Rozema-Dijst, E., Freijsen, A.H.J. and Huber, J.J.L. (1978). Population differentiation within *Festuca rubra* L. with regard to soil salinity and soil water. *Oecologia* **34**, 329–341.

Rozema, J., Dorel, F., Janissen, R.K., Lenssen, G.M., Broekman, R.A., Arp, W.J. and Drake, B.G. (1991). Effect of elevated atmospheric CO_2 on growth, photosynthesis and water relations of salt marsh grass species. *Aquat. Biol.* **38**, 45–55.

Sasse, F. (1976). *Ökophysiologische Untersuchungen der Serpentinvegetation in Frankreich, Italien, Österreich und Deutschland.* Dissertation, University of Münster.

Schleser, G.H. (1990). Investigations of the $\delta^{13}C$ pattern in leaves of *Fagus sylvatica* L. *J. Exp. Bot.* **41**, 565–572.

Schleser, G.H. (1991). Carbon isotope fractionation during CO_2 fixation by plants. In: *Modern Ecology. Basic and Applied Aspects*, pp. 603–622, eds. G. Esser and D. Overdieck. Amsterdam: Elsevier.

Schraudner, M., Ernst, D., Langebartels, C. and Sandermann, H. (1992). Biochemical plant responses to ozone. III. Activation of the defense-related proteins β-1,3-glucanase and chitinase in tobacco leaves. *Plant Physiol.* **99**, 1321–1328.

Schuster, W.S.F., Phillips, S.L., Sandquist, D.R. and Ehleringer, J.R. (1992). Heritability of carbon isotope discrimination in *Gutierrezia microcephala* (Asteraceae). *Am. J. Bot.* **79**, 216–221.

Schütte, F. (1957). Untersuchungen über die Populationsdynamik des Eichen-wicklers (*Tortrix viridiana* L.). *Z. Angew. Entomol.* **40**, 1–36.

Schwanitz, F. and Hahn, H. (1954). Genetisch-entwicklungsphysiologische Untersuchungen in Galmeipflanzen. *Z Bot.* **42**, 79–190.

Shinozaki, K. and Kira, T. (1956). Intraspecific competition among higher plants. VII. Logistic theory of the C–D effect. *J. Inst. Polytechnics, Osaka City Univ.* Ser. D7, 35–72.

Smedley, M.P., Dawson, T.E., Comstock, J.P., Donovan, L.A., Sherrill, D.E., Cook, C.S. and Ehleringer, J.R. (1991). Seasonal carbon isotope discrimination in a grassland community. *Oecologia* **85**, 314–320.

Stanton, M.L. (1984). Seed variation in wild radish: effect of seed size on components of seedling and adult fitness. *Ecology* **65**, 1105–1112.

Stauffer, B., Fischer, G., Neftel, A. and Oeschger, H. (1985). Increase of atmospheric methane recorded in antarctic ice core. *Science* **229**, 1386–1388.

Steponkus, P.L. (1990). Cold acclimation and freezing injury from a per-spective of the plasma membrane. In: *Environmental Injury to Plants*, pp. 1–16, ed. F. Katterman. San Diego, Calif.: Academic Press.

Strain, B.R. (1991). Possible genetic effects of continually increasing atmo-spheric CO_2. In: *Ecological Genetics and Air Pollution*, pp. 237–244, eds. G.E. Taylor, L.F. Pitelka and M.T. Clegg. New York: Springer Verlag.

Stuyfzand, P.J. (1991). *De Samenstelling van Regenwater Langs Hollands Kust.* Kiwa report SWE 91.010. Rijswijk: KIWA.

Tamm, C.O. (1968). The evolution of forest fertilization in European silviculture. In: *Forest Fertilization: Theory and Practice*, pp. 242–247. Muscle Shoals, Ala.: Tennessee Valley Authority.

Tax, M.H. (1989). *Atlas van Nederlandse Dagvlinders.* 's-Graveland: Vereniging tot Behoud van Natuurmonumenten in Nederland.

Taylor, G.E. and Murdy, W.H. (1975). Population differentiation of an annual plant species, *Geranium carolinianum* L., in response to sulphur dioxide. *Bot. Gaz.* **136**, 212–215.

Tilman, D. (1988). *Plant Strategies and the Dynamics and Structure of Plant Communities.* Princeton, NJ: Princeton University Press.

Tolsma, D.J. (1989). On the ecology of savanna ecosystems in south-eastern Botswana. Thesis, Vrije Universiteit, Amsterdam.

Van Andel, J., Rozijn, N.A.M.G., Ernst, W.H.O. and Nelissen, H.J.M.

(1986). Variability in growth and reproduction in F1-families of an *Erophila verna* population. *Oecologia* **69**, 79–85.

Van Baalen, J. (1982). Germination ecology and seed population dynamics of *Digitalis purpurea*. *Oecologia* **53**, 61–67.

Van Baalen, J. and Prins, E.G.M. (1983). Growth and reproduction of *Digitalis purpurea* in different stages of succession. *Oecologia* **58**, 84–91.

Van Breemen, N., Burrough, P.A., Velthorst, E.J., van Dobben, H.J., De Wit, T., Ridder, T.B. and Reijders, H.F.R. (1982). Soil acidification from atmospheric ammonium sulphate in forest canopy throughfall. *Nature* **299**, 548–550.

Van Capelleveen, H.E. (1985). *The Ecotoxicity of Zinc and Cadmium for Terrestrial Isopods*. Dissertation, Vrije Universiteit, Amsterdam.

Van der Eerden, L. (1992). Fertilizing effects of atmospheric ammonia on semi-natural vegetations. Thesis, Vrije Universiteit, Amsterdam.

Van der Meijden, R., Plate, C.L. and Weeda, E.J. (1989). *Atlas van de Nederlandse Flora. 3. Minder Zeldzame en Algemene Soorten*. Leiden: Rijksherbarium/Hortus Botanicus.

Van der Meijden, E., Klinkhamer, P.G.L., De Jong, T.J. and van Wijk, C.A.M. (1992). Metal population dynamics of biennial plants: how to exploit temporary habitats. *Acta Bot. Neerl.* **41**, 249–270.

Verkleij, J.A.C. and Prast, J.E. (1989). Cadmium tolerance and co-tolerance in *Silene vulgaris* (Moench) Garcke (= *S.cucubalus* (L.) Wib.). *New Phytol.* **111**, 637–645.

Verkleij, J.A.C., Bast-Cramer, W.B. and Levering, H. (1985). Effects of heavy metal stress on the genetic structure of populations of *Silene cucubalus*. In: *Structure and Functioning of Plant Populations*, pp. 355–365, eds. J. Haeck and J.W. Woldendorp. Amsterdam: North-Holland/Elsevier.

Von Liebig, J. (1840). *Chemistry in Agriculture and Physiology*. London: Taylor and Woltan.

Watkinson, A.R. (1984). Yield–density relationship: the influence of resource availability on growth and self-thinning in populations of *Vulpia fasciculata*. *Ann. Bot.* **53**, 469–482.

Woodward, F.J. (1987). Stomatal members are sensitive to increases in CO_2 from pre-industrial levels. *Nature* **327**, 617–618.

Wookey, P.A. and Ineson, P. (1991). Chemical changes in decomposing forest litter in response to atmospheric sulphur dioxide. *J. Soil Sci.* **42**, 615–628.

Wulff, A.R. and Alexander, H.M. (1985). Intraspecific variation in the response to CO_2 enrichment in seeds and seedlings of *Plantago lanceolata* L. *Oecologia* **66**, 458–460.

CHAPTER 3

Stress, competition, resource dynamics and vegetation processes

J. P. GRIME

3.1. Introduction

Many environmental factors temporarily or consistently constrain the growth, development and reproduction of plants. Viewed at the biochemical level or from the perspective of contemporary evolving populations, evolutionary responses to environmental stress present an apparently limitless array of adaptations. The challenge to ecologists is not to be mesmerized by this surface complexity but to classify it in a useful way that distinguishes primary recurring stress adaptations (relevant to general ecological theory) from particular stress adaptations (peculiar to local circumstances, taxa and populations). This plea for an assertion of priorities in the analysis of stress adaptations can be made as part of a broader argument (Grime, 1989) for the development and testing of high-level ecological generalizations. If the science of plant ecology is to become an effective contributor to contemporary debates about ecosystem sustainability and the impacts of land use and climatic change on the world's resources, we will require rapid progress towards the kinds of general theory which have already provided the unifying and predictive power of the physical sciences.

This chapter reviews progress in the classification of stress adaptations for ecological purposes. It is suggested that there are three fundamentally different responses to environmental stress and that these are an integral feature of three primary functional types, recognition of which provides a key to the understanding of plant community structure and vegetation dynamics. Crucial in this analysis is the role of stress adaptations in mechanisms of resource foraging, in competition and in the circulation of mineral nutrients within ecosystems.

3.2. Stress adaptations and primary functional types

How should we seek to establish a primary classification of stress adaptations in plants? One approach (Levitt, 1975) is to develop a framework based on the identity of the stress (low temperature, anoxia, moisture stress, low nitrogen, etc.). However, as Levitt recognized, this approach is not readily translated into a succinct typology of stress adaptations; this is because responses to particular stresses and their ecological consequences can vary according to the intensity of stress, the maturity of the plant, other circumstances prevailing in the habitat and, often most important of all, the evolutionary history and life-cycle of the species or population. This latter complication is amply illustrated by reference to the diversity of responses observed when plants differing in ecology are subjected to identical conditions of drought: some plants cease growth and tolerate desiccation, others show rapid changes in root:shoot allocation patterns, and still others (usually ephemerals) set seed early and die prematurely.

If stress responses vary according to evolutionary history, life-cycle and habitat, can such characteristics provide a basis for classifying and predicting stress adaptations? Are stress responses determined less by the chemical and physical nature of the stresses than by the kinds of plant experiencing them?

There is strong evidence suggesting that these questions should be answered in the affirmative, and Table 3.1 illustrates an early attempt to classify stress adaptations by reference to plant type.

The first type of stress adaptation recognized in Table 3.1 occurs in plants attuned to conditions of high productivity and low vegetation disturbance. Here the growing season is marked by the rapid accumulation of plant biomass above and below ground, and the fast-growing perennials that enjoy a selective advantage under such conditions quickly develop a dense leaf canopy and root network. Resources are rapidly intercepted by the plant and its neighbours and local stresses arise as depletion zones (shade above ground, mineral nutrient shortages around individual roots). Table 3.1 predicts that, in such circumstances, morphological plasticity will be the dominant form of stress response. This will be manifested as the attenuation of shoots and roots, and will have the effect of projecting the absorptive surfaces of the plant out of the depletion zones. Such responses are usually allied to short life-spans in individual leaves and roots, with the consequence that a continuous relocation of the effective leaf and root surfaces ('active foraging' *sensu* Grime, 1979) occurs throughout the growing season.

The second contingency in Table 3.1 considers the stress response likely to characterize plants (e.g. arable weeds) exploiting a productive but temporary habitat. Here, as first proposed by Salisbury (1942), stresses, whether imposed directly by the environment or by crowding by neighbours, are likely to be governed by a 'reproductive imperative' such that production of offspring tends to be maintained under stress, even at the cost of extreme vegetative stunting and accelerated mortality of the whole plant (Harper and Ogden, 1970; Hickman, 1975; Boot *et al.*, 1986).

The third form of stress adaptation proposed in Table 3.1 coincides with extremely low productivity arising from chronic mineral nutrient stress due

TABLE 3.1. *Morphogenetic responses to desiccation, shading, or mineral nutrient stress of competitive, stress-tolerant and ruderal plants and their ecological consequences in three types of habitat.*

Strategy	Response to stress	Consequences		
		Habitat 1*	Habitat 2†	Habitat 3‡
Competitive	Large and rapid changes in root:shoot ratio, leaf area and root surface area	Tendency to sustain high rates of uptake of water and mineral nutrients to maintain dry-matter production under stress and to succeed in competition	Tendency to exhaust reserves of water and/or mineral nutrients both in the rhizosphere and within the plant; aetiolation in response to shade increases susceptibility to fungal attack	Failure rapidly to produce seeds reduces chance of rehabilitation after disturbance
Stress tolerant	Changes in morphology slow and often small in magnitude	Overgrown by competitors	Conservative utilization of water, mineral nutrients and photosynthate allows survival over long periods in which little dry-matter production is possible	Failure rapidly to produce seeds reduces chance of rehabilitation after disturbance
Ruderal	Rapid curtailment of vegetative growth and diversion of resources into seed production	Overgrown by competitors	Chronically low seed production fails to compensate for high rate of mortality	Rapid production of seeds ensures rehabilitation after disturbance

*In the early successional stages of productive, undisturbed habitats. Stresses are mainly plant induced and coinciding with competition.
†In either continuously unproductive habitats (stresses being more or less constant and due to unfavourable climate and/or soil) or the late stages of succession in productive habitats.
‡In severely disturbed, potentially productive habitats. Stresses are either a prelude to disturbance (e.g. moisture stress preceding drought fatalities) or plant induced between periods of disturbance.

to absolute shortage (primary succession on immature soils) or sequestration and tight recycling within the vegetation (mature vegetation on poor soils). Under such conditions there is little seasonal change in biomass. Leaves and roots often have a functional life of several years, and resource capture is usually uncoupled from growth. Because of the slow turnover of plant parts, differentiating cells occupy a small proportion of the biomass, and morphogenetic changes are less likely to provide a viable mechanism of stress response. In these circumstances, where the same tissues experience a sequence of different climatic stresses as the seasons pass, the dominant form of stress response is likely to be cellular acclimation, whereby the functional characteristics and 'hardiness' of the tissues change rapidly through biochemical adjustments of membranes and organelles. These changes are reversible and can occur extremely rapidly in certain species (Hosakawa *et al.*, 1964; Mooney and West, 1964; Strain and Chase, 1966; Bjorkman, 1968; Mooney, 1972; Oechel and Collins, 1973; Larsen and Kershaw, 1975; Taylor and Pearcy, 1976).

3.3. Interaction of root and shoot responses to stress

The ecological classification of stress adaptations and responses outlined in Table 3.1 is supported by an extensive literature which has been reviewed elsewhere (Grime, 1977, 1985; Grime *et al.*, 1988). Before describing the role of these stress adaptations in vegetation processes, it is necessary to consider how the stress responses of roots and shoots will interact and vary in their interaction according to the ecological and evolutionary background of the plant.

3.3.1. Two hypotheses

In the majority of vascular plants, entirely different but essential resources are captured above and below ground and natural selection has resulted in apparently divergent paths of structural and functional specialization in shoots and roots. One of the tasks for plant ecologists is to progress beyond the beguiling simplicity of this generalization to an understanding of how root and shoot size, design, function and stress adaptations vary in relation to each other and between plants of contrasted ecology and distribution. Here there is no current consensus and two alternative schools of thought have emerged:

A. *The resource ratio hypothesis* (Newman, 1973, 1983; Iwasa and Rough-garden, 1984; Huston and Smith, 1987; Tilman, 1988, 1989). These authors propose that on an evolutionary time-scale, and within the life-span of an individual phenotype, trade-offs in allocation of captured resources between root and shoot will be of pivotal importance. Hence, where light is limiting shoots will be relatively large but allocation to roots will take precedence on dry or infertile soils.

B. *The root–shoot interdependence hypothesis* (Donald, 1958; Grime, 1973,

1979; Chapin, 1980; Field and Mooney, 1986; Aerts *et al.*, 1991; Campbell *et al.*, 1991). According to this interpretation short-term phenotypic adjustments in allocation between root and shoot will occur but, on an evolutionary time-scale, this trade-off will be severely constrained by the essential interdependence of root and shoot foraging activities.

Recently there have been several attempts to test these rival hypotheses experimentally. One feature of these studies has been a consistent falsification of hypothesis A (see for example Berendse and Elberse, 1989; Tilman and Cowan, 1989; Gleeson and Tilman, 1990; Olff *et al.*, 1990; Shipley and Peters, 1990; Aerts *et al.*, 1991; Campbell *et al.*, 1991). In two experimental studies (Campbell *et al.*, 1991; Campbell and Grime, 1992) and in a review paper (Grime, 1993) various threads of evidence, some of recent origin, others quite old and often neglected in contemporary models, are presented and found to be in strong support of hypothesis B.

In this section the objective is not to consider in exhaustive detail the evidence for interdependence and covariance between root and shoot function. After a brief restatement of some critical issues and evidence, the purpose is to examine the implications of hypothesis B for our understanding of how fundamental differences between plants in their responses to resource stress influence the capture, utilization and eventual release of resources and provide the primary mechanism controlling the spatial patterns and temporal dynamics of vegetation.

3.3.2. *Covariance between root and shoot in size, function and stress biology*

In laboratory experiments (e.g. Brouwer 1962a,b, 1963; Corré 1983a,b; Hunt and Nicholls, 1986), manipulation of mineral nutrient supply and irradiance has been shown to be capable of modifying root : shoot ratios to a profound extent, suggesting the existence of mechanisms that maintain a balance between photosynthesis and mineral nutrient capture. Therein lies the origin of hypothesis A, with its strong emphasis on trade-offs in allocation between root and shoot. It is pertinent to ask why such phenotypic plasticity evident in individual plants has not (as predicted by Bradshaw, 1965) been mirrored by genetic differences of a similar kind and detectable in interspecies comparisons in root : shoot ratio between plants of nutrient-poor soils and those of shaded habitats. Following Grime (1993), three explanations can be suggested for the weak explanatory power of hypothesis A:

1. The majority of habitats cannot be simply classified with respect to single limiting factors. Although shading, for example, is often a conspicuous feature of grasslands and woodlands, mineral nutrients also frequently limit plant production at particular sites.
2. Models such as those of Huston and Smith (1987) and Tilman (1988), and experiments such as those of Brouwer (1963) and Hunt and Nicholls (1986), involve circumstances in which resource depletion is imposed uniformly within the aerial or the rooting environment. These fail to simulate the 'patch' and 'pulse' phenomena which characterize resource

supply in natural environments. In particular, omission of the depletion zones that surround the root surfaces of fast-growing plants growing on fertile soils has led to serious underestimation of the expenditure of assimilate required to sustain the continuous process of root growth necessary to escape from the local but expanding zones of nutrient exhaustion (Bhat and Nye, 1973) which are an inescapable consequence of the physics of nutrient uptake (Nye and Tinker, 1977) from the rhizosphere.

3. There are biochemical limits to the trade-off between root and shoot. Autotrophy involves the assembly of components derived from *both* parts of the plant. Chemical analyses of plants reveal that the ratio of root- and shoot-derived elements remains relatively constant across a wide range of ecologies. Interdependence is further enforced by the carbon and energy demand of roots and the mineral nutrient demand of leaves. Scope for variation in root : shoot ratio across plant functional types is restricted by the fact that species with the capacity for high rates of photosynthesis and dry-matter production have higher concentrations of leaf nitrogen (Sharkey, 1985; Field and Mooney, 1986) and leaf phosphorus (Band and Grime, 1981) and in consequence are dependent on high rates of nutrient capture by the root system.

Evidence of strong covariance between root and shoot size and function has been obtained in a recent experiment (Campbell *et al.*, 1991) using two new techniques to assay the responses of the leaf canopies and root systems of a range of plants grown in isolation under standardized resource patchiness simulating aspects of the conditions experienced in a perennial community. Both techniques (Fig. 3.1) present patches of resource without the use of partitions or barriers, which could impede growth between patches. In both the root and the shoot assay, measurement was made of the partitioning of dry matter allocation between depleted and undepleted sectors imposed after an initial growth period in a uniform productive environment. The experiment involved eight common British herbaceous species, widely contrasted in morphology and ecology. To test the predictive value of the assays,

FIGURE 3.1 *Imposing standardized patches of resource depletion on growing shoot and root systems. Two new techniques have been developed to assay the 'resource-foraging' attributes of the leaf canopies and root systems of individual plants grown in isolation under standardized conditions simulating those experienced during competition. An important feature of both techniques is that plants are presented with standardized patches of resource depletion created without the use of partitions or barriers, which could impede growth between patches. (a, b) Light patches. (a) Section through a cone-shaped chamber for imposing partial shading on developing shoot systems. A transparent glass upper surface (G) is covered with filters (F) to produce standardized patches of shade. Fine struts (S) support leaves, and the chamber is supplied with compressed air (T). (b) View from above of the shading pattern: quadrants 1 and 3 are fully illuminated and quadrants 2 and 4 are shaded by filters. (c, d) Nutrient patches. (c) Section through a bowl used to*

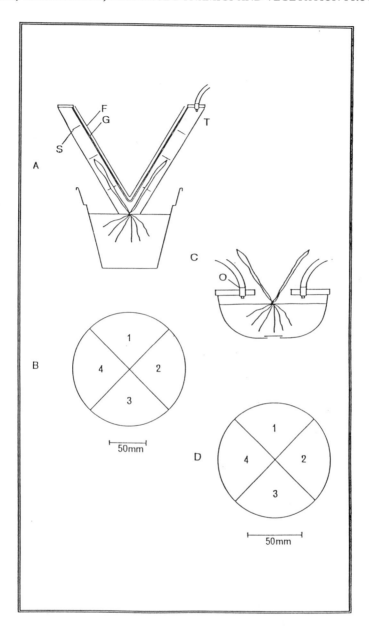

impose patches of nutrient depletion on developing root systems. Nutrient solution fed by peristaltic pumps was dripped continuously onto the surface at symmetrically arranged outlets (O). (d) View from above of the nutrient distribution pattern: quadrants 1 and 3 were supplied with nutrient-rich solution and quadrants 2 and 4 with nutrient-poor solution.

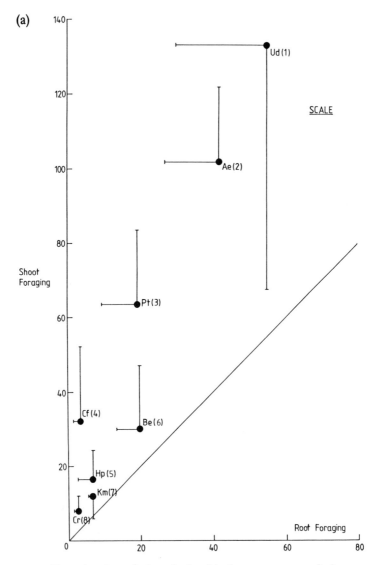

(a)

SCALE

Shoot Foraging

Root Foraging

FIGURE 3.2 *Examination of the relationship between root and shoot responses to resource heterogeneity in eight herbaceous species of contrasted ecology. A description of the methods used to expose the plants to resource patchiness is provided in Fig. 3.1. (a) The scales of foraging by the roots and shoots in the foraging assays are expressed as the respective increments of biomass (mg) to the undepleted quadrants. (b) The precision of foraging is estimated by calculating the respective increments as a percentage of the total increment to depleted and undepleted quadrants. Note that the increment to the unshaded quadrants by shoots can exceed 100% because of phototropic movements of shoots developed before introduction of the shaded quadrants. Ae, Arrhenatherum elatius; Be, Bromus erectus;*

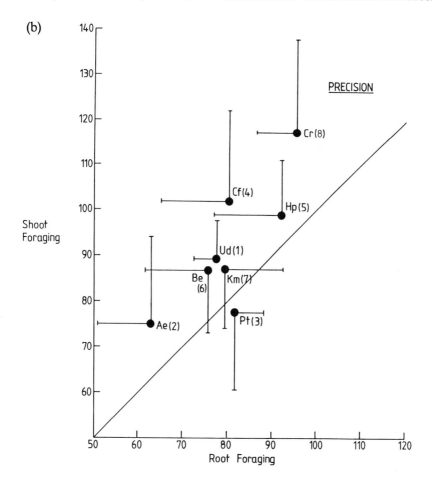

FIGURE 3.2 *Cont. Cf,* Cerastium fontanum; *Cr,* Campanula rotundifolia; *Hp,* Hypericum perforatum; *Km,* Koeleria macrantha; *Pt,* Poa trivialis; *Ud,* Urtica dioica. *Values in parentheses refer to the species ranking in a conventional competition experiment in which all eight species were grown together in an equiproportional mixture on fertile soil for 16 weeks.*

comparison was made with the status achieved in a conventional competition experiment in which the eight species were grown together in an equiproportional mixture under productive glasshouse conditions for 16 weeks. The results (Fig. 3.2) reveal a consistent relationship between the increment of dry matter to the undepleted sectors in both assays and the capacity for dominance in the competition experiment. It is also apparent that there is covariance between roots and shoots in their responses to resource patchiness. This relationship is maintained despite the consistent tendency for the scale and precision of leaf canopy adjustments to exceed those of the root system;

this, of course, is a predictable consequence of the relative freedom of movement of leaves and of the encasement of roots within a solid medium.

3.4. Resource capture in unproductive vegetation

Greatest uniformity in partitioning of captured resources between leaves and roots may be expected during the seedling phase in ephemerals and perennials of productive habitats. This is because these plants attain rapid seedling growth by immediate commitment of captured resources to new leaves and roots, and will therefore, at each point in time, tend to absorb and utilize carbon and mineral elements in proportions similar to those in which these elements occur in the plant tissues. The elemental composition of plant tissue would not be expected to exercise such a tight constraint upon root–shoot allometry in the slow-growing long-lived perennials of unproductive habitats; in these plants there may be substantial storage pools, resource capture is usually uncoupled from growth and, in the short term, there may be considerable independence between root and shoot functioning.

In contrast to the role of *spatial* patchiness of resources in productive habitats, there is evidence (Davy and Taylor, 1975; Gupta and Rorison, 1975; Taylor *et al.*, 1982) that *temporal* variation in resource capture is of key importance in unproductive habitats. This applies particularly to phosphorus and nitrogen, which tend to be sequestered in the living biomass of flowering plants and microorganisms or in litter as relicts of the antiherbivore defences that protect the living tissues of many of the plants of nutrient-limited vegetation (Grime 1988; Heal and Grime, 1991). Here opportunities for mineral nutrient capture may depend on disruptive events such as drying–wetting cycles and episodes of freeze and thaw. The frequency and duration of the pulses of nutrient release on infertile soils are suspected to be a critical determinant of plant success in unproductive vegetation (Grime *et al.*, 1986; Jackson and Caldwell, 1989). Soil microbial populations are capable of rapid response to sudden enrichment (Shields *et al.*, 1973; Ritz and Griffiths, 1987) and, in consequence, some pulses may be too infrequent and too short to be exploited effectively by the potentially dynamic but relatively short-lived roots of plants normally associated with fertile soils. In two recent experiments (Crick and Grime, 1987; Campbell and Grime, 1989), plants characteristic of fertile and infertile soils have been compared with respect to their ability to capture nitrogen from pulses of various durations. The results (Fig. 3.3) indicate that where pulses are short an advantage is enjoyed by slow-growing species of infertile soils; these plants appear to have low rates of tissue turnover and to have roots that remain functional despite chronic exposure to mineral nutrient stress.

3.5. The fate of captured resources

In this chapter it has been argued that plants of productive vegetation are capable of high rates of resource capture through foraging mechanisms that

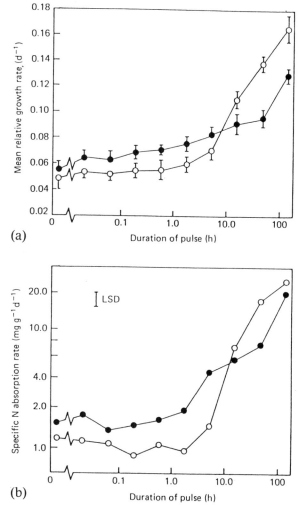

FIGURE 3.3 *(a) Mean relative growth rate of* Arrhenatherum elatius *(○) and* Festuca ovina *(●) plants exposed once every 6 days to pulses of nutrient enrichment of differing durations. Vertical bars are logarithmic s.d. (P > 0.05) for comparing means on a logarithmic scale. (b) Mean specific nitrogen absorption rate of* Arrhenatherum elatius *(○) and* Festuca ovina *(●) plants exposed once every 6 days to pulses of nutrient enrichment of differing duration. Means are shown ± 95% confidence limits.*

allow swift spatial readjustments of the leaf canopy and root system in response to the rapidly changing resource mosaics created by fast-growing plants. It is further argued that these mechanisms involve high rates of organ and tissue turnover and rapid release of captured carbon, energy and mineral nutrients.

More conservative mechanisms of resource capture and utilization are proposed for the plants of unproductive vegetation. Here it is suggested that the effect of natural selection has been to promote an uncoupling of resource capture from growth, to develop internal storage pools, to protect captured resources through antiherbivore defences and to develop 'sit and wait' foraging mechanisms involving pulse interception through long-lived surfaces.

It is immediately obvious that if such radically different mechanisms of resource capture, utilization and release distinguish the plants of productive and unproductive vegetation there will be profound implications for the control of vegetation dynamics and spatial patterns. These will now be examined.

3.6. Stress adaptations and vegetation processes

In preceding sections, it has been argued that three fundamentally different stress adaptations can be recognized in plants and that these not only help to define plant functional types but also have profound implications for the way in which resources are captured, utilized and released. Since resources are the essential currency of the functioning of communities and ecosystems it seems inevitable that eventually a typology of stress adaptations will find its place in general ecological theory.

Until recently, theories of vegetation succession and cyclical change in terrestrial ecosystems have been dominated by models relating to plant life-cycles, life-forms and mechanisms of regeneration (Clements, 1916; Egler, 1954; Drury and Nisbet, 1973; Connell and Slatyer, 1977; Whitmore, 1982; Finegan, 1984). However, implicit in all these models is the notion that competition for resources also plays an important part, and some ecologists have made explicit predictions concerning the role of resource dynamics in vegetation processes (Odum, 1969; Grime, 1977, 1987; Leps et al., 1982; Tilman, 1988; Redente et al., 1992; Colasanti and Grime, 1992).

In Fig. 3.4, various vegetation processes have been represented within the theoretical framework provided by the triangular array of primary functional types and associated stress adaptations. In each diagram the strategies and prevailing stress adaptations (*sensu* Table 3.1) of the dominant plants at particular points in time are indicated by the position of arrowed lines within the triangular model. The passage of time in years during succession is represented by numbers on each line, and shoot biomass at particular points is reflected in the size of the circles.

Fig. 3.4f depicts the course of secondary succession in a forest clearing situated on a moderately fertile soil in a temperate climate. Biomass development is initially rapid and there is a fairly swift replacement of species as the community experiences successive phases of competitive dominance by

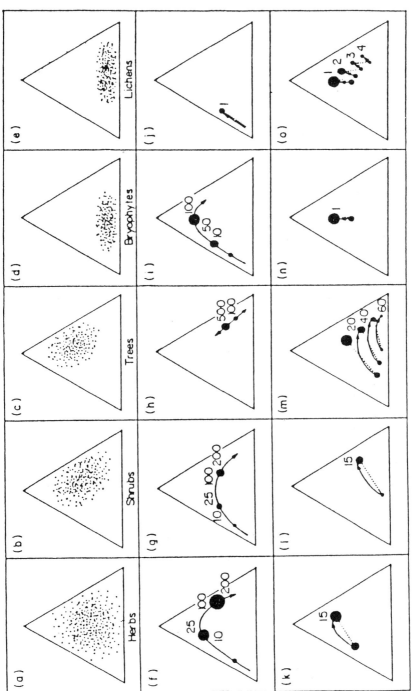

FIGURE 3.4 Models describing the strategic range of five life forms (a–e) and representing various successional phenomena (f–o). For a description of the succession diagrams see text.

rapidly growing herbs, shrubs and trees. At a later stage, however, the course of succession begins to deflect downwards towards the stress-tolerant corner of the triangular model. This process begins even during stages where the plant biomass is expanding appreciably, and reflects a change from vegetation exhibiting high rates of resource capture and loss with morphological plasticity characterizing the stress responses to one in which resources, particularly mineral nutrients, are efficiently retained in a biomass now mainly consisting of long-lived tissues in which cellular acclimation predominates as the stress response.

In Fig. 3.4g, secondary succession is described for a site of lower soil fertility. Here the course of events is essentially the same as that described in Fig. 3.4f except that the successional parabola is shallower and the plant biomass smaller as a consequence of the earlier onset of mineral nutrient limitation of the initial phase of competitive dominance.

Primary succession in a skeletal habitat such as rock outcrop is represented in Fig. 3.4h. In this case the initial colonists are stress tolerators of low biomass (lichens and bryophytes with well developed mechanisms of cellular acclimation). These plants occupy the site for a considerable period, giving way eventually to small slow-growing herbs and shrubs. This sequence coincides with the process of soil formation and provides an example of the facilitation model of vegetation succession (Connell and Slatyer, 1977).

In the examples so far considered, plant succession has been interpreted mainly as an interaction between resource availability and the characteristics of established plants. Greater sophistication can be introduced into the models by including circumstances where ineffective seed dispersal or failure in seedling establishment limits access of potential dominants into successional pathways. Fig. 3.4i, for example, describes the common situation (Niering and Goodwin, 1962; Kochummen and Ng, 1977) in which the development of a dense herbaceous cover before the arrival of the propagules of woody species prevents their establishment and strongly delays succession.

The range of models can be extended further to include loops representing vegetation responses to major perturbations. These include the truncation of succession by annual harvesting and fertilizer input in an arable field, a process which perpetuates occupation by ephemerals and sustains the reproductive imperative as the dominant stress response (Fig. 3.4j). Further examples include the cycles of vegetation change associated with coppicing *Fraxinus excelsior* woodland (Fig. 3.4k) or rotational burning of *Calluna vulgaris* moorland (Fig. 3.4l). Where the vegetation of moderately unproductive habitats is subjected to repeated cycles of destruction by burning, browsing or cropping (Fig. 3.4m), the declining mineral nutrient capital of the soil may be expected to bring about a series of arcs of progressively lower trajectory in successive cycles of vegetation recovery.

Models can also be drawn to describe vegetation processes and shifts in prevailing stress adaptations in agricultural grasslands. Fig. 3.4n describes a productive fertilized meadow in which the expansion of competitive dominants is restricted by annual mowing. In Fig. 3.4o we see the sequence of events where the meadow is not fertilized and there is a drift towards lower productivity and incursion by stress-tolerant species.

3.7. Conclusions

Although it remains necessary to investigate the stress adaptations of plants at all levels of organization from the ecosystem to the molecular, there is a specific need to identify major types of stress response which recur in association with particular evolutionary histories, life-cycles and habitats. Recognition of three fundamentally different stress responses contributes to the definition of primary functional types in plants and can provide useful insight into the capture, utilization and release of resources. This, in turn, is providing a basis for an understanding of the mechanisms whereby resource dynamics influence the structure and functioning of vegetation.

ACKNOWLEDGEMENTS

The research on which this paper is based was supported by the Natural Environment Research Council and draws on collaborative studies with colleagues at UCPE.

REFERENCES

Aerts, R., Boot, R.G.A. and van der Aart, P.J.M. (1991). The relation between above- and below-ground biomass allocation patterns and competitive ability. *Oecologia* **87**, 551–559.

Band, S.R. and Grime J.P. (1981). Chemical composition of leaves. *Annual Report 1981*, pp. 6–8. Sheffield: University of Sheffield, Unit of Comparative Plant Ecology (NERC).

Berendse, F. and Elberse, W.T. (1989). Competition and nutrient losses from the plant. In: *Causes and Consequences of Variation in Growth Rate and Productivity of Higher Plants*, pp. 269–284, eds. H. Lambers *et al*. The Hague: SPB Academic Publishing.

Bhat, K.K.S. and Nye, P.H. (1973). Diffusion of phosphate to plant roots in soil. 1. Quantitative autoradiography of the depletion zone. *Plant Soil* **38**, 161–175.

Bjorkman, O. (1968). Carboxydismutase activity in shade-adapted and sun-adapted species of higher plants. *Physiol. Plant.* **21**, 1–10.

Boot, R.G.A, Raynal, D.J. and Grime, J.P. (1986). A comparative study of the influence of drought stress on flowering in *Urtica dioica* and *U. urens*. *J. Ecol.* **74**, 485–495.

Bradshaw, A.D. (1965). Evolutionary significance of phenotypic plasticity in plants. *Adv. Genet.* **13**, 115–155.

Brouwer, R. (1962a). Distribution of dry matter in the plant. *Neth. J. Agric. Sci.* **10**, 361–376.

Brouwer, R. (1962b). Nutritive influences on the distribution of dry matter in the plant. *Neth. J. Agric. Sci.* **10**, 399–408.

Brouwer, R. (1963). Some aspects of the equilibrium between overground and underground plant parts. *Jaarboek Instituut voor Biologisch en Scheikundig Onderzoek van Landbouwgewassen*, pp. 31–39.

Campbell, B.D. and Grime, J.P. (1989). A comparative study of plant responsiveness to the duration of episodes of mineral nutrient enrichment. *New Phytol.* **112**, 261–267.

Campbell, B.D. and Grime, J.P. (1992). An experimental test of plant strategy theory. *Ecology* **73**, 15–29.

Campbell, B.D., Grime, J.P. and Mackey, J.M.L. (1991). A trade-off between scale and precision in resource foraging. *Oecologia* **87**, 532–538.

Chapin, F.S. III (1980). The mineral nutrition of wild plants. *Annu. Rev. Ecol. Syst.* **11**, 233–260.

Clements, F.E. (1916). *Plant Succession: An Analysis of the Development of Vegetation*. Washington, DC: Carnegie Institute.

Colasanti, R.L. and Grime, J.P. (1992). Resource dynamics and vegetation processes: a deterministic model using two dimensional cellular automation. *Functional. Ecol.* **7**, 169–176.

Connell, J.H. and Slatyer, R.O. (1977). Mechanisms of succession in natural communities and their role in community stability and organization. *Am. Naturalist* **111**, 1119–1145.

Corré, W.J. (1983a). Growth and morphogenesis of sun and shade plants. I. The influence of light intensity. *Acta Bot. Neerl.* **32**, 49–62.

Corré, W.J. (1983b). Growth and morphogenesis of sun and shade plants. III. The combined effects of light intensity and nutrient supply. *Acta Bot. Neerl.* **32**, 277–294.

Crick, J.C. and Grime, J.P. (1987). Morphological plasticity and mineral nutrient capture in two herbaceous species of contrasted ecology. *New Phytol.* **107**, 403–414.

Davy, A.J. and Taylor, K. (1975). Seasonal changes in the inorganic nutrient concentrations in *Deschampsia caespitosa* (L.) Beauv. in relation to its tolerance of contrasting soils in the Chiltern Hills. *J. Ecol.* **63**, 27–39.

Donald, C.M. (1958). The interaction of competition for light and for nutrients. *Aust. J. Agric. Res.* **9**, 421–432.

Drury, W.H. and Nisbet, I.C.T. (1973). Succession. *Journal of the Arnold Arboretum, Harvard University* **54**, 331–368.

Egler, F.E. (1954). Vegetation science concepts. I. Initial floristic composition, a factor in old-field vegetation development. *Vegetatio* **4**, 412–417.

Field, C. and Mooney, H.A. (1986). The photosynthesis–nitrogen relationship in world plants. In: *On the Economy of Plant Form and Function*, pp. 25–55, ed. T.V. Givnish. Cambridge: Cambridge University Press.

Finegan, B. (1984). Forest succession. *Nature* **312**, 109–114.

Gleeson, S.K. and Tilman, D. (1990). Allocation and the transient dynamics of succession on poor soils. *Ecology* **71**, 1144–1155.

Grime, J.P. (1973). Competitive exclusion in herbaceous vegetation. *Nature* **242**, 344–347.

Grime, J.P. (1977). Evidence for the existence of three primary strategies in plants and its relevance to ecological and evolutionary theory. *Am. Naturalist* **111**, 1169–1194.

Grime, J.P. (1979). *Plant Strategies and Vegetation Processes*. Chichester: Wiley.

Grime, J.P. (1985). The C–S–R model of primary plant strategies: origins, implications and tests. In: *Plant Evolutionary Biology*, eds. L.D. Gottlieb and S.K. Jain. London: Chapman & Hall.

Grime, J.P. (1987). Dominant and subordinate components of plant communities: implications for succession, stability and diversity. In: *Colonization, Succession and Stability*, pp. 413–428, eds. A.J. Gray, M.J. Crawley and P.J. Edwards. Oxford: Blackwell.

Grime, J.P. (1988). Fungal strategies in ecological perspective. *Proc. R. Soc. Edinb.* **B94**, 167–169.

Grime, J.P. (1989). The stress debate: symptom of impending synthesis? *Biol. J. Linn. Soc.* **37**, 3–17.

Grime, J.P. (1993). The role of plasticity in exploiting environmental heterogeneity. In Exploitation of Environmental Heterogeneity by Plants: Ecophysiological Processes Above and Below Ground, eds. M. Caldwell and R.W. Pearcy. San Diego, Academic Press.

Grime, J.P., Crick, J.C. and Rincon, J.E. (1986). The ecological significance of plasticity. In: *Plasticity in Plants*, pp. 5–19, eds. D.H. Jennings and A.J. Trewavas. Cambridge: Company of Biologists.

Grime, J.P., Hodgson, J.G. and Hunt, R. (1988). *Comparative Plant Ecology: A Functional Approach to Common British Species*. London: Unwin Hyman.

Gupta, P.L. and Rorison, I.H. (1975). Seasonal differences in the availability of nutrients down a podzolic profile. *J. Ecol.* **63**, 521–534.

Harper, J.L. and Ogden, J. (1970). The reproductive strategy of higher plants. I. The concept of strategy with special reference to *Senecio vulgaris L. J. Ecol.* **58**, 681–698.

Heal, O.W. and Grime, J.P. (1991). Comparative analysis of ecosystems: past lessons and future directions. In: *Comparative Analyses of Ecosystems: Patterns, Mechanisms and Theories*, pp. 7–23. New York: Springer-Verlag.

Hickman, J.C. (1975): Environmental unpredictability and plastic energy allocation strategies in the annual *Polygonum cascadense* (Polygonaceae). *J. Ecol.* **63**, 689–701.

Hosakawa, T., Odani, H. and Tagawa, H. (1964). Causality of the distribution of corticolous species in forests with special reference to the physiological approach. *Bryologist* **67**, 396–411.

Hunt, R. and Nicholls, A.O. (1986). Stress and the coarse control of root-shoot partitioning in herbaceous plants. *Oikos* **47**, 149–158.

Huston, M.A. and Smith, T.M. (1987). Plant succession: life history and competition. *Am. Naturalist* **130**, 168–198.

Iwasa, Y. and Roughgarden, J. (1984). Shoot/root balance of plants: optimal growth of a system with many vegetative organs. *Theor. Popul. Biol.* **25**, 78–104.

Jackson, R.B. and Caldwell, M.M. (1989). The timing and degree of root proliferation in fertile-soil microsites for three cold desert perennials. *Oecologia* **81**, 149–153.

Kochummen, K.M. and Ng, F.S.P. (1977). Arrest of tropical forest succession by dense under-storey vegetation. *Malaysian Forester* **40**, 61–78.

Larsen, D.W. and Kershaw, K.A. (1975). Acclimation in arctic lichens. *Nature* **254**, 421–423.

Leps, J.J., Osbornova-Kosinova, J. and Rejmanek, K. (1982). Community stability, complexity and species life-history strategies. *Vegetatio* **50**, 53–63.

Levitt, J. (1975). *Responses of Plants to Environmental Stresses*. p. 697. New York: Academic Press.

Mooney, H.A. (1972). The carbon balance of plants. *Annu. Rev. Ecol. Syst.* **3**, 315–346.

Mooney, H.A. and West, M. (1964). Photosynthetic acclimation of plants of diverse origin. *Am. J. Bot.* **51**, 825–827.

Newman, E.I. (1973). Competition and diversity in herbaceous vegetation. *Nature* **244**, 310.

Newman, E.I. (1983). Interactions between plants. Physiological plant ecology. III. Responses to the chemical and biological environment. In: *Encyclopedia of Plant Physiology*, New Series, Vol. **12C**, 697–710. Berlin: Springer-Verlag.

Niering, W.A. and Goodwin, R.H. (1962). Ecological studies in the Connecticut Arboretum Natural Area. I. Introduction and survey of vegetation types. *Ecology* **43**, 41–54.

Nye, P.H. and Tinker, P.B. (1977). *Solute Movement in the Soil–Root System*. Oxford: Blackwell Scientific.

Odum, E.P. (1969). The strategy of ecosystem development. *Science* **164**, 262–270.

Oechel, W.D. and Collins, N.J. (1973). Seasonal patterns of CO_2 exchange in bryophytes at Barrow, Alaska. In: *Primary Production and Production Processes*, eds. L.C. Bliss and F.E. Wielogolaski. Stockholm: Wenner–Gren Center.

Olff, H., Van Andel, J. and Bakker, J.P. (1990). Biomass and shoot/root allocation of five species from a grassland succession series at different combinations of light and nutrient supply. *Functional Ecol.* **4**, 193–200.

Redente, E.F., Friedlander, J.E. and McLendon, T. (1992). Response of early and late semiarid seral species to nitrogen and phosphorus gradients. *Plant Soil* **140**, 127–135.

Ritz, K. and Griffiths, B.S. (1987). Effects of carbon and nitrate additions to soil upon leaching of nitrate, microbial predators and nitrogen uptake by plants. *Plant Soil* **102**, 229–237.

Salisbury, E.J. (1942). *The Reproductive Capacity of Plants*, p. 244. London: Bell.

Sharkey, T.D. (1985). Photosynthesis in intact leaves of C_3 plants: physics, physiology and rate limitations. *Bot. Rev.* **51**, 53.

Shields, J.A., Paul, E.A. and Lowe, W.E. (1973). Turnover of microbial tissue in soil under field conditions. *Soil Biol. Biochem.* **5**, 753–764.

Shipley, B. and Peters, R.H. (1990). A test of the Tilman model of plant strategies: relative growth rate and biomass partitioning. *Am. Naturalist* **136**, 139–153.

Strain, B.R. and Chase, V.C. (1966). Effect of past and prevailing temperatures on the carbon dioxide exchange capacities of some woody desert perennials. *Ecology* **47**, 1043–1045.

Taylor, R.J. and Pearcy, R.W. (1976). Seasonal patterns in the CO_2 exchange characteristics of understorey plants from a deciduous forest. *Can. J. Bot.* **54**, 1094–1103.

Taylor, A.A., DeFelice, J. and Havill, D.C. (1982). Seasonal variation in nitrogen availability and utilization in an acidic and calcareous soil. *New Phytol.* **92**, 141–152.

Tilman, D. (1988). *Plant Strategies and the Structure and Dynamics of Plant Communities*. Princeton, NJ: Princeton University Press.

Tilman, D. (1989). Competition, nutrient reduction and the competitive neighbourhood of a bunchgrass. *Functional Ecol.* **3**, 215–219.

Tilman, D. and Cowan, M.L. (1989). Growth of old field herbs on a nitrogen gradient. *Functional Ecol.* **3**, 425–438.

Whitmore, T.C. (1982). On pattern and process in forests. In: *The Plant Community as a Working Mechanism*, pp. 45–60, ed. E.I. Newman. Oxford: Blackwell Scientific.

CHAPTER 4

What can models tell us?

M. REES and J.H. LAWTON

4.1. Introduction

Mathematical models are rather like democracy; they have many drawbacks, but nobody has invented anything better. Models force us to make our assumptions explicit. Then, given these assumptions, certain conclusions inevitably follow. One major reason for using models to understand the behaviour of, and to make predictions about, biological systems is that they are the only way, except intuition, that human beings have to explore the consequences of several interacting variables. Quantitative predictions about complex systems are impossible without some sort of explicit mathematical model. Yet there remains a remarkable undercurrent of hostility among some groups of ecologists to the use of models. The reasons for this hostility are many, but they include a belief, particularly by field workers, that models are too simple to capture anything close to the 'truth', and that only muddy-booted field biologists can fully appreciate the wonders of nature. We fundamentally disagree! In an unusual study, Caswell (1988) evaluated empirical and theoretical studies presented at a large, annual conference of ecologists. He counted how many potentially predictive factors were included in each study and found that mathematical ecologists included more factors in their models than field workers did in their experiments. It is the theoreticians who have the more complex view of nature.

Mathematical models are, however, simplified abstractions of real systems, and they can take many forms (Watt, 1968; Jeffers, 1982; Nisbet and Gurney, 1982; Curry and Feldman, 1987; Pacala, 1989). Three useful descriptors are realism, generality and precision (Levins, 1968). A model cannot simultaneously maximize realism, generality and precision, but must make compromises between them. For instance, the Lotka–Volterra competition equations found in any undergraduate ecology text are precise (they are formulated as equations, with specified parameters), general (they are broadly applicable to any pair of competing species), but not very realistic (it would be unwise to use them to predict the trajectories of two competing populations in the real world, or the time to extinction of the inferior competitor). But this lack of realism is not critical if the point of the exercise is to gain general understanding; in this case we learn from the equations that two competing species will coexist in stable equilibrium if, and only if, each species depresses its own population growth more than it depresses the growth of its competitor. Detailed computer simulations designed to make predictions about particular systems, for example in pest or fisheries management, are precise and realistic (often so much so that the model itself is too complicated to understand!) but, because of their striving to capture the realities of a particular system, they lack generality. A final class of models uses qualitative, often graphical representations of interactions (Y is a concave function of X) to create general, relatively realistic but imprecise descriptions of nature (e.g. Levins, 1968).

One of the most successful exercises in theoretical population biology has been the development of general, precise, analytically tractable models of the interactions between parasitic insects (predominantly in the orders Hymenoptera and Diptera, and usually referred to as parasitoids to distinguish them from 'real' parasites) and their hosts (many other kinds of insects). This work has a pedigree extending back nearly sixty years (Nicholson, 1933; Nicholson and Bailey, 1935). Because the biology of the interaction between hosts and parasitoids is relatively simple, theoreticians have been able to construct families of models that capture the essence of these interactions without becoming hopelessly complicated. Through these models we have learned a great deal not only about the roles of direct and delayed density dependence in population regulation (Nicholson's original interest), but also about such varied and fundamental problems as the importance of spatial and other types of heterogeneity in population dynamics (Pacala *et al.*, 1990; Hassell *et al.*, 1991), and the consequences of two or more species sharing natural enemies (Holt and Lawton, 1993). Emerging insights have also been used to try to improve success rates in practical biological pest control, and are currently guiding a wide range of experimental laboratory and field studies designed to test theoretical predictions.

Populations of terrestrial higher plants have more complicated biologies than insect host–parasitoid systems, and present theoreticians with a much greater challenge. Yet despite a shorter pedigree, and the difficulties inherent in modelling terrestrial plant populations, considerable progress has been made. Indeed, headway in understanding and predicting aspects of plant population dynamics in the field has arguably been at least as successful as the efforts of animal ecologists dealing with biologically simpler host–parasitoid systems. The complexities of modelling plant population dynamics have been reviewed by numerous authors; particularly useful are discussions in Harper (1977), Crawley (1983) and Pacala (1989). Drawing their combined insights together, the particular difficulties facing would-be modellers of plant population dynamics include the following:

1. There are many different kinds of plants, making it impossible to construct a basic 'all-purpose' model. For example, for work on the dynamics of annuals, rosette-forming biennials and most trees, individuals can be recognized, counted and their dynamics modelled much as zoologists model the dynamics of *Daphnia* or *Drosophila* populations. Clonal plants, for example grasses, cannot be treated in this way at all, and require either very detailed models of the population dynamics of individual modules (buds, leaves, etc., e.g. Harper (1977), Sackville Hamilton *et al.* (1987)) or models of changes in biomass, rather than numbers (Tilman and Wedin, 1991a).

2. Even where individuals can be identified, the plastic growth of plants means that reproductive individuals may vary enormously in size, while the fecundities of individuals of the same size may differ by several orders of magnitude. Animal ecologists sometimes have to confront similar problems (e.g. with fish), but can often ignore them (adult cabbage-white butterflies are all more or less the same size).

3. Insect hosts and their parasitoids operate on similar time-scales (they have roughly similar generation times), and when a parasitoid finds and attacks a host it usually kills it, generating a predictable number of parasitoids in the next generation. This all makes modelling rather easy. Unfortunately, plants and herbivores often have markedly different generation times, herbivores themselves come in many different guises, and being found and eaten by a single herbivore does not usually kill the plant (for further discussions see Caughley and Lawton (1981)).

4. With rare exceptions, plants cannot move. Interactions between plants are therefore local, and their dynamics are inherently spatial; average population densities often do not mean very much (although under special circumstances – see below – average densities may suffice). Explicitly spatial models are much harder to deal with mathematically than models that ignore spatial processes.

None of these difficulties is insurmountable, and a number of recent advances in modelling plant population biology have explicitly incorporated one or more of these phenomena to develop new insights about plant dynamics. We see several exciting lines of investigation opening up. One approach is to bypass the complexities, think hard about essential processes that must be crucial for all plant dynamics, and focus on these critical processes. Work by Tilman (Tilman, 1982, 1988; Tilman and Wedin, 1991b) provides an excellent example, in which the complexities of interspecific exploitation competition for resources (nutrients) are reduced to a single measure, R^*, defined as the level of the critical limiting resource at which population growth is just balanced by mortality. The species with the lowest R^* inevitably wins in competition. Another approach is to grapple with the complexity, and construct mathematically difficult models that may no longer be tractable analytically, but require computer simulation to reveal their behaviour, without being so complex as to be incomprehensible (e.g. Pacala, 1987). A third important development lies in obtaining better parameter estimates for plant population models, by employing more powerful data-analysis techniques than have hitherto been available; good parameter estimation has led to some remarkably accurate predictions of population behaviour (Pacala and Silander, 1990).

We will illustrate some of these developments with selected examples, linking the insights that are emerging to the general problem of predicting the effects of stress on plant population dynamics. But, first, we need to define what we mean by stress, and how we might expect stress to modify and influence plant population dynamics.

4.2. Incorporating stress into plant population models

Definitions of 'stress' in the botanical literature vary. Grime (1979) views stress as 'the external constraints which limit the rate of dry matter production of all or part of the vegetation' (see also Chapter 3). Alternatively, stress can be caused by any environmental factor (abiotic or biotic) which is liable to cause a 'potentially injurious strain in plants' (Levitt, 1980). Biotic stresses include competition, herbivory and disease; abiotic stresses embrace extreme temperatures, water availability, radiation, chemicals, etc. Grime's definition focuses on the ecological consequences; Levitt is more concerned with physiological mechanisms. Both emphasize that many natural environments are suboptimal for plant performance, a view that is shared by Chapin (1991).

In the context of plant population models, stresses reduce individual growth rates, size and fecundity, and increase mortality. They may also change interactions with pathogens and herbivores (which are not only stresses in their own right, but also respond to changes in the plant induced by stress) (see Chapters 17 and 18). For instance, various abiotic stresses commonly lead to an increase in the concentrations of soluble amino acids and carbohydrates in plant foliage (Chapin, 1991), which in turn may increase the palatability of foliage for insect herbivores, generating potentially damaging pest outbreaks (for further discussions of these ideas, see White (1984), Larsson (1989) and McQuate and Connor (1990)).

Given that almost anything in a plant's environment can be stressful, it is not our intention to explore the role of models in predicting the population dynamic consequences of particular stresses. Numerous papers in this book explore the details of what particular stresses do to particular plants. Nor are we going to review plant–herbivore or plant–pathogen models, which are reasonably well understood. Rather, we want to show how plant population dynamics respond to changes in such things as individual growth rates, size structure and fecundity; that is, to aspects of plant performance that might be changed by abiotic stresses. Our examples are illustrative, not exhaustive. They summarize significant recent advances in modelling plant populations and show what models could tell us if we knew how a particular stress could alter plant performance.

4.3. Dynamics of single-species populations

We start with the simplest case of an annual plant. Theoretical studies have demonstrated that the essential properties of a population of a single annual plant species can be understood by constructing Ricker curves which relate seed density in year t to that in year $t + 1$ (May and Oster, 1976). The slope of the relationship between seed densities in successive years, evaluated at the equilibrium point, determines whether the population will be characterized by a stable equilibrium point, population cycles or chaos (Fig. 4.1). If the slope is between 0 and 1 the equilibrium point is approached smoothly, if it is between 0 and -1 we have damped oscillations and finally when the slope is less than -1 we have population cycles and chaos, depending on how steep the slope is. For plants, this type of model structure is only appropriate for habitats when there are ample microsites available for colonization in every year, such as in sand dunes, salt marshes and around arable field margins (e.g. Symonides, 1979; Watkinson and Davy, 1985).

Consider the simplest case of a population of annual plants with no seed dormancy. To predict the number of seeds next year (and then from next year to the year after that, and hence the long-term dynamics of the population) we need to know four things:

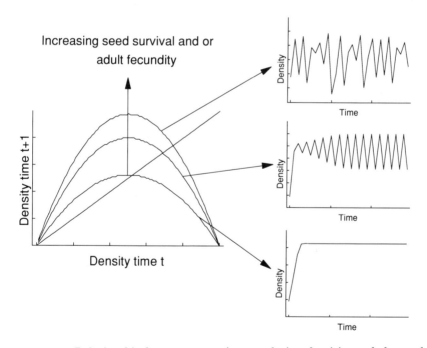

FIGURE 4.1 *Relationship between successive population densities and the resulting population dynamics. Note that in many plant populations the relationship is more or less flat topped, resulting in stable dynamics.*

1. How total biomass production is related to initial seed density
2. The distribution of biomass between individuals
3. The relationship between individual biomass and fecundity
4. The fraction of seeds that survive over winter

In the following paragraphs, we briefly review each of these phenomena to illustrate how they combine to yield simple models of annual plant population dynamics.

4.3.1. Biomass production and initial seed density

The relationship between total biomass and initial seed density is described by an empirical law – the law of constant yield – which states that total biomass production rapidly becomes independent of the density of seeds sown (Donald, 1951; Kira et al., 1953; Palmblad, 1968). Experimental studies demonstrate that the law of constant yield holds in a wide range of environments; for example, Donald (1951) sowed Bromus unioloides at a broad range of densities and at three nitrogen levels and found that, although the addition of nitrogen affects the total yield, in all cases the yield rapidly became independent of sowing density. Theoretical studies highlighted a problem, however, in that simple models, where the biomass of individual plants depends on a linear or exponentially decreasing function of the number of neighbours, often predict yield–density relationships that are humped and it is only special hyperbolic functions that result in constant yield (Pacala and Silander, 1985). However, by combining simulation studies and experimental work, Weiner and Thomas (1986) demonstrated that plant competition is often asymmetric, with large plants having a disproportionately adverse effect on small plants, on a weight-for-weight basis (Weiner, 1990). When the effects of competitive asymmetry are included into neighbourhood models, the law of constant yield arises naturally for most functions describing the effects of neighbours on individual biomass. Thus, asymmetry removes potential humps in yield–density relationships by allowing some plants to become large no matter what the sowing density (Pacala and Weiner, 1991).

4.3.2. Distribution of biomass among individuals

The distribution of biomass among individuals in a population depends on many factors such as differences in germination date, genetic make-up, abiotic environment, herbivore pressure and competitive interactions. Empirical studies have demonstrated that most plant populations consist of many small individuals and relatively few large ones (Obeid et al., 1967; Ogden, 1970). For example, in the high-density Linum populations studied by Obeid et al., 15% of the individuals make up 51% of the biomass. The mechanism underlying this observation is thought to be competition for light. Tall individuals shade short ones but not vice versa, and hence tall plants become taller, so generating positive feedback, which accentuates the differences between small and large individuals. Thus, the asymmetric competitive interactions discussed above are thought to be important in generating these positively skewed patterns in the distribution of biomass among individuals. In contrast, animal populations rarely show similar patterns of positive skew (Latto, 1992). Latto estimated that only 12% of bird populations and 8% of mammal populations studied possess positively skewed individual size distributions.

Positively skewed distributions of biomass among individuals have unexpected consequences for models of intraspecific competitive interactions between neighbours. In a detailed study of neighbourhood competitive interactions, Pacala and Silander (1990) determined the biomasses and number of competitors for over 12 000 plants. This allowed the radius over which competitive interactions took place, and the effect of the number of neighbours on individual biomass, to be determined. For both species studied, the competitive neighbourhood radii were less than 20 cm, demonstrating the local nature of plant interactions. However, surprisingly, given the local nature of the competitive interactions and the fact that the species were intraspecifically aggregated, and contrary to the expectations discussed on page 67, the spatial structure of the population turned out to be unimportant to the overall dynamics. Detailed analysis of population models indicated that this was a result of the spatial scale and level of aggregation being small and the enormous plasticity of plant performance resulting in total plant density rising rapidly to relatively high levels (Pacala and Silander, 1990).

4.3.3. Size–fecundity schedules

Plant size–fecundity schedules have recently received a lot of attention in the literature (Samson and Werk, 1986; Rees and Crawley, 1989; Klinkhamer *et al.*, 1992). Many plant species show, to a good approximation, linear relationships between reproductive and vegetative biomass (Samson and Werk, 1986; Rees and Crawley, 1989; see Fig. 4.2), although non-linear relationships have also been found (Klinkhamer *et al.*, 1992; Rees and Brown, 1992). Several studies have characterized size–fecundity schedules in different habitats where, presumably, the stresses that limit growth vary. These studies yield no consistent pattern; for example Kawano and Miyake's (1983) study of five annual foxtails demonstrated that average size differed between habitats, but the size–fecundity schedules were essentially the same. In contrast, other studies have found differences between slopes and intercepts in different habitats (Thompson *et al.*, 1991). This lack of consistent pattern is not surprising given the wide range of factors that could limit plant reproductive output. When the relationship is linear, attention focuses on whether there is a substantial threshold size for reproduction. Rees and Crawley (1989, 1991) argued that because plants are modular in construction the threshold size for reproduction will be small; on this basis they predicted that there would be a 50 : 50 split of positive and negative intercepts from regression studies of reproductive output against vegetative biomass. From a literature review they found that the split was in fact 60 : 40, broadly in agreement with the prediction. In contrast, the split for animal populations was 20 : 80 (Rees and Crawley, 1989, 1991); in other words, animals much more often have a critical size below which reproduction is impossible. If, to a good approximation, plants have no threshold size for reproduction,

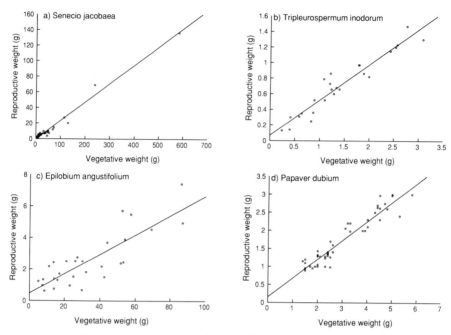

FIGURE 4.2 *Plant size–fecundity schedules. In all four cases the relationship is linear over a considerable range of plant biomass. None of the intercepts is significantly different from zero, suggesting that the plants that reproduce allocate a constant proportion of biomass to reproduction. Note, however, that in* Papaver *there is evidence that plants below 1.5 g fail to reproduce completely as they are too small to produce a single capsule.*

then total seed set is equal to $\Sigma_i \alpha B_i$ where α is the slope of the relationship between individual biomass, B_i, and seed set, which is of course $\alpha \Sigma_i B_i$. Hence, we only need to know the total biomass produced, and not the distribution of biomasses within the population, in order to predict the total seed set.

Using this simple fact, it follows that the law of constant yield for biomass will result in a constant yield for seed set. In contrast, if plants have substantial threshold sizes for reproduction, the proportion of total biomass that an individual allocates to reproduction will be an increasing function of its biomass. In this situation the distribution of individual biomasses, and so the spatial arrangement of neighbours and whether competition is asymmetric or not, becomes important in determining the total seed set and hence the long-term dynamics.

Consider the following two extremes:

1. All individuals are identical and competition is symmetric, so each plant has a biomass equal to the mean of the population. As population size

increases, so mean size decreases and a smaller and smaller proportion of total biomass is allocated to reproduction. As a result, total seed output decreases at high densities.

2. Individuals differ and competition is completely asymmetric, so there will be a group of large individuals, which together form a substantial proportion of the biomass produced. All these individuals will allocate approximately the same proportion of their biomass to reproduction and hence the relationship between total seed set and sowing density will again be flat topped.

Of course, real populations lie somewhere between these two extremes, although for the reasons outlined above most populations are probably closer to the latter than the former. Hence, if reproductive output is proportional to biomass or the threshold size for reproduction is large, but competition is primarily for light (resulting in asymmetric interactions), we would expect total seed set to become independent of sowing density. It is not difficult to see that this will tend to make the long-term dynamics very simple, with the population rapidly settling to a constant equilibrium, irrespective of initial conditions.

4.3.4. The proportion of seed that survives over winter

Having derived the relationship between initial sowing density and seed set, all we need to know to predict the number of seeds next year, from any initial seed density, is an estimate of the probability of seed survival over winter. There are two cases to consider:

1. The yield–density relationship is flat topped, resulting in stable population dynamics regardless of the probability of overwinter seed mortality, although the equilibrium population size increases as the fraction of seeds that survive over winter increases.
2. The yield–density relationship is humped; as the fraction of seeds that survives becomes smaller, the absolute value of the slope of the yield–density curve, evaluated at equilibrium, becomes shallower and the population dynamics become more and more stable (Fig. 4.1). In a similar fashion, decreasing the fertility of a habitat results in a decrease in average plant fecundity, which again makes the slope of the yield–density relationship shallower at equilibrium, making the population more stable.

These results can easily be extended to include the effects of dormancy and hence the formation of a seed bank. Inclusion of a seed bank is found to be stabilizing because even following a high-density reproductive crash there are plenty of seeds to recruit in the following year (Pacala, 1986). A good example of the biology described above is given by *Abutilon theophrasti* (Thrall *et al.*, 1989). This species has a reproductive threshold of 0.41 g and as a result its yield–density relationship is a decreasing function of seed density, which tends to destabilize the dynamics. However, it also forms a seed bank which is stabilizing and the interplay between these two factors determines the observed dynamics. Calibrated models predict that if *Abutilon* had no seed bank its population would perpetually oscillate. Inclusion of a seed bank stabilizes the population model, resulting in damped oscillations, which closely correspond to the dynamics observed in the field.

4.3.5. *Overview and the effects of stress*

Taken together, the theoretical and empirical results presented above lead to the simple prediction that annual plants should generally have very stable population dynamics in a constant environment; it is only species with large thresholds for reproduction, living in high-nutrient environments, with no seed bank and high seed survival over winter that will show population cycles. Consistent with these generalizations there are no well documented examples of cyclic annual plant populations (Rees and Crawley, 1991). Note, however, that the slow decay of litter can introduce a substantial time delay into systems dominated by perennial grasses, and this can result in oscillatory dynamics (Tilman and Wedin, 1991a).

Having established these general principles, we can now return to the effects of stress on plant population dynamics. These simple models make the counterintuitive prediction that the effects of stress on plant population dynamics will often be to *increase* population stability while lowering equilibrium population size. Stability is enhanced by increasing the size differential between dominant and subordinate plants, by reducing fecundity, and by reducing seed survival, any or all of which we might expect to see in stressed plants. Some effects of stress may, however, lead to a decrease in population stability, for example by reducing or eliminating the seed bank. In other words, intuition will be a poor guide for predicting the effects of particular stresses on plant population dynamics. The only way to make sensible predictions is to incorporate the effects into a model. These predictions will, of course, only be as good as the assumptions built into the model; but this is also true of intuitive predictions, derived from biological insight. The problem is that those who prefer insight to models rarely make their assumptions explicit.

4.4. Analysis

Modelling, data analysis to test model predictions, and parameter estimation are complementary processes. Accordingly, it is appropriate to say something, albeit briefly, about analysing census data.

Do analyses of plant census data support the prediction that many plant populations are characterized by stable dynamics? A common form of analysis is to regress population size in year t on that in year $t - 1$ and/or $t - 2$. If no significant relationship is found, some botanists have assumed that population sizes are decoupled, and hence that there are no density-dependent processes operating to stabilize the population. Both these conclusions are incorrect. We can demonstrate this by analysing a simple population model with two important features: parameters chosen so that the equilibrium point is stable and approached smoothly, and a per capita rate of increase which is a random variable that varies from year to year (because, for example, of variations in rainfall, drought or other stresses). A typical population trajectory produced from the model is shown in Fig. 4.3a. A regression of density at time t on density at time $t - 1$ shows no significant relationship (Fig. 4.3b). However, there is a strong, highly significant relationship between the logarithmic growth rate ($\log(N_{t+1}/N_t)$) and density. This latter analysis, which is the correct one to do, demonstrates that density-dependent processes are operating and the population is regulated. In fact we can go further because the relationship in Fig. 4.3c describes a population with a stable equilibrium point approached by a smooth trajectory. Hence, using an appropriate form of analysis, we can determine a great deal about the underlying population dynamics (Turchin and Taylor, 1992). The incorrect analysis (Fig. 4.3b) yields incorrect and misleading information.

The basic message here is simple. Understanding the response of a population to a change in the environment, for example a change in the level of some environmental stress, requires a model, and correct analysis of population data. The straightforward regression approach described above can be used to obtain a simple description of the dynamics. It is not a panacea (for a useful discussion of the pitfalls of this type of modelling see Crawley

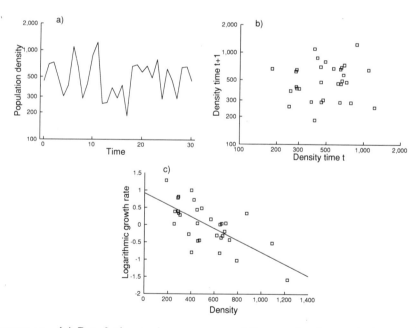

FIGURE 4.3 *(a) Population trajectory generated from a simple population model, see text for details. (b) Relationship between successive population densities. (c) Relationship between the logarithmic growth rate ($\log(N_{t+1}/N_t)$) and population density.*

(1983)), but a simple analysis can still provide useful information on the sensitivity of a population to perturbations and the return time to equilibrium.

4.5. Population persistence of annual plants in variable environments

Populations inhabit spatially and temporally variable environments. This obviously has important implications for population persistence. We now develop a simple model for an annual plant inhabiting an environment where the stresses that limit growth vary in both time and space. We initially consider only spatial variation. The condition for population persistence can be simply stated; we require that on average each seed replaces itself with more than one seed when population size is small. If this is the case, the population will increase when rare and eventually reach some density-dependent equilibrium. In contrast, if each seed replaces itself with less than a single seed, population size will decrease and the population will become extinct. By applying these intuitively obvious criteria it is relatively easy to explore how changes in the environment alter the likelihood of population persistence. Before we do so, however, note two things. First, population persistence is not the same as population stability (a population can persist without necessarily being stable). Secondly, although the models are derived with annual plants in mind, the qualitative insights apply to perennials. For perennials more complex stage-structured models must be constructed to predict persistence (e.g. Caswell, 1989), and the calculations are more complex algebraically, but they add nothing substantially new to the predictions.

The main pedagogic purpose of the sections that follow is to show how changing assumptions change predictions about population behaviour. It is exactly this property of models that worries sceptics. Such worries are fundamentally misguided. The whole point of modelling is to tell us what can happen, given certain conditions (i.e. assumptions). Good experimentalists also make sure that they have control over initial conditions.

Assume that the habitat consists of a large number of microsites. Some fraction $(1 - E)$ of these contain perennial plants, and any seeds from the annual that germinate at a site occupied by a perennial die before reproduction. Although this may seem unrealistic, it is often the case that recruitment from seed fails completely in the presence of perennial vegetation (Fenner, 1985; Rees and Brown, 1991; Rees and Long, 1992). We assume that some fraction d of the seeds die over winter and that in spring a fraction g of seeds germinate. Those that germinate in empty microsites produce F seeds and those that germinate in ones occupied by perennials produce none. Hence, the number of seeds next year, S_{t+1}, produced from the S_t seeds this year is

$$S_{t+1} = (1 - d)(1 - g)S_t + (1 - d)gEFS_t \qquad (4.1)$$

The first term is the seeds that survive but do not germinate (the seed bank), the second is the seeds produced by those seeds that survive and germinate in an empty microsite. Dividing S_{t+1} by S_t gives the finite rate of increase (λ) and for persistence we require $\lambda > 1$. Note that because we are only interested in the behaviour of the population when population size is small we have not included any density dependence into the model. Now, intuitively you might expect the formation of a seed bank, $g < 1$, to foster persistence of the annual. Intriguingly, this is not necessarily the case – the formation of a seed bank can drive the population to extinction (Rees and Long, 1992). This result occurs because delaying germination does not increase the probability of a seed germinating in an empty microsite and so cannot increase the finite rate of increase. In fact, there is a cost to forming a seed bank that results from the death of seeds before germination, which is why forming a seed bank makes persistence more difficult. The consideration of simple models focuses attention on the entire life-cycle, not on just, say, adult fecundity. It is therefore clear that, if we aim to understand the effects of a particular stress, we need to know how it affects seed mortality, germination, the fraction of sites suitable for colonization and average adult fecundity.

This simple model assumes that germination biology may be summarized by a single parameter, g. However, experimental work has demonstrated that the probability of a seed germinating depends on the presence of established plants (King, 1975; Gorski et al., 1977; Rice, 1985; Van Tooren and Pons, 1988; Rees and Brown, 1991). The main result from this broad body of experimental work is that the presence of established plants inhibits seed germination; this may occur through changes in the red to far red ratio of the incident radiation, the range of temperature fluctuations or a reduction in soil nitrogen levels. Where comparative data are available, the distribution of percentage inhibition appears to be of an all or nothing nature. Among the species studied by Gorski et al. (1977) 58% of the uncultivated species showed strong (80–100%) inhibition of germination by a leaf canopy, whereas only 10% of the cultivated species had a similar response, suggesting the importance of germination inhibition in natural populations.

It is therefore necessary to modify the previous model to allow different germination probabilities in empty microsites and in those occupied by perennials. If we assume that the probability of germinating in an empty microsite is g_u, the probability of germination in a microsite occupied by an established perennial plant is g_o, we obtain the following model:

$$S_{t+1} = (1 - d)\{1 - \Omega\}S_t + (1 - d)g_u EFS_t \qquad (4.2)$$

where $\Omega = g_u E + g_o(1 - E)$. For persistence, we require

$$(1 - d)\{1 - g_o(1 - E) + g_u E(F - 1)\} > 1 \qquad (4.3)$$

Comparison with the first model is facilitated by first setting $g = g_o(1 - E) + g_u E$ so that the fraction of seeds that germinates in each time interval is equal. (This ensures that the cost of forming a seed bank, as a result of seed mortality, is the same in both models.) Persistence then becomes easier whenever $g_u > g_o$. In other words, inhibition of germination by perennial plants promotes persistence.

The data collected by Gorski et al. (1977) suggest, to a rough approximation, that for the uncultivated species $g_o < 0.2 g_u$, assuming that g_o and g_u can be estimated by the probability of germination under a leaf canopy and in diffuse light respectively. Therefore, the germination biology described above is likely to be important in promoting persistence in many plant populations. It is worth repeating that it is not delaying germination *per se* that promotes persistence but the seed's germination response to the presence of established plants.

The potential for various stresses to change the ratio of g_u to g_o, and hence the likelihood of population persistence, seems not to have been explored, but it is clearly a possibility. For example, it is well know that nitrate can stimulate seed germination (Lehmann, 1909; Roberts and Smith, 1977) and also the growth of competitors, which can inhibit germination. This is also a good example of how models can suggest profitable lines for experimental investigation.

These simple models assume that all microsites not containing established perennials are equally suitable for growth. However, we would expect the presence of perennial neighbours to reduce microsite quality by shading or nutrient uptake. Several studies have demonstrated that the relationship between plant fecundity (F_c) and the weight or number of neighbours is a non-linear, decreasing function (Pacala and Silander, 1985; Goldberg, 1987; McConnaughay and Bazzaz, 1987; Miller and Werner, 1987). The relationship is often well described by the simple hyperbolic function

$$F_c = \frac{F}{1 + \alpha i} \tag{4.4}$$

or an exponential function

$$F_c = F \exp(-\alpha i) \tag{4.5}$$

where F is the fecundity of a plant with no neighbours, i is the number of perennial neighbours and α is a decay parameter. To determine the condition for persistence of an annual plant in such an environment, we must calculate its average fecundity. This may be approximated by using the Taylor series to expand F_c about the mean number of perennial neighbours and taking expectations, giving

$$\text{average fecundity} \approx F_c(\bar{i}) + \frac{\sigma_i^2 \, \mathrm{d}^2 F_c}{2 \, \mathrm{d}i^2} \bigg|_{\bar{i}} \tag{4.6}$$

where $\bar{i}$ and σ_i^2 are the mean and variance in the number of perennial neighbours, respectively. For the exponential function, this gives

$$\text{average fecundity} = F \exp(-\alpha \bar{i}) + \frac{F \alpha^2 \sigma_i^2}{2} \exp(-\alpha \bar{i}) \tag{4.7}$$

The first term on the right-hand side is the fecundity of a plant with the average number of perennial neighbours; the second term is positive and proportional to the variance in the number of neighbours, demonstrating that variance in microsite quality promotes persistence relative to the average environment. Thus, the spatial arrangement of perennial plants is important in determining persistence of annuals. Spatial aggregation of perennials (resulting in a large variance term) will promote persistence relative to a more even spatial distribution.

The results presented above assume that the fraction of sites available for colonization is constant from year to year. However, many annuals live in successional environments where the fraction of sites available for colonization, E, varies with time. In the simplest successional environment, virtually all microsites will be available for colonization after a large-scale disturbance ($E \approx 1$), whereas if there is no disturbance virtually all sites will be occupied by perennial plants ($E \approx 0$). Consider the case where a constant proportion of the seeds germinate. Earlier, for a constant environment, we showed that the formation of a seed bank made persistence more difficult. However, in a random environment, this simple result no longer holds. If the probability of germination is high, the population rapidly declines in years when there is no large-scale disturbance, resulting in extinction. A lower germination rate results in slower decay between disturbances, resulting in persistence. However, the germination rate cannot be too low, otherwise most seeds die before having a chance to germinate and this drives the population to extinction. Conversely, the presence of perennials often inhibits germination, and the incorporation of this into models strongly promotes persistence because seeds avoid germinating in the years between large-scale disturbances (Rees and Long, 1992).

For simple models it is possible to calculate the critical probability of large-scale disturbance, and so the critical time between disturbances, required for a population to persist (Rees and Long, 1992). This depends on average plant fecundity and the probabilities of seed germination and mortality, all of which can easily be estimated from experimental data. Using experimentally determined parameters, Rees and Long determined the effect of changes in mollusc herbivory, the ability of seeds to detect perennial vegetation, and seed mortality on the critical time between disturbances required for persistence. Experimental results indicated that mollusc herbivores reduced plant fecundity by approximately 30% (Rees and Brown, 1991), which most biologists would assume has a profound effect on the condition for persistence. Surprisingly, it has little effect on persistence compared with changes in the inhibitory effect of perennials on germination and the probability of seed mortality. For example, reducing average fecundity from 930 seeds to 612, as a result of mollusc herbivory, reduced the average time between disturbances required for persistence from 20 to 19 years. In contrast, the assumption that perennials have no effect on the probability of germination decreased the average time to 13 years, and reducing the probability of seed mortality from 0.2 to 0.1 increased the average time to 38 years, demonstrating the importance of understanding the processes that determine the rate of decay of the seed bank. These quantitative insights could only be obtained from the analysis of parameterized population models.

4.6. Concluding remarks

We have focused deliberately on insights that have recently emerged from new models in plant population biology. None of these models is explicitly directed at understanding or modelling the effects of stress on plant population dynamics. However, by now it ought to be obvious that to understand the population dynamic consequences of any stress we need to know how it affects critical stages in the life-cycle and performance of individual plants. It should also be obvious that, just because stress reduces some aspect of plant performance, it does not necessarily mean that populations will decline, disappear or become less stable.

Modelling forces one to consider the entire life-cycle and also to take a long-term view. The difference between short-term and long-term effects has led to several debates in the literature; for example, Grime and Tilman's debate on competition (Tilman, 1987; Thompson and Grime, 1988). As far as stress is concerned, short-term and long-term views can be diametrically opposed. Consider an annual plant growing in an environment where growth is nitrogen limited. Addition of nitrogen results in increased plant size, and the environment is now less stressful. However, in the long term, addition of nitrogen results in increased growth of perennial grass, which prevents successful colonization from seed, resulting in a decrease in population size, and so the environment is now more stressful. To a population ecologist or conservation manager the long-term view would seem more appropriate.

There would appear to be considerable potential in the field of plant stress to explore the consequences for plant population dynamics of various environmental stresses. The important things to measure are not only how a particular stress changes growth rates, fecundity, seed survival, etc., but also the frequency and duration of the stress, and the spatial and temporal variance in plant responses. The modelling tools now exist; it is up to botanists interested in predicting the effects of stress on plant populations to use them to develop predictive models, and to guide future field experiments.

REFERENCES

Caswell, H. (1988). Theory and models in ecology: a different perspective. *Bull. Ecol. Soc. Am.* **69**, 102–108.

Caswell, H. (1989). *Matrix Population Models*. Sunderland, Mass.: Sinauer Associates.

Caughley, G. and Lawton, J.H. (1981). Plant–herbivore systems: In: *Theoretical Ecology. Principles and Applications*, pp. 132–166, ed. R.M. May. Oxford: Blackwell Scientific.

Chapin, F.S. III (1991). Integrated response of plants to stress: a centralized system of physiological responses. *Biosci.* **41**, 29–36.

Crawley, M.J. (1983). *Herbivory. The Dynamics of Animal–Plant Interactions*. Oxford: Blackwell Scientific.

Curry, G.L. and Feldman R.M. (1987). *Mathematical Foundations of Population Biology*. College Station, Texas: Texas A&M University Press.

Donald, C.M. (1951). Competition among pasture plants. I. Intra-specific competition among annual pasture plants. *Aust. J. Agric. Res.* **2**, 355–376.

Fenner, M. (1985). *Seed Ecology*. London: Chapman & Hall.

Goldberg, D.E. (1987). Neighbourhood competition in an old-field plant community. *Ecology* **68**, 1211–1223.

Gorski, T., Gorska, K. and Nowicki, J. (1977). Germination of seeds of various herbaceous species under leaf canopy. *Flora Batava* **166**, 249–259.

Grime, J.P. (1979). *Plant Strategies and Vegetation Processes*. Chichester: Wiley.

Harper, J.L. (1977). *Population Biology of Plants*. London: Academic Press.

Hassell, M.P., May, R.M., Pacala, S.W. and Chesson, P.L. (1991). The persistence of host–parasitoid associations in patchy environments. 1. A general criterion. *Am. Naturalist* **138**, 568–583.

Holt, R.D. and Lawton, J.H. (1993). Apparent competition and enemy-free space in insect host–parasitoid communities. *Am. Naturalist* **138**, in press.

Jeffers, J.N.R. (1982). *Modelling. Outline Studies in Ecology*. London: Chapman & Hall.

Kawano, S. and Miyake S. (1983). The productive and reproductive biology of flowering plants. X. Reproductive energy allocation and propagule output of five congeners of the genus *Setatrai* (Gramineae). *Oecologia* **57**, 6–13.

King, T.J. (1975). Inhibition of seed germination under leaf canopies in *Arenaria serpyllifolia, Veronica arvensis* and *Cerastium holosteoides*. *New Phytol.* **75**, 87–90.

Kira, T., Ogawa, H. and Shinozaki, K. (1953). Intraspecific competition among higher plants. I. Competition–density–yield inter-relationship in regularly dispersed populations. *J. Inst. Polytechnics, Osaka City University* **4**, 1–16.

Klinkhamer, P.G.L., Meelis, E., de Jong, T.J. and Weiner J. (1992). On the analysis of size-dependent reproductive output in plants. *Functional Ecol.* **6**, 308–316.

Larsson, S. (1989). Stressful times for the plant stress–insect performance hypothesis. *Oikos* **56**, 277–283.

Latto, J. (1992). The differentiation of animal body weights. *Functional Ecol.* **6**, 386–395.

Lehmann, E. (1909). Zur Keimungsphysiologie und -biologie von *Ranuculus sceleratus* und einigen anderen Samen. *Ber. D. Bot. Ges.* **27**, 476–494.

Levins, R. (1968). *Evolution in a Changing Environment. Some Theoretical Explorations*. Princeton, NJ: Princeton University Press.

Levitt, J. (1980). *Responses of Plants to Environmental Stresses*. Vols I & II. San Diego, Calif.: Academic Press.

May, R.M. and Oster, G.F. (1976). Bifurcations and dynamic complexity in simple ecological models. *Am. Naturalist* **110**, 573–599.

McConnaughay, K.D.M. and Bazzaz, F.A. (1987). The relationship between gap size and performance of several colonizing annuals. *Ecology* **68**, 411–416.

McQuate, G.T. and Connor, E.F. (1990). Insect responses to plant water deficits. I. Effects of water deficits in soybean plants on the feeding preference of Mexican bean beetle larvae. *Ecol. Entomol.* **15**, 419–431.

Miller, T.E. and Werner, P.A. (1987). Competitive effects and responses between plant species in a first-year old-field community. *Ecology* **68**, 1201–1210.

Nicholson, A.J. (1933). The balance of animal populations. *J. Animal Ecol.* **2**, 132–178.

Nicholson, A.J. and Bailey, V.A. (1935). The balance of animal populations. Part 1. *Proc. Zool. Soc. Lond.* **1935**, 551–598.

Nisbet, R.M. and Gurney, W.S.C. (1982). *Modelling Fluctuating Populations.* Chichester: Wiley.

Obeid, M., Machin, D. and Harper, J.L. (1967). Influence of density on plant to plant variations in fiber flax, *Linum usitatissimum. Crop Sci.* **7**, 471–473.

Ogden, J. (1970). Plant population structure and productivity. *Proc. NZ Ecol. Soc.* **17**, 1–9.

Pacala, S.W. (1986). Neighbourhood models of plant population dynamics. 4. Single-species and multispecies models of annuals with dormant seeds. *Am. Naturalist* **128**, 859–878.

Pacala, S.W. (1987). Neighbourhood models of plant population dynamics. 3. Models with spatial heterogeneity in the physical environment. *Theor. Popul. Biol.* **31**, 359–392.

Pacala, S.W. (1989). Plant population dynamic theory. In: *Perspectives in Ecological Theory,* pp. 54–67, eds. J. Roughgarden, R.M. May and S.A. Levin. Princeton, NJ: Princeton University Press.

Pacala, S.W. and Silander, J.A. Jr. (1985). Neighbourhood models of plant population dynamics. I. Single-species models of annuals. *Am. Naturalist* **125**, 385–411.

Pacala, S.W. and Silander, J.A. Jr. (1990). Field tests of neighborhood population dynamic models of two annual weed species. *Ecol. Monogr.* **60**, 113–134.

Pacala, S.W. and Weiner, J. (1991). Effects of competitive asymmetry on a local density model of plant interference. *J. Theor. Biol.* **149**, 165–179.

Pacala, S.W., Hassell, M.P. and May, R.M. (1990). Host–parasitoid associations in patchy environments. *Nature* **344**, 150–153.

Palmblad, I.G. (1968). Competition in experimental populations of weeds with emphasis on the regulation of population size. *Ecology* **49**, 26–34.

Rees, M. and Brown, V.K. (1991). The effects of established plants on recruitment in the annual forb *Sinapis arvensis. Oecologia* **87**, 58–62.

Rees, M. and Brown, V.K. (1992). Interactions between invertebrate herbivores and plant competition. *J. Ecol.* **80**, 353–360.

Rees, M. and Crawley, M.J. (1989). Growth, reproduction and population dynamics. *Functional Ecol.* **3**, 645–653.

Rees, M. and Crawley, M.J. (1991). Do plant populations cycle? *Functional Ecol.* **5**, 580–582.

Rees, M. and Long, M.J. (1992). Germination biology and the ecology of annual plants. *Am. Naturalist* **139**, 484–508.

Rice, K.J. (1985). Responses of *Erodium* to varying microsites: the role of germination cueing. *Ecology* **66**, 1651–1657.

Roberts, E.H. and Smith, R.D. (1977). Dormancy and the pentose phosphate pathway. In: *The Physiology and Biochemistry of Seed Dormancy and Ger-*

mination, pp. 385–411, ed. A.A. Khan. Amsterdam: Elsevier Biomedical Press.

Sackville Hamilton, N.R., Schmidt, B. and Harper, J.L. (1987). Life history concepts and the population biology of clonal organisms. *Proc. R. Soc. B* **232**, 35–57.

Samson, D.A. and Werk, K.S. (1986). Size-dependent effects in the analysis of reproductive effort in plants. *Am. Naturalist* **126**, 667–680.

Symonides, E. (1979). The structure and population dynamics of psammophytes on inland dunes. II. Loose-sod populations. *Ekol. Polska* **27**, 191–234.

Thompson, K. and Grime, J.P. (1988). Competition reconsidered: a reply to Tilman. *Functional Ecol.* **2**, 114–116.

Thompson, B.K., Weiner, J. and Warwick, S.I. (1991). Size-dependent reproductive output in agricultural weeds. *Can. J. Bot.* **69**, 442–446.

Thrall, P.H., Pacala, S.W. and Silander, J.A. (1989). Oscillatory dynamics in populations of an annual weed species *Abutilon theophrasti. J. Ecol.* **77**, 1135–1149.

Tilman, D. (1982). *Resource Competition and Community Structure.* Princeton, NJ: Princeton University Press.

Tilman, D. (1987). On the meaning and the mechanisms of competitive superiority. *Functional Ecol.* **1**, 304–315.

Tilman, D. (1988). *Plant Strategies and the Dynamics and Structure of Plant Communities.* Princeton, NJ: Princeton University Press.

Tilman, D. and Wedin, D. (1991a). Oscillations and chaos in the dynamics of a perennial grass. *Nature* **353**, 653–655.

Tilman, D. and Wedin, D. (1991b). Dynamics of nitrogen competition between successional grasses. *Ecology* **72**, 1038–1049.

Turchin, P. and Taylor, A.D. (1992). Complex dynamics in ecological time series. *Ecology* **73**, 289–305.

Van Tooren, B.F. and Pons, T.L. (1988). Effects of temperature and light on the germination in chalk grassland species. *Functional Ecol.* **2**, 303–311.

Watkinson, A.R. and Davy, A.J. (1985). Population biology of salt marsh and sand dune annuals. *Vegetatio* **62**, 487–497.

Watt, K.E.F. (1968). *Ecology and Resource Management.* New York: McGraw-Hill.

Weiner, J. (1990). Asymmetric competition in plant populations. *Trends Ecol. Evol.* **5**, 360–364.

Weiner, J. and Thomas, S.C. (1986). Size variability and competition in plant monocultures. *Oikos* **47**, 211–222.

White, T.C.R. (1984). The abundance of invertebrate herbivores in relation to the availabity of nitrogen in stressed food plants. *Oecologia* **63**, 90–105.

CHAPTER 5

Understanding photosynthetic adaptation to changing climate

D. W. LAWLOR and A. J. KEYS

5.1. Introduction

Photosynthesis, and its response and adaptation to climate change, are important because the carbon economy of plants is a major link between the atmosphere, geosphere and biosphere. The atmospheric carbon dioxide pool is small and rapidly exchanges with other carbon pools, including those of the biosphere because of the dominant role of photosynthesis. Photosynthesis is also the basis of all ecosystem processes, including those on which human society depends. Thus the way in which photosynthesis behaves in the face of rapid and substantial changes in atmospheric chemistry and expected changes in climate is of more than academic interest (Jäger and Ferguson, 1991). The rate of photosynthesis, per unit area of both leaf and ground surface, depends on suitable environmental conditions; greatest photosynthetic productivity occurs on areas of the earth's surface with mean temperatures between 5 and 25 °C, and where water and nutrients are available. Photosynthetic rate responds to environmental conditions on a time-scale ranging from almost instantaneous, in the case of light, to the slow response to seasonal changes in temperature and radiation at high latitudes. Changes in the rate of photosynthesis caused by the environment are quickly reflected in the carbon dioxide content of the atmosphere (Watson *et al.*, 1990; Roeckner, 1992). The rate of photosynthesis also largely determines net primary productivity, which influences the production of biomass and thereby terrestrial soil carbon pools and deposition of carbon in wood and peat, all of which have short-to-medium rates of turnover, as well as in coal and natural gas which are much more slowly recycled (Schlesinger, 1991). Alterations in the global atmospheric composition (e.g. increasing carbon dioxide) and climate (e.g. rising temperature) may increase rates of photosynthesis and thus ameliorate the changes in global conditions, whereas reduced rates of photosynthesis may exacerbate the climate change. The response of photosynthesis to altered atmospheric carbon dioxide level is therefore of importance for its potential effects on global conditions and also for its consequences on the biosphere.

The subject of photosynthetic adaptation to climate change has been reviewed by Stitt (1991) and Bowes (1991) from biochemical viewpoints and by Long (1991) with physiological emphasis. Aspects of plant responses to potential global climate change have been discussed by Lawlor and Mitchell (1991), Lawlor (1991) and Mousseau and Saugier (1992). The following analysis considers the concepts and information available at the interface between biochemical processes and physiological functions from which plant photosynthetic responses to changing environment may be assessed, and suggests aspects of plant function which require more attention.

5.2. The basis of climate change

Climate change is the process of alteration in long-term average weather conditions in a particular geographical area (Schlesinger, 1991). Climate change may encompass all features of the environment that constitute the climate, for example temperature, humidity, cloud cover, rainfall, snow and drought, as well as the occurrence of extreme meteorological events. As these features of the environment are interrelated, e.g. cloud cover with rain, and also affect other aspects of the environment, e.g. cloud determines the solar radiation reaching the earth's surface, change in one may have significant, indeed – as far as the biosphere is concerned – profound, consequences for global conditions (Jäger and Ferguson, 1991). In addition to the changes in meteorological parameters of climate, the term has come to include changes in the composition of the earth's atmosphere. Strictly, the change in climate is separate from alterations in atmospheric composition and chemistry, although the former is related to or even a consequence of, the latter. The term 'environmental change' expresses both aspects.

Current information about the composition of the atmosphere and changes that have occurred over the last two centuries is very precise (Roeckner, 1992). The evidence that over the period from about 1750 to the present the concentration of carbon dioxide in the atmosphere has risen worldwide is irrefutable (Boden et al., 1990), increasing from some 280 to 355 cm^3 m^{-3}; the rate of increase is now approximately 1.5 cm^3 m^{-3} $year^{-1}$, with evident acceleration (Watson et al., 1990). Predictions of future atmospheric composition are also well founded (Jäger and Ferguson, 1991); at present rates of carbon dioxide emissions and with the projected world population growth, the atmospheric carbon dioxide level is expected to reach 700 cm^3 m^{-3} by the end of the 21st century. Other trace gases are also increasing in the atmosphere: methane, oxides of nitrogen and, more recently, the chlorinated fluorocarbons (CFCs). The increases are solely ascribable to the burgeoning human population of the planet, accompanied by increasing industrialization and the associated increase in use of fossil fuels as energy sources. In addition, large quantities of carbon dioxide are released into the atmosphere by oxidation of accumulated biomass and soil organic carbon due to destruction of forests and alteration in land use (Watson *et al.*, 1990).

The consequences of changing atmospheric chemistry for the long-term stability of the earth's climate are not yet established unequivocally. Carbon dioxide, methane and CFCs absorb long-wave infrared radiation and thus 'trap' energy emitted from the earth's surface. On a molar basis, methane and CFCs are more efficient absorbers of infrared radiation than carbon dioxide. The trapped energy warms the atmosphere; this is the basis of the 'greenhouse effect' (Watson *et al.*, 1990; Jäger and Ferguson, 1991). As a consequence, the temperatures of the atmosphere, oceans and continents will rise. Current global circulation models (GCMs), which calculate global energy balances based on atmospheric physics and the content of infrared-absorbing gases (particularly the effects of increasing carbon dioxide levels) predict that temperatures will increase by up to 4°C for a doubling of carbon dioxide (Watson *et al.*, 1990). However, the energy balance of the globe largely depends on the atmospheric water vapour content, which strongly absorbs infrared radiation and is at very high concentration, so that relatively small changes in water vapour would have profound effects on global energy balance. Therefore, quantitation of the warming, and its rate, are subject to considerable uncertainty. The principal complication is the behaviour of the global energy balance, which depends on the amount of solar radiation reflected back into space, the albedo, and hence on clouds. Even a modest change in cloud as a consequence of temperature change may be significant.

Evidence for changing global temperature is persuasive but not yet conclusive. Compared with the period 1950–1969 (taken as a reference period), and corrected for many potential sources of error in the temperature record, the last 20 years have seen an increase of about 0.5 °C globally. The period before 1950 was some 0.3 °C cooler than the baseline period. There is no significant correlation between the rise in carbon dioxide level and temperature. The evidence of increased temperatures may be obscured by the large thermal inertia of the oceans, counteractive effects of change in the solar 'constant' and random variation in the data (Jäger and Ferguson, 1991).

Changes in temperature across the globe are not expected to be uniform. GCMs suggest that warming will be greatest at high latitudes and smallest at the equator. Rainfall patterns may also alter, with less rain in mid-latitudes but more at high latitudes and possibly at the equator. However, global atmospheric carbon dioxide composition should remain very uniform as mixing is rapid. The carbon dioxide level does fluctuate, both in the short term and on the small scale (e.g. in vegetation during the daily cycle of radiation) and seasonally on a large scale related to variation in carbon dioxide exchange by vegetation as a consequence of temperature-dependent growth cycles in higher, essentially northern, latitudes. However, the fluctuations are small compared with the background (5 as opposed to $350 \, cm^3 \, m^{-3}$).

As global atmospheric carbon dioxide and other gases are increasing and will certainly continue to do so for the next century unless major unforeseen changes in human population occur (Lashof and Tirpak, 1990), global climate (in the strict sense) is very likely to change, with warmer temperatures, especially at higher latitudes. More specific changes in other conditions, e.g.

rainfall and other forms of precipitation, cloudiness and extreme weather, on a geographically local scale, are not predictable.

5.3. Effects of carbon dioxide and climate change on photosynthesis

Given the probability of environmental change, what will be the effects on the composition and productivity of the earth's vegetation? Will climate change constitute a 'stress' on plants, and if so how and what methods may be employed to ameliorate it? Here we address the question of photosynthetic response, as it underlies all other ecosystem processes.

Atmospheric carbon dioxide is the substrate for photosynthetic carbon assimilation by plants (Bowes, 1991; Stitt, 1991). The photosynthetic rate of most plants with the C3 mode of carbon metabolism, which constitute the major part of the earth's vegetation in virtually all habitats, is not saturated by the current carbon dioxide concentration (Fig. 5.1). When the carbon

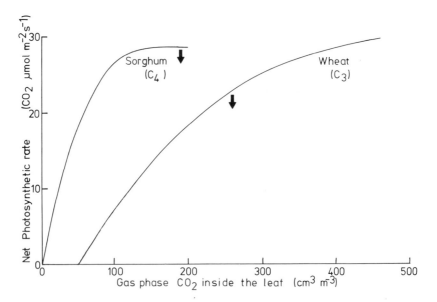

FIGURE 5.1 *Relation between photosynthetic rate, A, and the intercellular carbon dioxide concentration, C_i, for typical C3 and C4 plant leaves at optimum temperatures. In the current atmosphere and with stomatal conductance, g_s, of 0.2–0.4 μmol m^{-2} s^{-1} in C3 leaves, C_i is substantially smaller than that required for saturation of photosynthesis. In contrast, C4 photosynthesis is saturated, despite g_s of 0.1–0.2 μmol m^{-2} s^{-1}. With increasing atmospheric carbon dioxide levels, C_a, C3 assimilation approaches saturation; as g_s decreases with exposure to elevated carbon dioxide levels, C_i does not increase in proportion to C_a, but at 700 cm^3 m^{-3} A is saturated.*

dioxide level is increased from 350 to 700 cm^3 m^{-3}, the rate of photosynthesis increases, if light and other essential factors are adequate, by about 50% over a period of minutes and may remain large for many hours to days (Cure and Acock, 1986). However, over longer periods, the rate of photosynthesis decreases in many studies (Sage et al., 1989; Besford, 1993). The changes in photosynthesis are not uniform or consistent within or between species, nor are they related simply to particular environmental conditions. In contrast to C3 plants, C4 species respond by about 10% when the current level of carbon dioxide is doubled.

The instantaneous response of the rate of photosynthetic carbon assimilation by C3 plants to increased carbon dioxide supply can be partly explained by the catalytic properties of the carboxylating enzyme ribulose bisphosphate carboxylase–oxygenase (Rubisco) (Laing et al., 1974; Bowes, 1991; Stitt, 1991). This enzyme catalyses the first step in conversion of inorganic carbon to organic form. Oxygen competes with carbon dioxide – the oxygenase activity of this bifunctional enzyme – and gives rise to photorespiration, which is approximately 20–30% of net photosynthesis at temperatures of 20 °C, and thus decreases carboxylation efficiency (Laing et al., 1974; Jordan and Ogren, 1984; Bowes, 1991). Elevated carbon dioxide provides more substrate for carboxylation and reduces the competition by oxygen, thus decreasing photorespiration and stimulating gross photosynthesis. Therefore, the increase in atmospheric carbon dioxide, C_a, is expected to stimulate photosynthesis, particularly of C3 plants, and their global production of carbon biomass; this is confirmed in most experiments. Averaged over many species and growth conditions, doubling the carbon dioxide concentration from about 300 to 600 cm^3 m^{-3} increased dry matter production by 33% (Kimball, 1983). The response to carbon dioxide decreases as the absolute concentration rises, as expected, because photosynthesis becomes saturated at a carbon dioxide concentration within the leaf intercellular spaces, C_i, of about 500 cm^3 m^{-3} i.e. at atmospheric concentrations of about 600–700 cm^3 m^{-3} (Bunce, 1992). The precise concentration depends markedly on temperature; the advantages of elevated levels of carbon dioxide are greater at high temperatures.

The much smaller response of C4 plants to a concentration of carbon dioxide double the current ambient (Fig. 5.1) is a consequence of the carbon dioxide concentrating mechanism, which results in more efficient assimilation than that of C3 plants, especially in warmer conditions and in bright light. The mechanism is based on the enzyme phosphoenol pyruvate carboxylase (PEPc), which has a high affinity for carbon dioxide and can assimilate it effectively from the small concentration in the atmosphere. This PEPc carbon dioxide pump provides Rubisco, the carboxylating enzyme of the photosynthetic carbon reduction (Calvin) cycle, with a high concentration of carbon dioxide. Thus, C4 assimilation is largely saturated at current ambient atmospheric carbon dioxide concentrations and assimilation and dry matter production are increased little by elevated carbon dioxide (Edwards et al., 1985).

In addition to the effects on photosynthetic metabolism, there is considerable and consistent evidence for many C3 and C4 plants that double current ambient carbon dioxide markedly and persistently decreases stomatal aperture, g_s, resulting in smaller conductance (about 40%) to carbon dioxide and water vapour transport (Valle et al., 1985; Mansfield et al., 1990; Eamus, 1991; Morison, 1993). For a given atmospheric water vapour pressure and a particular leaf temperature (both of which determine the gradient of water vapour pressure between evaporating surfaces in the leaf mesophyll and the bulk air) the flux of water vapour from the leaf to air (transpiration) decreases in elevated carbon dioxide. However, because of the interactions between leaf energy balance, evaporation rates, humidity and wind speed, the exact consequences for photosynthetic response to climate change are difficult to predict (Eamus, 1991; Lawlor and Mitchell, 1991). The benefits to photosynthesis and productivity of improving the plant water balance have been demonstrated (Gifford, 1979; Morison, 1993). Another effect of elevated carbon dioxide levels, which has been widely observed but is not well understood, is a decrease in the rate of dark respiration in many species (Amthor, 1991); the significance of reducing assimilate consumption to productivity has not been evaluated.

There is a wealth of experimentation on the effects of atmospheric carbon dioxide on plants, albeit predominantly on annual crop plants, particularly of warm climates such as the continental USA in summer, and with marked bias towards soybean and cotton (Lawlor and Mitchell, 1991). The knowledge that carbon dioxide stimulates photosynthesis and productivity on an experimental scale should permit the prediction that elevated levels will increase the productivity of vegetation. In addition, decreased water loss should be an added benefit, reducing water stress. From this viewpoint, elevated carbon dioxide is beneficial; crops and natural vegetation will produce more dry matter (Kimball, 1983; Cure and Acock, 1986) and reproductive and harvestable yield. C3 plants will benefit from additional carbon dioxide much more than C4 plants.

However, the effects of long exposure to elevated carbon dioxide on plant productivity and ecosystem responses cannot be extrapolated from the short-term effects on assimilation, nor from the response of annual crops (Cure and Acock, 1986). The marked initial stimulation of photosynthesis on exposure to elevated carbon dioxide levels is often not persistent; there is much literature showing that prolonged exposure results, in many cases, in decreased photosynthetic capacity (Sage et al., 1989; Bunce, 1992; Besford, 1993). The phenomenon is called acclimation and is illustrated in Fig. 5.2. The decreased assimilation with time explains the discrepancy between the stimulation of assimilation (an average of 50%) caused by short-term exposure compared with the increase of production (33%) resulting from long-term exposure to elevated carbon dioxide levels (Kimball, 1983). Thus, other factors contribute to the regulation or acclimation of photosynthetic rate to the plant's carbon balance.

The acclimation of photosynthesis is of considerable importance. It has been reported in many dicotyledonous plants, e.g. bean (Sage et al., 1989) and is particularly marked in tomato (Besford, 1993); fewer studies have been made on the response of monocotyledonous plants. However, there are few generalities to guide assessment of the phenomenon; suggestions (Lawlor, 1991) that decreased capacity for assimilation is related to determinate growth patterns remain to be established. In this vein, perennial species, particularly trees, appear not to be as subject to 'negative acclimation' as annuals and

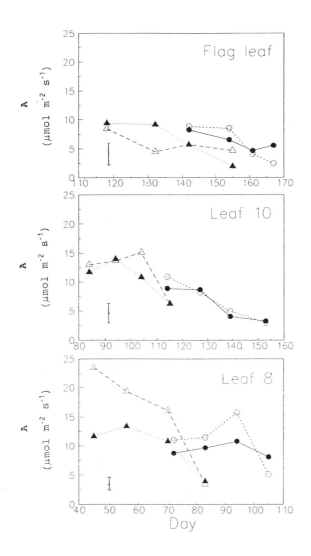

FIGURE 5.2 *Acclimation of photosynthesis in winter wheat. A crop of winter wheat (var. Mercia) was grown in carbon dioxide at 350 cm³ m⁻³ at ambient temperature,* ● *or at ambient +4 °C,* ▲, *or in 700 cm³ m⁻³ carbon dioxide at ambient temperature,* ○, *or at ambient +4 °C,* △, *from sowing in October. The rates of photosynthesis were measured under a photosynthetic radiation flux of 400 μmol m⁻² s⁻¹ at the carbon dioxide concentration under which the plants grew, on leaves developed as the crop aged. Leaf 8 was formed during late February, leaf 10 in late March to early April and the flag leaf in early May at Rothamsted. Early formed leaves grown in elevated carbon dioxide levels had high rates as of assimilation, as expected, but there is progressive decrease in the rate as leaves age and also in the later formed leaves; this indicates loss of ability to assimilate carbon dioxide. It correlates with a decrease in the chlorophyll and soluble protein content of leaves as they aged and as they formed later in plant development. (E. Delgado, S. Driscoll and D. W. Lawlor, unpublished data.)*

herbaceous plants. Acclimation means that the longer term response of plants to elevated carbon dioxide levels may be smaller than predicted from experiments involving brief exposure. Not only must the response of photosynthesis and productivity to elevated carbon dioxide levels be assessed, but also the response to increasing temperature and other aspects of potential climate change, e.g. to warmer and more frequent extreme (hot, dry) environments. As with the understanding of responses to elevated carbon dioxide levels, there is a vast literature describing and analysing photosynthetic responses to temperature, mineral nutrition, light, water supply, humidity and wind. However, no comprehensive model exists which enables the response to future temperature regimens to be assessed (Long, 1991). An added complication is that the analyses of temperature, drought and other responses have largely been made in the current carbon dioxide concentrations or smaller. Enough is known of the response to combinations of conditions during development and growth of plants to caution against simple one-factor analysis. Although there is a dearth of information on the interactions under 'realistic' growing conditions, e.g. light and temperature regimens corresponding to environments under which plants normally grow, the problem is now being addressed in many climate change programmes.

The effects of environmental changes in temperature and increasing carbon dioxide on photosynthesis and plant production, and particularly their interaction with other environmental factors (e.g. nutrition, water supply), are still largely speculative. Interactions will be of the greatest importance; an example is the interaction between elevated carbon dioxide levels, transpiration and drought (Gifford, 1979; Eamus, 1991; Morison, 1993) mediated via decreased g_s. Smaller g_s reduces evaporation from the leaves and hence evaporative cooling, and so the leaves warm, thus tending to increase water loss. However, under many conditions, the balance between g_s and leaf temperature is favourable and less water is used under elevated carbon dioxide conditions, slowing the onset of soil drying and water stress. If the plant becomes drought stressed, the elevated C_i may protect against photoinhibition and oxidative damage, e.g. superoxide and peroxides. However, despite the advantages conferred by elevated carbon dioxide levels to photosynthesis, stress will decrease it and the production of dry matter; the question of response thus becomes a quantitative one. Models that incorporate these semiquantitative concepts into rigorous quantitative relationships are of only limited accuracy. Thus predictions of the likely effects of climate change are, at best, semiquantitative. Better models, using existing knowledge, are essential to identify the gaps in understanding, to quantify them and to assess the long-term effects of changing environment.

5.4. Adaptation and acclimation

The terms adaptation and acclimation, as applied to plant responses to their environment, are synonymous; the latter seems more accepted to denote long-term permanent change in plants, but implies no direction. Metabolic regulation is the complementary short-term response. Photosynthetic acclimation occurs when photosynthetic rate, measured under standard conditions (e.g. 'normal' atmospheric carbon dioxide level and temperature), is altered by growth of the plant or exposure of the leaf to different conditions and the change is associated with altered tissue composition ('machinery') or functions of the machinery or parts of it. Acclimation may be positive (increased photosynthesis, greater content or enhanced activity of components of the machinery) or negative (loss of assimilation rate, machinery, etc.). The rate at which acclimation occurs may vary but the process is essentially long term (taking days or weeks), irreversible under those conditions and related to changes in tissue composition as well as activity of components (Bowes, 1991; Lawlor, 1991; Bunce, 1992). Regulation may be said to occur if exposure of the plant or leaf to novel conditions induces rapid change (occurring over minutes or hours) in the photosynthetic rate measured under standard conditions, or alters the activity of the machinery or, less likely, amounts of machinery. Regulation is seen as readily reversible on return to the standard conditions; it may be positive ('up regulation') or negative ('down regulation'). Acclimation of photosynthetic rate via change in composition is equivalent to 'coarse control' of metabolic systems, regulation to 'fine control' (Stitt, 1991).

5.5. Photosynthetic system

To analyse the response of photosynthesis to carbon dioxide and climate change factors, a simplified picture (model) of the system is useful (Fig. 5.3). Regulation of photosynthetic rate (A) and acclimation may occur by changes in many parts of it; some known and potential alterations under conditions of climatic change are now considered.

(a) Carbon dioxide supply to Rubisco

The best attested change is the decreased stomatal conductance (g_s) (Mansfield *et al.*, 1990) with exposure of leaves to elevated carbon dioxide levels and, conversely, increases as the carbon dioxide content decreased. The factors regulating the mechanism responsible for determining g_s in relation to light, atmospheric humidity and C_a and C_i are unclear (Mansfield *et al.*, 1990). Stomatal opening depends on generation of large turgor pressure in the guard cells. Light induces this by stimulating accumulation of potassium ions, malate and chloride in guard cell vacuoles; the ions decrease the osmotic potential and the resultant water influx raises the turgor pressure. There is a requirement for energy, as ATP, for the transport of K^+; ATP is probably produced by the light reactions of the guard cell thylakoids. A source of organic carbon for the synthesis of malate is also required. Guard cell chloroplasts do not contain sufficient Rubisco for carbon dioxide assimilation by the photosynthetic carbon reduction cycle. Rather, PEPc catalyses carbon dioxide assimilation, with phosphoenol pyruvate (PEP) as substrate, giving oxaloacetate, which is reduced to malate. Starch is abundant in guard cells and is converted in the light to PEP, thus providing a ready source of malate. With bright light the energy supply in guard cell chloroplasts will be greater than in dim light, thus supporting stomatal opening (Schnabl, 1992). However,

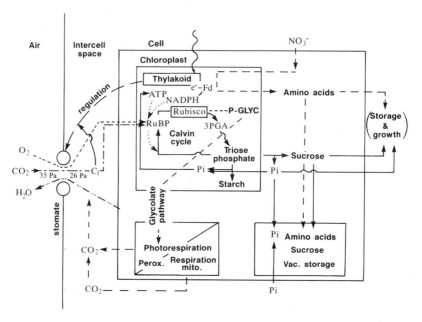

FIGURE 5.3 *Highly schematic representation of the photosynthetic system of a C3 plant. The path of carbon assimilation includes the supply of carbon dioxide via the stomata and transport across the mesophyll cell walls, membranes, etc., carboxylation of ribulose bisphosphate (RuBP) catalysed by Rubisco, the regeneration of RuBP by the photosynthetic carbon reduction (PCR) cycle, transport of the assimilation products from the chloroplast to the cytosol, where sucrose is synthesized, and translocation of sucrose to storage or growing organs. The process is driven by light capture in the thylakoid membranes, which results in electrons flowing from water to $NADP^+$, forming the reductant NADPH. Also ATP is synthesized by light energy driven processes. Both NADPH and ATP are required for regeneration of RuBP by the PCR cycle. The carbon flux via the photo-respiratory pathway is indicated.*

this mechanism does not readily explain the response of stomata to carbon dioxide. Low carbon dioxide concentration should reduce its assimilation by PEPc and hence decrease malate and starch content, lowering turgor pressure and decreasing g_s. Elevated carbon dioxide levels should increase starch and malate production and increase g_s. These effects are counter to the evidence. An alternative regulatory mechanism must be considered, perhaps with carbon dioxide assimilation serving to inhibit the opening mechanism. Such a system is suggested, based on the role of ATP in guard cell opening. Guard cell ATP content, some 30% of which is derived from photophosphorylation in the light (with oxidative phosphorylation by mitochondria providing the remainder) (Schnabl, 1992), is the main controlling factor which responds to C_i. Elevated carbon dioxide levels would stimulate carbon dioxide assimilation by the guard cell or tissue, thus competing for ATP with ion transport. Inadequate ATP would decrease the K^+ flux and content in guard cells. As PEPc is stimulated by potassium ions, its activity would be lower and less malate would be produced. Mitochondrial oxidation of malate would also decrease, and with it oxidative phosphorylation. Hence elevated carbon dioxide levels would decrease g_s. Such a (speculative) mechanism would also serve as a regulator of C_i: in bright light, low C_i means a low rate of photosynthesis and adequate ATP for ion transport and stomatal opening. This mechanism would be part of the control of g_s and would be integrated with other controlling factors, e.g. leaf water content and atmospheric humidity. Thus, several factors that may change in future climates determine stomatal responses.

Plants may also acclimate morphologically to elevated carbon dioxide levels, with decreased stomatal frequency (Woodward and Bazzaz, 1988). The mechanism of this long-term adaptation is not understood, but must involve genetic changes. Acclimation of the plant in the transport of carbon dioxide from intercellular spaces to the chloroplast stroma might also be a limitation to photosynthesis. This is not a purely physical process, as the rates of interconversion of carbon dioxide and bicarbonate ions are affected by the enzyme carbonic anhydrase (CA). CA activity decreases in leaves of tomato and lettuce plants grown in elevated carbon dioxide levels (Porter and Grodzinski, 1986; Katayama, 1990), so the equilibrium rate may be slowed. However, Beeson and Graham (1991) found no such increase in rose leaves. The reported reduction may not be important as the enzyme is greatly in excess and drastic decrease would be needed to affect assimilation (Hatch and Burnell, 1990). The mechanisms of such effects are not known, but must involve changes in the regulation of gene expression and selection in populations.

Despite smaller g_s in elevated carbon dioxide levels compared with normal air, the carbon dioxide concentration predicted for the end of the next century still substantially increases assimilation (Fig. 5.1). Most importantly, for given leaf temperature and atmospheric humidity, the substantial and consistent decrease in water loss in short-term and long-term experiments (Eamus, 1991) will improve the plant water balance and improve photosynthesis, resulting in greater water use efficiency.

(b) Light reactions, electron transport and photophosphorylation
Development of the thylakoid system responsible for light harvesting and associated processes is very light dependent, with many plant species showing considerable flexibility in response to light (facultative sun or shade types), although others are more genetically determinate, obligate sun or shade plants. However, there is no such well established response to carbon dioxide or to temperature. Elevated carbon dioxide levels decrease the content of chlorophyll per unit leaf area in some situations (see later), as do warm temperatures (Lawlor *et al.*, 1987a–c, 1988). This may decrease efficiency of light utilization but, because of the much greater photosynthetic rate

due to higher carbon dioxide concentrations and faster rate of enzyme reactions in warmer conditions, total photosynthesis may still be enhanced. Studies of the *in vivo* thylakoid reactions in leaves grown with elevated carbon dioxide levels show no substantial differences from plants grown in normal air (Delgado and Lawlor, unpublished observations on winter wheat grown under simulated future climate conditions; also Professor N. R. Baker, personal communication). Therefore, the basic energy capture and transduction mechanisms will remain constant. Increased photosynthesis may decrease the energy load on thylakoids and reduce the possibility of photoinhibition; the carotenoid protective mechanisms may then decrease in importance.

The amount of light energy available under climate change conditions is crucial to both photosynthesis and production. If solar radiation incident on vegetation were to decrease as a result of increased cloud cover (Lawlor and Mitchell, 1991), both would decrease. In many environments, productivity of complex vegetation (in contrast to photosynthesis of individual leaves) is a linear function of intercepted radiation (Monteith, 1977; Chapter 1). As more carbon is assimilated per unit of radiation captured, the efficiency of energy use will increase with elevated carbon dioxide level. However, warmer temperatures increase respiration and hence may decrease energy use efficiency. The net effect will depend on the response (possibly a decrease; Amthor, 1991) of respiration to elevated carbon dioxide level and to temperature.

(c) Carboxylation: Rubisco activity and the photosynthetic carbon reduction cycle

Interpretation of the photosynthetic response to carbon dioxide and temperature depends largely on the characteristics of the carboxylation reactions and specifically on the characteristics of the most abundant protein in the world, Rubisco (Stitt, 1991; Bowes, 1991; Long, 1991). This photosynthetic carbon reduction (PCR) cycle enzyme is responsible, in all photosynthetic organisms, for net accumulation of carbon with a net energy gain although, as mentioned earlier, PEPc enhances the efficiency of C4 plants. Rubisco catalyses carboxylation of ribulose 1,5-bisphosphate (RuBP), forming two molecules of 3-phosphoglyceric acid (PGA); this is used to regenerate RuBP by the PCR cycle or to produce carbohydrates. Oxygen competes with carbon dioxide for RuBP at the same active site on the enzyme; in the reaction, one molecule of PGA and one of phosphoglycollate (PG) is produced (Laing *et al.*, 1974; Jordan and Ogren, 1984). The PGA is consumed as described; PG is metabolized by the glycollate pathway, resulting in carbon dioxide release by photorespiration (PR). Kinetic analysis provides a way of calculating rates of carboxylation and oxygenation if carbon dioxide and oxygen are each regarded as competitive inhibitors of the other and the enzymic mechanism involves RuBP binding to Rubisco before the dissolved gaseous substrates bind. Such kinetic analysis has provided the basis of several mechanistic models of leaf photosynthesis (Farquhar *et al.*, 1980), which have been used to estimate the effects of carbon dioxide and temperature on photosynthesis. Rather than present the equations describing Rubisco function, we give calculated rates of carboxylation and oxygenation that would be catalysed at three different temperatures (Fig. 5.4), assuming that RuBP is in excess and that the effective oxygen concentrations in the chloroplast stroma are the same as in water in equilibrium with air containing $0.21 \, m^3 \, O_2 \, m^{-3}$ at the chosen temperatures. The calculations are based on kinetic constants for spinach Rubisco (Jordan and Ogren, 1984). At 10°C the rate of oxygenation at $350 \, cm^3 \, CO_2 \, m^{-3}$ is lower as a percentage of carboxylation than at the higher temperatures, but the shape of the curve shows that carboxylation is saturated at lower carbon dioxide concentration. This is shown more formally in Table 5.1 as the values of 'apparent' Michaelis constants $K_{m(app)}$ at the three temperatures; at 30°C half saturation of the carboxylase reaction is achieved at well above the current ambient concentration of carbon dioxide.

The significance of such information can be appreciated only when consideration is given to the extent to which elevated carbon dioxide diminishes photorespiration (oxygenation) as opposed to increasing carboxylation. The stoichiometry of the photorespiratory metabolic pathway (Farquhar et al., 1980) shows that one molecule of carbon dioxide is released in the glycine-to-serine conversion for every two molecules of RuBP oxygenated. If the increase in atmospheric carbon dioxide from the current 350 to $700 \, cm^3 \, m^{-3}$ changes the concentration in the chloroplast stroma from the equivalent of 200 to $400 \, cm^3 \, m^{-3}$ in the gas phase, we can derive the relative contributions of decreased oxygenase activity (PR) and increased saturation of carboxylation to

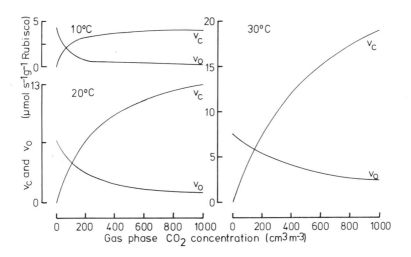

FIGURE 5.4 *Predictions of rates of carboxylation (v_c) and oxygenation (v_o) of ribulose 1,5-bisphosphate (RuBP) by ribulose bisphosphate carboxylase-oxygenase at 10, 20 and 30°C, with saturating RuBP. The oxygen concentration in solution is taken as that in equilibrium with air containing $0.21 \, m^3 \, O_2 \, m^{-3}$, and carbon dioxide concentration as that in solution in equilibrium with the gas-phase carbon dioxide shown. Rates were calculated using the kinetic constants for spinach Rubisco reported by Jordan and Ogren (1984).*

TABLE 5.1. *Effect of doubling atmospheric carbon dioxide on rates of carboxylation (v_c) and oxygenation (v_o), net carbon gain due to decreased photorespiration, and apparent affinity for carbon dioxide, $K_{m(app)}$ at three temperatures predicted from Rubisco kinetic parameters.*

Temperature (°C)	v_o (μmol s^{-1} g^{-1} Rubisco)		v_c (μmol s^{-1} g^{-1} Rubisco)		Increase in carbon gain because of decreased oxygenation (photorespiration) (%)	$K_{m(app)}$ (cm^3 m^{-3})
	200 cm^3 m^{-3} internal CO$_2$	400 cm^3 m^{-3} internal CO$_2$	200 cm^3 m^{-3} internal CO$_2$	400 cm^3 m^{-3} internal CO$_2$		
10	0.85	0.47	3.32	3.75	30.6	56.2
20	3.45	2.28	6.83	9.75	16.7	297
30	5.27	4.05	7.33	11.90	11.8	664

The calculations assume that changing ambient carbon dioxide from 350 to 700 cm^3 m^{-3} will change leaf internal gas phase carbon dioxide concentration from 200 to 400 cm^3 m^{-3}. Oxygen in the leaf is assumed to be that in water in equilibrium with air containing 0.21 m^3 O$_2$m^{-3}. Carbon dioxide evolved as a consequence of the oxygenase reaction is assumed to be produced at half the predicted rate of oxygenation. The $K_{m(app)}$ was calculated from the data for v_c in Fig. 5.1 using a computer program.

net photosynthesis from Fig. 5.4 (Table 5.1). Although oxygenation is faster at warmer than at cooler temperatures, its rate is little increased by temperature rises of the order predicted for climate change over the next 100 years (4°C), whereas the potential increase in carboxylation due to elevated carbon dioxide level is substantial. Thus, the observed stimulation of photosynthesis resulting from elevated carbon dioxide and changed temperature are explicable at the biochemical level in terms of Rubisco characteristics (Farquhar et al., 1980; Long, 1991).

The effects may not depend only on the Rubisco kinetics observed in vitro. In vivo, the amount of Rubisco may decrease (as mentioned earlier) as the tissue acclimates. The possible reasons for this are considered later. In addition, Rubisco is regulated in vivo; the degree of activation may decrease if RuBP supply falls and inhibitors bind to the active sites. Carboxyarabinitol 1-phosphate (CA1P) is an inhibitor, derived from carbohydrate metabolism, which is present in many species in darkness. Increased tissue carbohydrates with elevated carbon dioxide level could stimulate its production. The binding of CA1P is regulated by Rubisco activase, a protein that requires ATP (Streusand and Portis, 1987) and thus functions in the light to remove CA1P inhibition (Robinson and Portis, 1988). If Rubisco activase does not increase in activity in photosynthetic tissues grown in elevated carbon dioxide concentration, the specific activity of Rubisco could be reduced in vivo.

Reduction of the oxygenase reaction and thereby increased availability of RuBP for carboxylation increases net carbon dioxide assimilation substantially. Although Rubisco could be limited by the supply of RuBP, this is unlikely: the light reactions seem little affected by elevated carbon, dioxide, and there is no evidence of impaired PCR cycle activity. However, RuBP supply is determined by light and this will not increase in a changed climate, indeed the opposite is likely if cloud cover increases: RuBP will ultimately limit the capacity of photosynthesis to respond to elevated carbon dioxide. Of course, decreased synthesis of any components of the ATP/NADPH synthesizing mechanism of the PCR cycle (including Rubisco) or of systems delivering carbon dioxide to Rubisco, e.g. carbonic anhydrase, would limit the response. Currently, there is little biochemical or physiological evidence to suggest that such changes would be more than second order.

When models based on the Rubisco mechanism are used to predict the photosynthetic response of leaves to carbon dioxide and temperature (Farquhar et al., 1980; Long, 1991), the advantage of elevated carbon dioxide levels, particularly at warmer temperatures, becomes apparent. However, the models are not as successful in predicting the effects of cool temperatures and have not been applied to the interactions of carbon dioxide with nutrient supply. Under these conditions, phosphate transport between cell compartments or the amount of phosphate available may restrict the response and thus act as the limiting factor (Leegood and Furbank, 1986).

5.5.1. Carbon and nitrogen fluxes

The increased net assimilation in elevated carbon dioxide results in accumulation of carbohydrates, such as monosaccharides, disaccharides (particularly sucrose) and storage polymers, e.g. starch and fructans, in leaves and other organs (Cure and Acock, 1986; Stitt, 1991; Lawlor, 1991). Presumably all cellular pools increase, although storage compartments (e.g. vacuoles) would be expected to accumulate them earlier and to much greater concentration than other compartments. The ability to maintain the correct conditions for metabolic processes in particular compartments is essential for efficient system function, for example control of sugar concentrations is necessary to allow proper gene expression (Sheen, 1990; see later). Alterations in the amounts and proportions of other substrate and regulatory metabolites in particular compartments (e.g. amino acids, ATP) could affect response to carbon dioxide. Effective regulatory mechanisms will be required to prevent damage to metabolism. The fact that photosynthesis and production do respond to elevated carbon dioxide levels shows that homeostasis can be maintained, although the limits of adjustment are not understood for any plant under a range of conditions.

5.5.2. Photorespiratory fluxes

Rubisco kinetics show (Table 5.1) that the carbon flux in the glycollate pathway (Wallsgrove et. al., 1992) will decrease both absolutely and relative to the total carbon flux. The prediction from Table 5.1 is that increasing the carbon dioxide from 350 to $700 \, cm^3 \, m^{-3}$ would decrease the total flux of carbon through the glycollate pathway by 34% at 20°C and by 23% at 30°C. As a consequence, net carbon gain would be increased by 17 and 12% (Table 5.1) in the cooler and warmer temperatures, respectively. Although the carbon flux from the chloroplast to the cytosol through the glycollate pathway would be decreased, the absolute and relative amounts of total assimilate leaving the chloroplast as triose phosphate via the triose phosphate–inorganic phosphate (P_i) translocator in exchange for P_i from the cytosol must increase. In addition, accumulation of assimilate in the chloroplast stroma stimulates starch deposition there and may impair photosynthesis (Stitt, 1991), particularly when nutrient deficiency or cool temperatures limit growth. These fluxes are yet to be quantified in relation to elevated carbon dioxide levels and altered temperatures. A possible effect of such large changes in photorespiratory fluxes and compartmentation with elevated carbon dioxide level is a marked alteration in the content and proportions of carbohydrates and their relation to other metabolites (e.g. amino acids) in cytosol compared with chloroplasts. The implications of such changes for metabolism are only now becoming apparent, as with the effects of sugars on gene expression (see later).

Plants grown in elevated carbon dioxide conditions have a greater content of carbohydrates in tissues and therefore a smaller content of nitrogen in dry matter and also reduced content of soluble amino acids per unit of water (Lawlor, unpublished data), indicating their decreased availability for synthetic processes, e.g. protein production. The glycollate pathway is associated with a very active nitrogen metabolism (Wallsgrove et al., 1992). Ammonia released in the mitochondria during the decarboxylation of glycine and the formation of serine, is reassimilated by the glutamine synthetase (GS)–glutamate synthase (GOGAT) cycle in chloroplasts. In addition there is active transamination in

PR metabolism. A decrease of 20–35% in the flux of carbon through the glycollate pathway is likely to decrease the total pools and proportions of different amino acids in the cytosol. Also, the glycine-to-serine conversion is linked to reduction of NADH and potentially ATP synthesis in mitochondria (Moore *et al.*, 1992) and to consumption of reduced pyridine nucleotides in the peroxisomes; both could be decreased in elevated carbon dioxide, thus altering the adenylate and redox balance in cell compartments, although the quantitative aspects of the processes are not established (Moore *et al.*, 1992).

Although changes to the PR nitrogen cycle may affect amino acid proportions, net nitrogen gain by plants depends on reduction of nitrate and incorporation of ammonia into amino acids, both of which are linked to photosynthetic energy and carbon metabolism; this has been shown clearly in algae (Turpin, 1992). If the availability of photosynthetically derived reductant in the form of reduced ferredoxin decreases as a consequence of competition by carbon dioxide assimilation, the reduction of nitrate to ammonia could decline. Similarly, deficiency of ATP could inhibit the GS/GOGAT cycle of ammonia assimilation, reducing amino acid synthesis. The observed decrease in total amino acid content per unit fresh mass of plants grown in elevated carbon dioxide may be related to the reduced rate of nitrogen assimilation and could result in changes in cell structure and composition; it is important to know whether such effects occur and, if so, the mechanisms.

Increased consumption of light-generated reductant by assimilation in elevated carbon dioxide conditions should reduce the risk of photoinhibition to the photosynthetic apparatus and thus protect against excess light damage and also against the stresses that induce it. These effects have been observed in the marsh plant *Scirpus olneyii* in bright light (Long and Drake, 1991), suggesting its ecological importance.

5.5.3. *Photosynthesis, sources and sinks, and environment*

Temperature has marked effects on photosynthesis and photorespiration, as expected from the characteristics of Rubisco (Table 5.1 and Fig. 5.4). High temperature increases PR relativé to photosynthesis (Lawlor *et al.*, 1987b). However, the effects on tissue components are complex; in particular conditions (Lawlor *et al.*, 1987c) soluble protein and activity of some enzymes (e.g. fructose bisphosphatase, nitrate reductase, carbonic anhydrase) decreased whereas others increased (e.g. glycollate oxidase) and some remain unaffected (Rubisco, glutamine synthetase, serine- and glutamate-glyoxalate aminotransferases). The effects of increased temperature on metabolites are also complex but generally amino acids decrease markedly, sucrose and starch less so and RuBP and PGA are little altered in wheat leaves (Lawlor *et al.*, 1987a). Warmer temperatures stimulate growth and demand for assimilates, altering the assimilate supply (source) to demand (sink) balance (Lawlor *et al.*, 1988). Such interactions have not been examined for plants grown over long periods in a range of carbon dioxide concentrations combined with a range of temperatures. The projected increase in global temperature is relatively small compared with current ranges in particular environments, but may be crucial to the efficiency and competitiveness of vegetation. Information on the metabolic roles of temperature is important to assess the consequences of climate change on photosynthesis, plant growth and production.

Accumulation of assimilates in tissues of plants grown in elevated carbon dioxide conditions shows clearly that sink demand is less than supply (Stitt, 1991; Lawlor, 1991). The P_i transporter in the chloroplast membrane or the sucrose transporter at the plasmalemma could, theoretically, restrict the much increased fluxes out of the chloroplast and leaf, but this is unlikely given their capacity (Stitt, 1991). Sink demand is much more important, with warm temperatures accelerating rates of development, organ growth and dark respiration, and reducing assimilate accumulation (Lawlor et al., 1988). An example is the effect on winter wheat of current ambient temperatures for the UK and those 4°C above the ambient, combined with 350 and 700 cm^3 CO_2 m^{-3}. Dry matter production of the crop was increased by 15% by elevated carbon dioxide at ambient temperatures, whereas temperatures 4°C above ambient in 350 cm^3 CO_2 m^{-3} decreased growth by a similar proportion; the reduction in growth due to warmer conditions was compensated by the elevated carbon dioxide (Mitchell and Lawlor, unpublished data).

5.5.4. Nutrition and climate change

Nutrient supply substantially affects photosynthetic responses to environment, in determining both the development of photosynthetic structures and their function. The former is illustrated by the requirement for nitrogen during leaf development for synthesis of proteins and of pigments (Lawlor et al., 1987a). Inadequate nitrogen prevents growth of leaves as well as of sink/ storage organs, so that leaves that are formed may overproduce carbohydrates (Lawlor et al., 1987a,c, 1988); elevated carbon dioxide level sexacerbate this. The reduced nitrogen content in dry matter of plants grown in higher concentrations of carbon dioxide is a consequence of greater carbon accumulation in structural (cell walls, vascular tissue) and storage materials (carbohydrates), relative to the 'living' proteinaceous components ('dilution') (Lawlor, 1991). However, the larger biomass resulting from extra carbon dioxide may be achieved with about the same or even slightly more total nitrogen than in plants grown at current carbon dioxide levels (Mitchell and Lawlor, unpublished data). Phosphate supply is also of direct relevance to photosynthetic adaptation. Inadequate P_i results in deficiency of ATP and slows the transport of triose phosphates out of the chloroplast on the translocator. Add to this the inhibition of organ growth and it is clear that P_i supply may greatly affect the response to elevated carbon dioxide. Both P_i and nitrogen deficiency lead to carbohydrate accumulation, and this will be increased by elevated carbon dioxide levels. When carbon is made more available, deficiencies of other nutrient elements become apparent. Research beyond that directed to nitrogen and phosphorus supply is needed.

5.5.5. Feedback regulation of photosynthesis

The consequences of carbohydrate accumulation in photosynthetic cells are complex. Generally, inhibition of photosynthesis may take some time (many hours to days) to appear. There is no consensus about the mechanism of regulation (fine control) nor about the mechanisms of acclimation (coarse control). Reduction of available phosphate due to decreased import to the stroma and phosphorus-containing metabolite accumulation is likely. Rates of particular enzyme reactions may be slowed by accumulation of products (Stitt, 1991; Bowes, 1991) in the PCR cycle, e.g. phosphoglycerate kinase is regulated by triose phosphate and by AMP and ADP, which also regulate ribulose-5-phosphate kinase. Also, Rubisco activase is inhibited by ADP and stimulated by ATP, thus Rubisco activity is related to energy charge and CA1P (see earlier), production of which is possibly linked to carbohydrate status (Robinson and Portis, 1988). If assimilation is reduced by such mechanisms, utilization of captured light energy could decrease, resulting in photo-inhibitory damage and loss of photosynthetic capacity. Depending on the extent of the damage, the age of the leaf, its physiological state, etc., the effect might be reversible (i.e. regulatory) or not (i.e. acclimatory). The importance of such mechanisms remains to be assessed in relation to climate change.

The evidence that the accumulation of carbohydrates in cells regulates gene expression via sensitive promoters (Sheen, 1990) is potentially of considerable importance *in vivo*. The mechanism would explain decreased Rubisco and loss of pigment in elevated carbon dioxide conditions. Regulation of one gene may trigger a cascade of responses in other parts of metabolism, including synthesis of particular enzymes involved in senescence. This is very likely in photosynthesis, where many components of the mechanism are synthesized and also degraded in a regulated, coordinated manner, e.g. loss of thylakoids and Rubisco during ageing and senescence of leaves. The phenomenon of acclimation is thus not a *sine qua non* of the response to elevated carbon dioxide levels; rather it depends on the quantitative responses of other parts of the system (e.g. storage and growth) to variable environmental factors (e.g. nutrient supply, temperature) and is a regulatory mechanism that may become more important for plants growing in a changing climate.

5.6. Conclusions

Photosynthetic adaptation to climate change is, like climate change itself, subject to considerable uncertainty and thus open to speculation. Knowledge of the photosynthetic mechanism is sufficiently advanced to allow some predictions, based largely on models incorporating Rubisco enzyme kinetics, which match the experimental observations of increased photosynthesis and dry matter production. Some features of the changes in assimilate pools are understood, but details of how different metabolic processes interact, source–sink relations and the influence of changing atmospheric carbon dioxide and temperature, and of nutrition, are largely speculative, requiring experimental test. Models of the processes producing and consuming assimilates are, like our understanding of the mechanisms themselves, inadequate, yet necessary for quantitative assessment of the photosynthetic and other plant responses to climate change.

REFERENCES

Amthor, J.S. (1991). Respiration in a future, higher-CO_2 world. *Plant Cell Environ.* **14**, 13–20.

Beeson, R.C. and Graham, M.E.D. (1991). CO_2 enrichment of greenhouse roses affects neither Rubisco nor carbonic-anhydrase activities. *J. Am. Soc. Hort. Sci.* **116**, 1040–1045.

Besford, R.T. (1993). Photosynthetic acclimation in tomato plants grown in high CO_2. In: *CO_2 and Biosphere*, pp. 441–448, eds. J. Rozema, H. Lambers, S.C. van de Geijn and M.L. Cambridge. Dordrecht: Kluwer.

Boden, T.A., Kanciruk, P. and Farrell, M.P. (1990). *Trends '90, a Compendium of Data on Global Change*. Oak Ridge, Tenn.: Carbon Dioxide Information Analysis Center.

Bowes, G. (1991). Growth at elevated CO_2: photosynthetic responses mediated through Rubisco. *Plant Cell Environ.* **14**, 795–806.

Bunce, J.A. (1992). Light, temperature and nutrients as factors in photosynthetic adjustment to an elevated concentration of carbon dioxide. *Physiol. Plant.* **86**, 173–179.

Cure, J.D. and Acock, B. (1986). Crop responses to carbon dioxide doubling: a literature survey. *Agric. Forest Meteorol.* **38**, 127–145.

Eamus, D. (1991). The interaction of rising CO_2 and temperatures with water use efficiency. *Plant Cell Environ.* **14**, 843–852.

Edwards, G.E., Ku, M.S.B. and Monson, R.K. (1985). C4 photosynthesis and its regulation. In: *Topics in Photosynthesis*, Vol. 6, pp. 287–328, ed. J. Barber. Amsterdam: Elsevier.

Farquhar, G.D., von Caemmerer, S. and Berry, J.A. (1980). A biochemical model of photosynthetic CO_2 assimilation in leaves of C_3 species. *Planta* **149**, 78–90.

Gifford, R.M. (1979). Growth and yield of CO_2-enriched wheat under water limited conditions. *Aust. J. Plant Physiol.* **6**, 367–378.

Hatch, M.D. and Burnell, J.N. (1990). Carbonic anhydrase in leaves and its role in the first step of C4 photosynthesis. *Plant Physiol.* **93**, 825–828.

Jäger, J. and Ferguson, H.L., eds. (1991). *Climate Change: Science, Impacts and Policy*. Proceedings of the 2nd World Climate Conference. Cambridge: Cambridge University Press.

Jordan, D.B. and Ogren, W.L. (1984). The CO_2 specificity of ribulose 1,5-bisphosphate concentration, pH and temperature. *Planta* **161**, 308–313.

Katayama, M. (1990). Effect of high concentration of carbon dioxide in the atmosphere on carbonic anhydrase activity in plants. *Bull. Univ. Osaka Prep. Ser. B.* **42**, 71–75.

Kimball, B.A. (1983). Carbon dioxide and agricultural yield: an assemblage and analysis of 430 prior observations. *Agron. J.* **75**, 779–782.

Laing, W.A., Ogren, W.L. and Hageman, R.H. (1974). Regulation of soybean net photosynthetic CO_2 fixation by the interaction of CO_2, O_2 and ribulose 1,5-diphosphate carboxylase. *Plant Physiol.* **54**, 678–685.

Lashof, D.A. and Tirpak, D.A. (1990). *Policy Options for Stabilizing Global Climate*. New York: Hemisphere Publishing Corporation.

Lawlor, D.W. (1991). Response of plants to elevated carbon dioxide: the role of photosynthesis, sink demand and environmental stresses. In: *Global Climatic Changes on Photosynthesis and Plant Productivity*, pp. 431–445, eds. Abrol, Y.P. *et al.* Proceedings of the Indo-US Workshop, New Delhi, India.

Lawlor, D.W. and Mitchell, R.A.C. (1991). The effects of increasing CO_2 on crop photosynthesis and productivity: a review of field studies. *Plant Cell Environ.* **14**, 807–818.

Lawlor, D.W., Boyle, F.A., Young, A.T., Kendall, A.C. and Keys, A.J. (1987a). Nitrate nutrition and temperature effects on wheat: soluble components of leaves and carbon fluxes to amino acids and sucrose. *J. Exp. Bot.* **38**, 1091–1103.

Lawlor, D.W., Boyle, F.A., Young, A.T., Keys, A.J. and Kendall, A.C. (1987b). Nitrate nutrition and temperature effects on wheat: photosynthesis and photorespiration of leaves. *J. Exp. Bot.* **38**, 393–408.

Lawlor, D.W., Boyle, F.A., Kendall, A.C. and Keys, A.J. (1987c). Nitrate nutrition and temperature effects on wheat: enzyme composition, nitrate and total amino acid content of leaves. *J. Exp. Bot.* **38**, 378–392.

Lawlor, D.W., Boyle, F.A., Keys, A.J., Kendall, A.C. and Young, A.T. (1988). Nitrate nutrition and temperature effects on wheat: a synthesis of plant growth and nitrogen uptake in relation to metabolic and physiological processes. *J. Exp. Bot.* **39**, 329–343.

Leegood, R.C. and Furbank, R.T. (1986). Stimulation of photosynthesis by 2% oxygen at low temperatures is restored by phosphate. *Planta* **168**, 84–94.

Long, S.P. (1991). Modification of the response of photosynthetic productivity to rising temperature by atmospheric CO_2 concentrations: has its importance been underestimated? *Plant Cell Environ.* **14**, 729–739.

Long, S.P. and Drake, B.G. (1991). Effects of long-term elevation of CO_2 concentration in the field on the quantum yield of photosynthesis in the C3 sedge *Scirpus olneyi*. *Plant Physiol.* **96**, 221–226.

Mansfield, T.A., Hetherington, A.M. and Atkinson, C.J. (1990). Some current aspects of stomatal physiology. *Annu. Rev. Plant Physiol. Plant Mol. Biol.* **41**, 55–75.

Monteith, J.L. (1977). Climate and the efficiency of crop production in Britain. *Phil. Trans. R. Soc. Lond.* B **281**, 277–294.

Moore, A.L., Siedow, J.N., Fricaud, A.C., Vojnikov, V., Walters, A.J. and Whitehouse, D.G. (1992). Regulation of mitochondrial respiratory activity in photosynthetic systems. In: *Plant Organelles*, pp. 189–210, ed. A.K. Tobin. Cambridge: Cambridge University Press.

Morison, J.I.L. (1993). Response of plants to CO_2 under water limited conditions. In: *CO_2 and Biosphere*, pp. 193–209, eds. J. Rozema, H. Lambers, S.C. van de Geijn and M.L. Cambridge. Dordrecht: Kluwer.

Mousseau, M. and Saugier, B. (1992). The direct effect of increased CO_2 on gas exchange and growth of forest tree species. *J. Exp. Bot.* **43**, 1121–1130.

Porter, M. and Grodzinski, B. (1986). Acclimation to high CO_2 in bean carbonic anhydrase and ribulose 1,5-bisphosphate carboxylase. *Plant Physiol.* **74**, 413–416.

Robinson, S.P. and Portis, A.R. (1988). Release of the nocturnal inhibitor, carboxyarabinitol-1-phosphate from ribulose bisphosphate carboxylase/oxygenase by Rubisco activase. *FEBS Lett.* **233**, 413–416.

Roeckner, E. (1992). Past, present and future levels of greenhouse gases in the atmosphere and model projections of related climatic changes. *J. Exp. Bot.* **43**, 1097–1109.

Sage, R.F., Sharkey, T.D. and Seemann, J.R. (1989). Acclimation of photosynthesis to elevated CO_2 in five C3 species. *Plant Physiol.* **59**, 590–596.

Schlesinger, W.H. (1991). *Biogeochemistry: An Analysis of Global Change.* San Diego, Calif.: Academic Press.

Schnabl, H. (1992). Metabolic interactions of organelles in guard cells. In: *Plant Organelles*, pp. 265–279, ed. A.K. Tobin. Cambridge: Cambridge University Press.

Sheen, J. (1990). Metabolic repression of transcription in higher plants. *Plant Cell* **2**, 1027–1038.

Stitt, M. (1991). Rising CO_2 levels and their potential significance for carbon flow in photosynthetic cells. *Plant Cell Environ.* **14**, 741–762.

Streusand, V.J. and Portis, A.R. (1987). Rubisco activase mediates ATP-dependent activation of ribulose bisphosphate carboxylase oxygenase. *Plant Physiol.* **85**, 152–154.

Turpin, D.H. (1992). Metabolic interactions during photosynthetic and respiratory nitrogen assimilation in a green alga. In: *Plant Organelles*, pp. 49–78, ed. A.K. Tobin. Cambridge: Cambridge University Press.

Valle, R., Mishoe, J.W., Jones, J.W. and Allen, L.H. (1985). Transpiration rate and water use efficiency of soybean leaves adapted to different CO_2 environments. *Crop Sci.* **25**, 477–482.

Wallsgrove, R.M., Baron, A.C. and Tobin, A.K. (1992). Carbon and nitrogen cycling between organelles during photorespiration. In: *Plant Organelles*, pp. 79–96, ed. A.K. Tobin. Cambridge: Cambridge University Press.

Watson, R.T., Rodhe, H., Oescheger, H. and Siegenthaler, U. (1990). Greenhouse gases and aerosols. In: *Climate Change: The IPCC Scientific Assessment*, pp. 1–40. Cambridge: Cambridge University Press.

Woodward, F.I. and Bazzaz, F.A. (1988). The response of stomatal density to CO_2 partial pressure. *J. Exp. Bot.* **39**, 1771–1781.

Whole plant responses

CHAPTER 6

Temperature stress

C. J. POLLOCK, C. F. EAGLES, C. J. HOWARTH,
P. H. D. SCHÜNMANN and J. L. STODDART

6.1. Introduction

Temperature, moisture and light are the major determinants of plant growth. Each can be limiting in specific circumstances and, frequently, their influences on growth or survival are exerted coincidentally. In higher latitudes, temperature and moisture limitations usually have seasonally separated impacts but, in many subtropical areas, simultaneous stresses are frequently imposed. It is a generalization, however, that the more intense survival demands are often primarily related to temperature, with exposure to opposite extremes evoking differing cellular responses.

Thermal responses in biological systems occur with varying temporal characteristics. They can be apparently instantaneous or progressive over a time period and, with the exception of situations where transgression of the tolerance limits is sufficiently severe to provoke catastrophe, the effects are normally reversible. It is also evident that many species have evolved genetic strategies to confer protection against extreme events, either by metabolic alterations or by anticipatory physiological changes regulated by environmental cues correlated with temperature change. An example of the latter is photoperiodically induced leaf drop in deciduous temperate tree species.

Long-term acclimation leads to progressive changes in cell structure and metabolism, which reduce sensitivity to thermal damage. These changes involve altered gene expression and are typically adaptations to slow-rate sustained trends in the temperature environment. Such adaptation, or 'hardening', is freely reversible when conditions return to a more favourable range.

In contrast, heat-shock responses are protective measures against potentially lethal rapid-rate upward departures from optimal temperatures. Here again gene expression is invoked but is accompanied by shutdown of the pre-existing protein synthesis. Heat-shock genome elements code for polypeptides acting as protein conformation protectants, closely related to chaperonins. Only very short exposures to high temperatures are required for the induction of heat-shock proteins, and synthesis ceases when the stress is relieved.

Finally, and fundamentally, there are the short-term cellular adjustments to temperature change. These may be at the thermodynamic level in respect of enzyme kinetics or may reflect higher order interactive events in organelles or other subcellular structures. Such events are not normally viewed as stress related, but they clearly set the basic response limits of the organism and involve genetically determined constraints which protect against damage. Assessments of short-term thermal sensitivities for basic physiological processes (photosynthesis, respiration, growth) show clearly that growth processes have the highest Q_{10} values, usually by at least an order of magnitude (Pollock et al., 1983). The kinetics of rapid effects of temperature change on growth do not provide compelling evidence for de novo gene expression but indicate action sites concerned with the modulation of cell-wall rheology.

To modify the thermal stress tolerance, or growth kinetics, of crop species, at either end of the temperature range, it will be necessary to understand the details of subcellular control and the genetic basis of the appropriate response types. Furthermore, these will correlate with other critical metabolic and developmental processes. The following sections of this chapter deal with acclimation, heat-shock and instantaneous effects at the whole plant and cellular levels, and have the ultimate objective of providing such a manipulative framework.

6.2. Low-temperature acclimation

Long-term exposure to low temperature often produces modifications in plant form as a result of alterations in the patterns of cell division and extension. Such changes have been described during overwintering and during prolonged exposure of plants to low temperatures in controlled environment facilities (Dale, 1965; Eagles, 1967; Haycock, 1981; Huner et al., 1981; Thomas, 1983; Krol et al., 1984; Pollock et al., 1984). These responses are frequently characterized by prostrate plants with shorter leaf sheaths and laminae. Marked genotypic differences in the degree of such responses have been described for a range of temperate species (Cooper, 1964; Robson and Jewiss, 1968; Ostgard and Eagles, 1971; Eagles and Othman, 1981). Eagles (1967) showed that the morphological changes induced by growth at 5 °C, particularly in short-day conditions, were much more extreme in northern European populations of Dactylis than those observed in plants of Mediterranean origin. The patterns of plant response may be modified by interactions between low temperature and other environmental, or developmental, factors (Thomas and Norris, 1977; Parsons and Robson, 1980; Kemp et al., 1989; Hay, 1990).

Exposure of temperate plants to declining temperatures in the autumn is not only associated with reductions in growth and altered plant form, but also with the development of increased ability to tolerate subsequent exposure to freezing temperatures. Winter survival under field conditions will often depend on the ability to tolerate a wider range of environmental stresses, including rapidly fluctuating temperatures, low light intensities, desiccation, wind, snow and ice cover, and disease. The relative importance of these factors depends on local climatic conditions as well as management practices, although across sites and years low temperature is probably the most important single factor causing winter damage.

When considering effects of suboptimal temperatures on plant processes, it is important to distinguish between limitations attributable to reversible reduction in the rates of certain key processes and those imposed by injury, where injury can be defined as an inability to restore normal function, following removal of the causal constraint. Cold-induced injury occurs over a wide range of temperatures, dependent on both genotype and environment. This has led to classification of plant species into three groups. Chilling-sensitive plants suffer injury at positive temperatures, usually below 10–15 °C; freezing-sensitive plants are damaged by exposure to any temperature below 0 °C and freezing-resistant plants are able to survive subzero temperatures down to a limit which is characteristic of the genotype (Larcher, 1981). The present consideration of long-term acclimation will concentrate on freezing-resistant plants, using temperate grasses as a case study.

Survival of the stresses mentioned previously frequently depends on the development of resistance or tolerance through a process of hardening. The ability to cold harden and the resultant cold tolerance are essential attributes for successful exploitation of a perennial or overwintering crop in a northern temperate climate. This ability to harden may limit the extent to which enhanced low-temperature growth can be exploited, because a strong negative correlation invariably exists between growth at low temperatures and acquired freezing tolerance (Cooper, 1964). In a more extensive study of 86 perennial ryegrass accessions, Humphreys (1989) derived relationships between 14 winter-related characters which showed good freezing tolerance to be associated with low autumn growth and with maintenance of a high water soluble carbohydrate content through the winter. Factors which reduce carbohydrate reserves, such as autumn disease or disease under snow cover, autumn growth forced by high nitrogen applications, late cutting or uncompensated respiratory loss through growth at low light intensities, can all seriously reduce winter hardiness. These adverse effects of autumn managements on ability to cold harden have also been reported by Charles et al. (1975) and Hides (1978a,b).

The physiological basis of these agronomic responses has been studied in controlled environment conditions using a number of laboratory-based techniques (Lawrence et al., 1973; Fuller and Eagles, 1978; Larsen, 1978). Two major factors influence the extent of cold hardening achieved by a genotype: the temperature and the duration of acclimation. The hardening process is relatively slow, requiring several days' exposure to appropriate temperatures for completion. Using the artificial freezing technique described by Fuller and Eagles (1978), we find that temperate grasses and cereals have an LT_{50} (lethal temperature for 50% kill) of about -6 °C for unhardened plants grown at 15 °C. During the first 2 days' acclimation at 2°C, the change in LT_{50} can range from 0.3 °C for cv. Bulwark, a susceptible winter oat, to more than 2.5 °C for cv. Engmo, a resistant timothy, whereas cultivars of

Lolium perenne show intermediate hardening. With longer durations of exposure to 2 °C, an asymptote is normally approached, with LT_{50} values ranging from about -9 °C for the susceptible winter oat cultivar to -20 °C for the hardy Norwegian timothy cultivar and -12 to -17 °C for cultivars of *L. perenne*.

With this potential to acquire hardiness it is reasonable to predict that winter damage should not occur in relatively mild maritime environments. However, this is not the situation; winter damage is frequently recorded and may even be more severe in warmer lowland sites than in more stable upland environments (Thomas and Norris, 1979). It is apparent that factors other than hardening ability can determine winter damage and, in the widely fluctuating temperatures typical of a maritime environment, the ability to maintain acquired hardiness at elevated temperatures may be critical to survival. It is well established that long-term acclimations are normally reversible, although there is genetic variation in the time scale and extent of this reversal.

Genotypic differences in temperature thresholds for hardening and dehardening have been proposed previously (Habeshaw and Swift, 1978; Fuller and Eagles, 1981). Recently, hardening and dehardening experiments conducted in a more comprehensive range of temperature environments in a thermal-gradient tunnel (Eagles, 1989) have reinforced the view that hardy cultivars harden more effectively at warmer temperatures than susceptible cultivars, which require temperatures below 4 °C for effective hardening. Similarly, hardy cultivars are much better able to maintain hardiness acquired at 2 °C when temperatures are raised than susceptible cultivars, which deharden significantly even at 4 °C.

Further evidence of the genetic variation in these responses to temperature was provided when hardening and dehardening were analysed as quantitative processes dependent on temperature and time (Gay and Eagles, 1991). The model was based on the time courses of the processes changing exponentially to an asymptote which was logistically related to temperature. Parameters of biological interest, such as the initial rates of hardening and dehardening for a given temperature and the percentage of the process completed in a given time, were derived and compared for cultivars of contrasting hardiness (Table 6.1). Initial rates of hardening at 4 °C were almost five times faster for a hardy than for a susceptible cultivar of *L. perenne*, and a much greater proportion of the potential hardening was also completed after 14 days. In contrast, the initial rates of dehardening at 4 °C were about two times faster for the susceptible than for the hardy cultivar, whereas a greater proportion of the dehardening process was completed after 14 days for the susceptible cultivar. Genotypic differences in these parameters may determine the seasonal fluctuations in cold tolerance recorded in a maritime environment. Earlier development of hardiness in the autumn for hardier cultivars and the loss of hardiness of susceptible cultivars during mid-winter are examples of such temperature responses. The implications of earlier development of hardiness for poor autumn growth should not be overlooked.

Whole-plant responses have dominated studies of cold tolerance because of the agronomic or breeding context in which they have been conducted.

TABLE 6.1. *Derived parameters for cold hardening and dehardening in* Lolium perenne.

	Premo (hardy)	Grasslands Ruanui (susceptible)
Initial rate of hardening at 4°C (°C day^{-1})	−0.673 (0.073)	−0.138 (0.019)
Initial rate of dehardening at 4°C (°C day^{-1})	0.204 (0.041)	0.440 (0.092)
Completion of hardening in 14 days (%)	75.3 (5.4)	34.1 (12.6)
Completion of dehardening in 14 days (%)	81.5 (4.8)	97.6 (1.3)

Values in parentheses are standard errors.

More detailed studies of the method of recovery after freezing in grasses and cereals have pointed to tissue-specific hardening (Eagles *et al.*, 1993). Several experiments have shown, somewhat surprisingly, that the main apex does not harden. Regardless of the LT_{50} recorded for plant survival the apical meristem is killed at about −6 °C for cultivars of *L. perenne* (Fig. 6.1), *L. multiflorum*, *Avena sativa* and *Phleum pratense*. The change in whole-plant LT_{50} with hardening is a function of the plant's increased ability to produce viable regrowth from tiller sites which, at the time of freezing, are not apparent as differentiated tissues. The basis of this recovery has not yet been established,

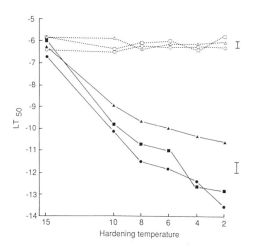

FIGURE 6.1. *Cold tolerance* (LT_{50}) *of main apex (. . . . . .) and whole plant* (——) *of three cultivars of* Lolium perenne *hardened at five temperatures (°C).*

although alternative hypotheses are currently being investigated.

With this wide range of genetic variation in cold tolerance responses it should be possible to improve this character by conventional breeding techniques. In fact, significant improvements in LT_{50} can be achieved by selection of survivors from rigorous cold tests followed by a crossing programme, but undesirable growth patterns often result. Modern molecular techniques may offer an alternative approach for improvement of cold tolerance. Changes in the products of gene expression during hardening have been shown in barley (Hughes and Pearce, 1988; Dunn et al., 1990) and oats (Quigley, 1991). Caution may be necessary in the selection of tissue for analysis in view of tissue-specific hardening responses.

6.3. Heat-shock proteins and rapidly induced thermotolerance

Rapid temperature changes, particularly those towards the upper end of the adaptation range for individual plant species, can produce dramatic changes in the patterns of gene expression. These changes occur concomitantly with the induction of enhanced thermotolerance and there is strong, albeit circumstantial, evidence for a causal link between the two processes.

Protein synthesis, in general, is a very thermosensitive metabolic process and changes in temperature can affect both the amount and type of polypeptides produced. If the growth temperature of a plant is increased beyond a particular point, a sudden decline in the ability to synthesize protein occurs. Concomitantly, seedling growth ceases. However, tolerance of both protein synthesis and seedling growth to a previously lethal high temperature can be induced by prior short exposure to a sublethal high temperature. During such a pretreatment the synthesis of a specific set of proteins, the heat-shock proteins (HSPs), occurs from messenger ribonucleic acid (mRNA) that is newly transcribed in response to high temperature (Fig. 6.2). The synthesis of the normal cellular proteins is reduced. This process is detectable within minutes of the onset of stress and is found in all plant and animal species so far studied (Schlesinger et al., 1982; Ougham and Howarth, 1988). In different species, the response is induced at different temperatures; this temperature is related to the optimal growth temperature for that organism. For example, in the tropical cereals sorghum and millet, HSPs are induced at 45 °C; in the temperate grass *Lolium temulentum*, HSPs are induced at 35 °C; and in the snow fungus *Fusarium nivale* (growing at 12 °C) HSPs are induced at 25 °C. In each case these are temperatures that the organisms might expect to encounter. In general, the heat-shock response is induced by a 10 °C rise from the optimal growth temperature. There are a number of major families of HSPs, classified on the basis of molecular weight: high molecular weight HSPs (> 80 000), the HSP70 family (i.e. HSPs of molecular weight approximately 70 000) and low molecular weight HSPs (14 000–30 000). Other proteins that are heat inducible include ubiquitin, a 76 amino-acid protein. Two-dimensional electrophoresis emphasizes the complex nature of the changes that occur during heat shock (Fig. 6.3), with the HSP70

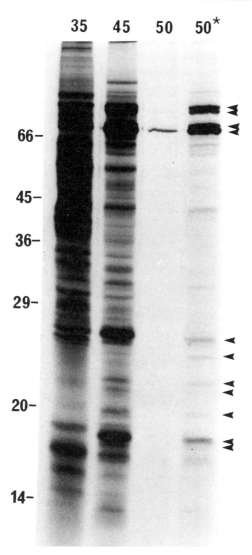

FIGURE 6.2. *Electrophoretic separation of proteins synthesized by 2-day-old millet seedlings, grown at 35 °C, during a 2 h treatment in the presence of a radioactive amino acid at the temperatures shown. Radiolabelled proteins visualized by autoradiography. *Seedlings exposed to 45 °C for 1 h prior to treatment. ▶, Major heat-shock proteins.*

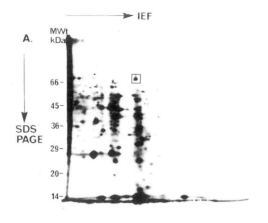

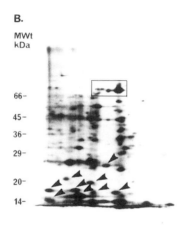

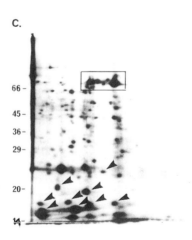

FIGURE 6.3. *Two-dimensional electrophoresis of* in vitro *translation products of total RNA extracted from sorghum seedlings after treatment at 35 °C (A), 35 °C, 3 h + 100 μM arsenite (B), or 45 °C, 3 h (C). Proteins were separated in the first*

family of proteins resolving into at least 20 distinct spots (Wilkinson *et al.*, 1990).

The heat-shock response is controlled at both the transcriptional and the translational level. In almost all tissues, induction of HSP synthesis requires transcriptional activation of their genes (Lindquist, 1981). A *cis*-acting DNA sequence, the heat-shock element (HSE), has been found to be necessary for heat-induced transcription (Pelham, 1985). The induction of *HS* gene expression is mediated by the binding of a *trans*-acting transcriptional activator, the heat shock factor (HSF), to the HSE (Bienz and Pelham, 1987). Not only is there transcriptional activation of *HS* genes, but transcriptional repression of most previously active genes (Schöffl *et al.*, 1987). Repression of normal protein synthesis is also under translational control, exerted over the normal cellular mRNAs. These are mostly not degraded during heat shock but persist in a translatable form (Lindquist, 1981). On recovery from heat shock, these pre-existing mRNAs are translated (Storti *et al.*, 1980). If RNA prepared from heat-shocked tissue is translated *in vitro*, the labelled products include not only heat-shock proteins but also those proteins synthesized *in vivo* at normal temperatures.

The heat-shock response is not only very rapidly induced but is a transient phenomenon; a prolonged high temperature treatment does not result in a continuation of the response. In soybean, continuous heat-shock at 40 °C results in a return to the normal pattern of protein synthesis by 6 h, although HSP mRNAs persist for at least 9 h. However, run-off transcription from nuclei extracted from heat-shocked tissue indicated that active transcription of the *HS* genes occurs only during the first 2 h at 40 °C (Kimpel *et al.*, 1990). When soybeans are returned to the control temperature of 28 °C, HSP mRNA rapidly declines. HSP mRNAs are apparently more stable at heat-shock temperatures. The response of protein synthesis to long-term heat shock differs between species. Necchi *et al.* (1987) found that, in a number of cereal species, heat shock of longer than 6 h resulted in the appearance of a new set of proteins, which they termed 'late' HSPs. Prolonged heat shock in maize also resulted in the appearance of a novel set of proteins unrelated to HSPs (Cooper and Ho, 1983). A long-term heat treatment is likely to impose many secondary consequences. This is due to indirect effects on metabolism caused by alterations in assimilation and carbon allocation, in addition to coping with the immediate problem of a sudden short temperature rise.

It is hypothesized that HSPs and other proteins induced under stress conditions enable the plant to make biochemical and structural changes that permit them to endure the stress. One of the reasons for implicating HSPs in thermotolerance is the correlation between their synthesis and the development of the ability for survival at a previously non-permissive temperature

---

dimension by isoelectric focusing (IEF) and in the second dimension by sodium dodecyl sulphate–polyacrylamide gel electrophoresis (SDS-PAGE), and radio-labelled proteins were visualized by autoradiography. Major low molecular weight proteins are indicated by ▶, and HSP70 family is delineated by □.

(Fig. 6.2). Treatments other than heat shock are also known to induce the synthesis of HSPs. If the synthesis of HSPs does result directly in thermotolerance, one might expect that HSP synthesis, as a result of a stress other than heat shock, would also induce thermotolerance. This approach has been used to examine the effect of various chemical treatments on protein synthesis and the development of thermotolerance in sorghum and millet seedlings. In sorghum seedlings, treatment with sodium arsenite or sodium malonate (both inhibitors of respiration) at normal growing temperatures resulted in the synthesis of a subset of HSPs, although 'normal' protein synthesis was not affected. In millet seedlings, little alteration in the proteins synthesized was detected. Interestingly, treatment with arsenite and malonate induced thermotolerance in both sorghum and millet seedlings. Additionally, HSPs were synthesized at a previously non-permissive temperature if such a pretreatment was given (Howarth, 1990a). Thus both arsenite and malonate induced thermotolerance of growth and of protein synthesis with little synthesis of HSPs during the pretreatment with these chemicals. Extraction of RNA from arsenite-treated seedlings, however, indicated that translatable RNA for HSPs is present in both sorghum and millet at normal growing temperatures, so it is possible that it is not translated *in vivo* until an increase of temperature is also detected. Two-dimensional electrophoresis indicates the remarkable similarity between the translation products of RNA from heat-shocked and arsenite-treated seedlings (Fig. 6.3). Thus the expression of HSP genes and the development of thermotolerance are both induced by arsenite, and this provides further evidence for their involvement in thermotolerance.

The induction of thermotolerance is dependent on the temperature of the inductive conditions. For example, a high pretreatment temperature, during which HSPs are not actually synthesized although their gene expression is initiated, requires a recovery period, during which HSPs are synthesized, before thermotolerance is induced (Howarth, 1991). Thermotolerance develops in parallel to the recovery of protein synthesis and the synthesis of HSPs. However, a lower inductive temperature induces thermotolerance immediately. This is because protein synthesis continues and HSPs are synthesized during the inductive period. Li and Hahn (1980) proposed that there are two stages in the development of thermotolerance: an initial trigger and the subsequent development of thermotolerance. During moderate pretreatments, these processes occur simultaneously. At higher temperatures, only the trigger occurs and development of thermotolerance takes place once the tissue is returned to its normal growth temperature. These two stages can also be analysed in protein synthesis terms, the trigger resulting in gene transcription, and the development process being translation of those mRNAs. The different time-scales in the development of thermotolerance that result from the difference in the severity of the inductive pretreatment explain why different laboratories often obtain contrasting results and make contrasting conclusions. Not only does the relative severity of pretreatments differ but also of the subsequent treatment, both in actual temperature and in duration. The physiological state of the tissue at the outset of the experiment can be critical. For young seedlings, the condition of the seed and indeed the maturation of that seed on the parent plant can affect its subsequent thermosensitivity (Kiegl and Mullen, 1986; Howarth, 1989, 1990b). When attempting to interpret the consequences of a given treatment to a tissue it is important to consider its physiology in addition to biochemical or molecular analyses.

Survival of high-temperature conditions also requires that the tissue can recover to its normal cellular functioning once heat shock has ceased. The recovery of normal protein synthesis following heat shock is also dependent on the severity of that heat shock. Even when normal protein synthesis has resumed, once heat-shock conditions are no longer prevalent, HSPs persist in tissue for many hours following their synthesis (Howarth, 1991; Chen *et al.*, 1990). The intracellular localization of HSPs may alter during recovery from heat shock. In *Drosophila*, HSPs concentrate in the nuclei during heat shock, but on return to normal growth conditions they become distributed throughout the cytoplasm (Velazquez and Lindquist, 1984). A subsequent heat shock results again in the concentration of HSPs in the nucleus. Chen *et al.* (1990), using an antibody to HSP21 in pea, demonstrated that it was still present 3 days after the initial shock. This may be due to the involvement of HSPs in the repair processes. Additionally, it may be to enable survival in a future heat shock. In sorghum and pearl millet seedlings it was found that, despite the presence of HSPs after 24 h of recovery, thermotolerance did not persist (Howarth, 1991). Repeat heat shock, during which HSPs are again synthesized from newly transcribed mRNAs, returns the tissue to its thermotolerant state. This suggests that *de novo* synthesis of HSPs is required for thermotolerance or that the acquisition of thermotolerance was not related to the presence of HSPs. It may be the free HSP concentration that is important. If, for example, HSP70 binds to thermally denatured proteins, this will affect its effective concentration. The synthesis of HSPs is transient and is not concerned with long-term heat hardening of tissue, during which complex readjustments in metabolism occur.

6.4. Heat-shock protein synthesis in the natural environment

Although temperature changes in the field can be very rapid (Fig. 6.4), particularly in tropical areas, plants are not exposed to a temperature shock as such but to a relatively gradual increase in temperature. Field conditions can be mimicked in the laboratory and it is found that the heat-shock response is not only induced by a sudden increase in temperature (Altschuler and Mascarenhas, 1985; Chen *et al.*, 1990; Howarth, 1991). Moreover, thermotolerance is also induced by a gradual increase in temperature. The upper limit for both protein synthesis and survival when the temperature is increased gradually is much higher than for a sudden increase, as the acclimation process actually occurs during the change in temperature. HSP gene expression has also been detected in field-grown plants (Kimpel and Key, 1985; Burke *et al.*, 1985).

Plants in the field are often exposed to high temperature conditions on a daily basis (Fig. 6.4), occurring for a few hours at midday and recurring

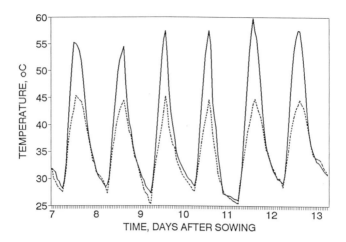

FIGURE 6.4. *Diurnal temperature variation in Rajasthan, India, during the growing season (June 1990). Temperature was measured using thermocouples throughout the day. Full seedling emergence occurred by 6 days after sowing. ——, Soil surface; . . . , air.*

each day, and this can have a detrimental effect. Experiments with pearl millet seedlings have shown that, although seedlings may be able to synthesize HSPs on their first exposure to high temperature, on subsequent days this ability is lost (Howarth, 1991). It is possible that the seedlings have not recovered from the consequences of a preceding heat shock by the time they are exposed to further high-temperature conditions.

In the natural environment, high temperatures are often found in combination with limited water availability. This in itself can affect plant temperature, as the plant may be unable to cool itself effectively by transpiration. Burke *et al.* (1985) found that HSPs were expressed to a much greater extent in cotton grown in non-irrigated fields where leaf temperatures were 10 °C higher than in irrigated fields. Seedlings in very young stages of development may be particularly vulnerable as they have not developed a full transpiration stream and an effective cooling system.

In the field there can also be a spatial distribution of stress. A seedling emerging from the soil is likely to be exposed to a wide range of temperatures. This is particularly evident in tropical areas where, at midday, a seedling may be exposed to a root zone temperature of 40 °C, a soil surface temperature of 60 °C and an air temperature of 45 °C. Within the diurnal cycle, these temperatures may drop to 25, 20 and 21 °C respectively (Fig. 6.4). For a grass or cereal seedling which has its shoot meristem situated near the soil surface, this has considerable consequences for the growth and survival of the seedling (Peacock, 1975). HSPs may be important in the survival of such wide fluctuations in temperature by stabilizing this meristematic tissue.

Thus, the heat-shock response is not confined to sudden temperature changes in the laboratory. The shock protein syndrome may well be an important part of the acclimatory strategies of higher plants.

6.5. Rapid responses to temperature change

Chemical reactions show an immediate and reproducible response to temperature change. It is generally accepted that this response is independent of measurement temperature, i.e. Q_{10} values are constant (Pollock, 1990). The existing chemical machinery of plant cells also responds rapidly to temperature change, although the effects are more complex. Q_{10} values for both biochemical reactions and physiological processes depend strongly on measurement temperature, leading to threshold values for specific responses. In addition, genetic variability in response is common. At the biochemical level, for example, alterations in starch–sucrose balance in potato tubers stored at low temperature can be attributed to the differential cold lability of key glycolytic enzymes (ap Rees et al., 1988). Whole-plant responses are equally dramatic, although the specific chemical basis may not be fully understood.

At the whole-plant level, cold is known to slow the growth of plants. In an attempt to understand the cause of this response, the temperature sensitivity of various processes following the transfer of grass plants from 20 °C (close to their growth optimum) to 5 and 2 °C were measured (Table 6.2; Pollock et al., 1983). In the grass Lolium temulentum the processes of carbon fixation were found to be much less sensitive to cold than was growth itself, even though the growth reduction observed was freely reversible if plants were subsequently returned to 20 °C (Table 6.2). This reversible cessation of growth at low positive temperatures is characteristic of chilling-resistant plants and can be observed in the field during the winter when a few warm days immediately lead to a flush of grass growth, which ceases when the next frost occurs.

TABLE 6.2. *Response of primary processes to temperature in seedlings of Lolium temulentum.*

Parameter	Rate at 20 °C	Rate at 5 °C	Rate at 2 °C	Q_{10} 2–20 °C	Q_{10} 2–5 °C
Relative leaf extension rate (d^{-1})	0.70	0.11	0.02	7.2	293.7
Relative growth rate (d^{-1})	0.064	0.025	0.008	3.2	44.6
Relative root growth rate (d^{-1})	0.069	0.030	0.015	2.3	10.1
Photosynthetic capacity $(\mu l\ O_2 h^{-1} mg^{-1}$ chlorophyll)	220	40	24	3.4	5.3
Whole plant relative growth rate (d^{-1})					
Control plants	0.09	–	–	–	–
following 41 days at 2 °C	0.08	–	–	–	–

For the study of growth, grasses and cereals are particularly well suited. Leaves elongate in an essentially linear fashion from a basal extension zone immediately above the site of insertion of the leaf. Increase in length is much easier to measure than is the increase in area of a growing dicot leaf, and auxanometers have been used by botanists for many years to amplify and quantitate this response. Auxanometers were used to measure how growth rate changed immediately following transfer of whole seedlings from 20 °C to either 5 or 2 °C (Thomas and Stoddart, 1984). Cold reduced the rate of growth within 30 min of transfer (the resolution limit of the auxanometers), and the new reduced growth rate remained stable throughout the subsequent growth of that leaf. The rapidity of this response suggested that cold was directly affecting the processes of leaf extension rather than being sensed elsewhere in the plant and the stimulus transmitted chemically to the extension zone. This suggestion was confirmed by studies using localized cooling and an even more sensitive auxanometer capable of measuring leaf length increments as small as $10 \mu m$, equivalent to approximately 20 s of growth at 20 °C by seedlings of *L. temulentum* (Stoddart *et al.*, 1986). Imposition of localized cooling anywhere on the seedling except the basal extension zone had no immediate effect on growth; cooling the extension zone, on the other hand, produced an immediate effect with no discernible lag phase. The effect of chilling the meristem was freely reversible and confirmed that cold acts directly on the extension zone to affect the processes involved in the increase in cell size. Temperature change may also affect the rates of cell division (Francis and Barlow, 1988), but the speed of the responses are such that the supply of divided, yet unextended, cells would be essentially unchanged throughout the experiment.

Cell extension is dependent on the interplay of two factors: the rigidity of the cell wall and the turgor pressure developed by the protoplasm inside it. If the turgor pressure exceeds the value at which the cell wall becomes plastic, cell size will increase. Measurement of turgor pressure in extending leaf cells of *L. temulentum* shows that it is largely unaffected by temperature change (Table 6.3), strongly suggesting that cold reduces extension growth by increasing the rigidity of the primary cell walls (Thomas *et al.*, 1989). This rigidity is controlled by a range of chemical and physical interactions between various classes of wall polymer, and by the actions of specific

TABLE 6.3. *Effect of temperature on extension rate and mesophyll cell turgor during leaf growth in* Lolium temulentum.

Measurement temperature (°C)	Growth rate ($\mu m \ min^{-1}$)	Turgor pressure (MPa)
18	25	0.50
15	10	0.50
10	8	0.52
5	2	0.51
2	0	0.49

enzymes. Some confirmation that temperature change is both sensed and transduced into altered extension rates by alterations in the physical properties of cell walls within the extension zone has been obtained using a stature mutant of barley known as *slender* (Pollock *et al.*, 1992). The mutant, which is the product of a single recessive gene, continues extension growth to $-7\,°C$, whereas the normal form ceases growth at $+4\,°C$ (Stoddart and Lloyd, 1986).

Turgor–growth rate–temperature interactions of the two forms can be compared directly, as can the rheological properties of the extension zone (using the extensiometer described by Van Volkenburgh *et al.*, 1983). Extension zone turgor and rates of leaf extension were measured over a range of temperatures using both genotypes (Pollock *et al.*, 1990). As in the case of *L. temulentum*, turgor was independent of temperature whereas extension rate declined during cooling (Table 6.4). Differences in extension rate between normal and *slender* were not associated with significant differences in turgor, supporting the hypothesis that wall properties determine the gross temperature response. Evidence consistent with this suggestion was obtained by measuring wall extensibility. Measurements were made of the displacement produced per unit load applied and the plastic component was assessed by calculating the hysteresis between strain and relaxation curves. This is equivalent to the work lost as heat and would be zero for a perfectly elastic solid. Comparisons between mature leaves of both normal and mutant genotypes showed no significant differences, with both behaving like rigid elastic solids (Table 6.5). In contrast, extending leaves showed marked plastic deformation, with those from the *slender* genotype being noticeably less rigid. The differences were lost when seedlings were killed before measurement (Table 6.5). Normal seedlings appeared to become more plastic during this treatment, suggesting that some constraining character present in living cells of normal leaves had been lost in the mutant, resulting in greater extensibility. It is suggested that, among other things, this difference could be manifested in an altered temperature response (Pollock *et al.*, 1990).

TABLE 6.4. *Extension and mesophyll cell turgor in leaves of normal and slender barley.*

Meristem temperature (°C)	Growth rate (μm min^{-1})		Turgor (MPa)	
	Normal	*Slender*	Normal	*Slender*
20	11.2 (3.2)	29.0 (3.8)	0.65 (0.07)	0.69 (0.05)
14	6.5 (4.4)	21.8 (1.4)	0.68 (0.05)	0.71 (0.05)
10	4.1 (2.0)	11.3 (1.9)	0.62 (0.11)	0.65 (0.10)
4	2.5 (1.4)	8.4 (0.8)	0.65 (0.06)	0.69 (0.09)
2	1.7 (2.8)	6.4 (2.2)	0.63 (0.06)	0.71 (0.08)
0	No detectable growth	3.8 (1.6)	0.63 (0.09)	0.68 (0.11)

Values in parentheses are standard deviations.

TABLE 6.5. *Physical properties of extending and mature leaves from normal and* slender *seedling of* Hordeum vulgare.

Tissue	Condition	Genotype	Extensibility (μm extended g^{-1} applied load)	Load/relaxation hysteresis under 10 g applied load (μJ)
Mature leaf	Living	Normal	25 (3)	1.1 (0.3)
		slender	32 (4)	1.2 (0.3)
Extending leaf	Living	Normal	116 (12)	3.3 (0.5)
		slender	330 (28)	8.4 (0.9)
Extending leaf	Killed	Normal	470 (31)	7.2 (0.6)
		slender	480 (39)	7.9 (0.7)

Data are means of four determinations. Values in parentheses are standard deviations.

Recently, studies have concentrated on establishing whether the biophysical and physiological differences between genotypes described above are detectable at the chemical or enzymological level. Orthodox chemical analysis of wall carbohydrates prepared from within the extension zone showed no significant differences between genotypes. Both normal and mutant material showed a monosaccharide distribution characteristic of monocot primary cell walls (McNeil et al., 1980; Pollock et al., 1990). However, the dynamic state of living extending cells, where chemical linkages are undergoing rapid turnover (Labavitch, 1981), might mean that measurements of this kind would be unable to detect differences which could profoundly affect cell extensibility.

Although measurement of wall turnover is complex, and there is no good coherent model for the process, the availability of the *slender* mutant simplifies the experimental approach, since any differences from the normal form must be presumed to have some connection, albeit tenuous, with the primary genetic lesion. Thus, three physiological indices which relate to the processes of wall metabolism were all shown to differ in the mutant form. Autolytic release of sugars was examined during incubation of buffered wall extracts. This technique has been used previously to study auxin-mediated changes in metabolism in the extension zones of oat (Heyn, 1986). In our study, the normal genotypes showed higher levels of autolysis, mainly attributed to the release of arabinose and arabinose-rich oligosaccharides, suggesting that the mutation altered the interaction between specific chemical structures and the enzymes that act on them (Pollock et al., 1990). Further differences were detected when radioactivity was measured in the component monosaccharides of the cell walls within the extension zone following administration of very short pulses of $^{14}CO_2$ to the autotrophic tips of the growing first leaf. Material extracted at the end of a 5 min pulse showed more radioactivity in acid-hydrolysable glucose in preparations from normal seedlings than in preparations from the *slender* genotype. As the length of the chase period was increased, the differences became smaller and radioactivity appeared in other sugars such as arabinose, xylose and fucose (Pollock *et al.*, 1990). Finally, measurement of release of sugars following treatment with purified hydrolytic enzymes showed differences in the release of uronic acids between the two genotypes. Gross chemical analysis showed that both genotypes contained about 12% total uronic acids, and so differences in enzymatic release can be attributed to differences in accessibility of the material to the purified enzymes. Again this is consistent with the association of the mutation with altered wall architecture (Pollock *et al.*, 1990).

As yet, our studies have been able to provide only supporting evidence for the central hypothesis concerning the perception and transduction of temperature change. In addition, we have considered only the reversible changes in wall properties associated with temporary reduction in the rate of cell extension. The irreversible cessation of leaf growth involves increases in wall crosslinking, which can be presumed to be qualitatively different from the cold-sensitive wall loosening discussed above. The timing of irreversible growth cessation may itself be sensitive to temperature, possibly acting via alterations in the patterns of gene expression, and will thus strongly influence both the final form and the duration of leaf growth.

6.6. Molecular studies of the *slender* lesion

Comparisons of the rate of leaf extension suggest that the observed increase in overall leaf size in *slender* is due not so much to a greater rate of expansion at any one point but, rather, that the cells continue to extend over a longer period and thus further from the leaf base (Ougham *et al.*,1993). To enable molecular biological comparisons of this extending region to be made, cell-free translation products of mRNA extracted from leaf sections of normal and *slender* seedlings were analysed, and physiologically equivalent parts of the leaf identified. cDNA libraries were constructed from leaf base tissue and screened by differential gene hybridization. This led to the isolation of three clones showing altered patterns of expression in the *slender* mutant. Sequence analysis has so far identified one clone as the mRNA encoding NADPH-protochlorophyllide oxidoreductase (PCOR; Schultz *et al.*, 1989). The second clone is thought to be involved in cell wall metabolism but, as yet, shows no homology with any published sequences. The clone shows higher levels of expression in the *slender* mutant and strong expression in both the leaf sheath (coleoptile) and roots. When seedlings are grown at 4 °C, levels of the transcript are considerably more reduced in normal than they are in *slender*. The third clone again has stronger expression in *slender* and is of unknown function, being only weakly expressed in the leaf sheath and not at all in the roots. Expression of this clone is unaffected by temperature (unpublished observations).

The PCOR enzyme is nuclear encoded. It catalyses the light-dependant conversion of protochlorophyllide to chlorophyllide and thus controls a key step in chlorophyll biosynthesis (Apel *et al.*, 1980). The concentrations of both the protein and mRNA are light regulated and greatly elevated within dark grown (etiolated) tissue, but decline rapidly on illumination (Santel

A

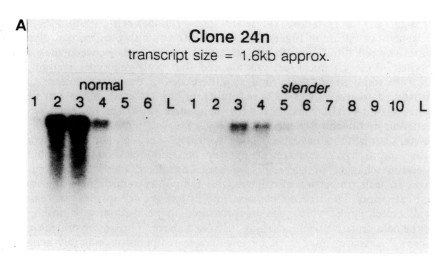

Clone 24n
transcript size = 1.6kb approx.

normal   slender
1 2 3 4 5 6 L 1 2 3 4 5 6 7 8 9 10 L

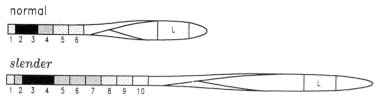

normal

1 2 3 4 5 6   L

slender

1 2 3 4 5 6 7 8 9 10   L

Basal two sections are 2.5mm, remainder of sections are 5mm, except 10mm for the fully emerged leaf section (L).

B

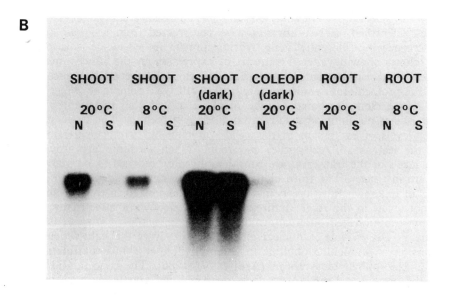

SHOOT	SHOOT	SHOOT (dark)	COLEOP (dark)	ROOT	ROOT
20°C	8°C	20°C	20°C	20°C	8°C
N S	N S	N S	N S	N S	N S

and Apel, 1981; Apel, 1981). *Slender* seedlings grown at 20 °C and 16 h daylight show a greatly reduced level of PCOR mRNA compared with the wild type (Fig. 6.5). Whereas dark-grown leaf tissue contains considerably more of the transcript, *slender* barley again has a lower level than normal. As *slender* seedlings are capable of producing high levels of PCOR mRNA (as shown in etiolated seedlings), it would appear that the relationship between the observed *slender* phenotype and regulation of the transcript level is complex. An investigation into the amount of chlorophyll *a* and *b* present within young seedlings illustrated that, although fully emerged leaf tissue (sections L in Fig. 6.5) of normal and *slender* seedlings were comparable, levels throughout emerging (and predominantly expanding) leaf tissue were considerably lower in *slender* (unpublished observations). From our results it is tempting to speculate that (at least in barley) photosynthesis and extension growth are to some extent antagonistic processes.

6.7. Conclusions

This review has concentrated on the mechanisms that assist acclimation of plants to the extremes of their current environments. What must be emphasized, however, is the diversity of response that exists within the plant kingdom. Although individual species may only survive within quite narrow temperature ranges, the potential for altering these ranges is vast. Detailed characterization of the genetic systems involved in tolerance and development of techniques to transfer such systems across taxanomic boundaries generates the potential to use existing variation to alleviate many of the problems associated with anthropogenic climate change.

REFERENCES

Altschuler, M. and Mascarenhas, J.P. (1985). Transcription and translation of heat-shock and normal proteins in seedlings and developing seeds of soybean exposed to a gradual temperature increase. *Plant Mol. Biol.* **5**, 291–297.

Apel, K. (1981). The protochlorophyllide holochrome of barley (*Hordeum vulgare* L.). Phytochrome-induced decrease of translatable mRNA coding

FIGURE 6.5. *(A) Comparative levels of protochlorophyllide oxidoreductase (PCOR) mRNA (transcript size 1.6 kb) in successive sections of extending normal and* slender *barley seedlings. Sections used, showing the increased leaf length of the* slender *seedling. Regions where the transcript is present are indicated by intensity of shading proportional to the relative message level. (From Ougham et al., 1993.) (B) Expression of PCOR in total leaf tissue (shoots), leaf sheaths (coleoptiles) and roots of normal and* slender *seedlings. Plants were grown at either 20 or 8 °C with either 16 h light or complete darkness.*

for the NAPDH-protochlorophyllide oxidoreductase. *Eur. J. Biochem.* **120**, 89–93.

Apel, K., Santel, H.-J., Redlinger, T.E. and Falk, H. (1980). The protochlorophyllide holochrome of barley (*Hordeum vulgare* L.): isolation and characterization of the NADPH-protochlorophyllide oxidoreductase. *Eur. J. Biochem.* **111**, 251–258.

ap Rees, T., Burrell, M.M., Entwistle, T.G., Hammond, J.B.W., Kirk, D. and Kruger, N.J. (1988). Effects of low temperature on the respiratory metabolism of carbohydrates by plants. In: *Plants and Temperature*, Symposium of the Society for Experimental Biology pp. 377–393, eds. S.P. Long and F.J. Woodward. Cambridge: Company of Biologists.

Bienz, M. and Pelham, H.R.B. (1987). Mechanisms of heat-shock gene activation in higher eukaryotes. *Adv. Genet.* **24**, 31–72.

Burke, J.J., Hatfield, J.L., Klein, R.R. and Mullet, J.E. (1985). Accumulation of heat-shock proteins in field-grown cotton. *Plant Physiol.* **78**, 394–398.

Charles, A.H., England, F. and Thomson, A.J. (1975). The effect of nitrogen application and autumn management on autumn growth of *Lolium perenne* L. at Aberystwyth, Edinburgh and Cambridge. 1. Spaced plants. *J. Br. Grassland Soc.* **30**, 315–325.

Chen, Q., Lauzon, L.M., Derocher, A.E. and Vierling, E. (1990). Accumulation, stability and localization of a major chloroplast heat-shock protein. *J. Cell Biol.* **110**, 1873–1883.

Cooper, J.P. (1964). Climatic variation in forage grasses. I. Leaf development in climatic races of *Lolium* and *Dactylis*. *J. Appl. Ecol.* **1**, 45–61.

Cooper, P. and Ho, T.H.D. (1983). Heat-shock proteins in maize. *Plant Physiol.* **71**, 215–222.

Dale, J.E. (1965). Leaf growth in *Phaseolus vulgaris*. 2. Temperature effects and the light factor. *Ann. Bot.* **29**, 293–308.

Dunn, M.A., Hughes, M.A., Pearce, R.S. and Jack, P.L. (1990). Molecular characterisation of a barley gene induced by cold treatment. *J. Exp. Bot.* **41**, 1405–1413.

Eagles, C.F. (1967). The effect of temperature on vegetative growth in climatic races of *Dactylis glomerata* in controlled environments. *Ann. Bot.* **31**, 31–39.

Eagles, C.F. (1989). Temperature-induced changes in cold tolerance of *Lolium perenne*. *J. Agric. Sci.* **113**, 339–347.

Eagles, C.F. and Othman, O.B. (1981). Growth at low temperature and cold hardiness in white clover. In: *Plant Physiology and Herbage Production*, Occasional Symposium No. 13, pp. 109–113, ed. C.E. Wright. Hurley, UK: British Grassland Society.

Eagles, C.F., Williams, J. and Louis, D.V. (1993). Recovery after freezing in *Avena sativa*, *Lolium perenne* and *Lolium multiflorum*. *New Phytol.* in press.

Francis, D. and Barlow, P.W. (1988). Temperature and the cell cycle. In: *Plants and Temperature*, Symposium of the Society for Experimental Biology, pp. 181–201, eds. S.P. Long and F.J. Woodward. Cambridge: Company of Biologists.

Fuller, M.P. and Eagles, C.F. (1978). A seedling test for cold-hardiness in *Lolium perenne*. *J. Agric. Sci.* **91**, 217–222.

Fuller, M.P. and Eagles, C.F. (1981). Effect of temperature on cold dehardening of *Lolium perenne* L. seedlings. *J. Agric. Sci.* **96**, 55–59.

Gay, A.P. and Eagles, C.F. (1991). Quantitative analysis of cold hardening and dehardening in *Lolium*. *Ann. Bot.* **67**, 339–345.

Habeshaw, D. and Swift, G. (1978). Frost damage and the winter hardiness of Italian ryegrass varieties. *Technical Note*, East of Scotland College of Agriculture. No. 180C, 6 pp.

Hay, R.K.M. (1990). The influence of photoperiod on the dry-matter production of grasses and cereals. *New Phytol.* **116**, 233–254.

Haycock, R. (1981). Environmental limitations to spring production in white clover. *Plant Physiology and Herbage Production*, Occasional Symposium No. 13, pp. 119–123, ed. C.E. Wright. Hurley, UK: British Grassland Society.

Heyn, A.N.J. (1986). A gas chromatographic analysis of the release of arabinose from coleoptile cell walls under the influence of auxin and dextranase preparation. *Plant Sci.* **45**, 77–82.

Hides, D.H. (1978a). Winter hardiness in *Lolium multiflorum* Lam. I. The effect of nitrogen fertilizer and autumn cutting managements in the field. *J. Br. Grassland Soc.* **33**, 99–105.

Hides, D.H. (1978b). Winter hardiness in *Lolium multiflorum* Lam. II. Low temperature responses following defoliation and nitrogen application. *J. Br. Grassland Soc.* **33**, 175–179.

Howarth, C.J. (1989). Heat-shock proteins in *Sorghum bicolor* and *Pennisetum americanum*. I. Genotypic and developmental variation during seed germination. *Plant Cell Environ.* **12**, 471–477.

Howarth, C.J. (1990a). Heat-shock proteins in sorghum and pearl millet: ethanol, sodium arsenite, sodium malonate and the development of thermotolerance. *J. Exp. Bot.* **41**, 877–883.

Howarth, C.J. (1990b). Heat-shock proteins in *Sorghum bicolor* and *Pennisetum americanum*. II. Stored RNA in sorghum seed and its relationship to heat shock protein synthesis during germination. *Plant Cell Environ.* **13**, 57–64.

Howarth, C.J. (1991). Molecular responses of plants to an increased incidence of heat-shock. *Plant Cell Environ.* **14**, 831–841.

Hughes, M.A. and Pearce, R.S. (1988). Low temperature treatment of barley plants causes altered gene expression in shoot meristems. *J. Exp. Bot.* **39**, 1461–1467.

Humphreys, M.O. (1989). Assessment of perennial ryegrass (*Lolium perenne* L.) for breeding. 11. Components of winter hardiness. *Euphytica* **41**, 99–106.

Huner, N.P.A., Palta, J.P., Li, P.H. and Carter, J.V. (1981). Anatomical changes in leaves of Puma rye in response to growth at cold-hardening temperatures *Bot. Gaz.* **142**, 55–62.

Kemp, D.R., Eagles, C.F. and Humphreys, M.O. (1989). Leaf growth and apex development of perennial ryegrass during winter and spring. *Ann. Bot.* **63**, 349–355.

Kiegl, P.J. and Mullen, R.E. (1986). Changes in soybean seed quality from high temperature during seed fill and maturation. *Crop Sci.* **26**, 1212–1216.

Kimpel, J.A. and Key, J.L. (1985). Presence of heat-shock mRNAs in field grown soybeans. *Plant Physiol.* **79**, 672–678.

Kimpel, J.A., Nagao, R.T., Goekjian, V. and Key, J.L. (1990). Regulation of the heat-shock response in soybean seedlings. *Plant Physiol.* **94**, 988–995.

Krol, M., Griffith, M. and Huner, N.P.A. (1984). An appropriate physiological control for environmental temperature studies: comparative growth response of winter rye. *Can. J. Bot.* **62**, 1062–1068.

Larcher, W. (1981). Effects of low temperature stress and frost injury on productivity. In: *Physiological Processes Limiting Plant Productivity*, pp. 253–269, ed. C.B. Johnson. London: Butterworths.

Larsen, A. (1978). Freezing tolerance in grasses: methods for testing in controlled environments. *Meldingen fra Norges Landbrukhogskole* **57**, 56.

Labavitch, J.M. (1981). Cell wall turnover in plant development. *Annu. Rev. Plant Physiol.* **32**, 385–406.

Lawrence, T., Cooper, J.P. and Breese, E.L. (1973). Cold tolerance and winter hardiness in *Lolium perenne* L. II. Influence of light and temperature during growth and hardening. *J. Agric. Sci.* **80**, 341–348.

Li, G.C. and Hahn, G.M. (1980). A proposed operational model of thermotolerance based on effects of nutrients and initial treatment temperature. *Cancer Res.* **40**, 4501–4508.

Lindquist, S. (1981). Regulation of protein synthesis during heat-shock. *Nature* **293**, 311–314.

McNeil, M., Darvill, A.G. and Albersheim, P. (1980). Structure of plant cell walls. X. Rhamnogalacturonan I, a structurally complex pectic polysaccharide in the walls of suspension-cultured sycamore cells. *Plant Physiol.* **66**, 1128–34.

Necchi, A., Pogna, N.E. and Mapelli, S. (1987). Early and late heat-shock proteins in wheats and other cereal species. *Plant Physiol.* **84**, 1378–1390.

Ostgard, O. and Eagles, C.F. (1971). Variation in growth and development in natural populations of *Dactylis glomerata* from Norway and Portugal. I. Leaf development and tillering. *J. Appl. Ecol.* **8**, 383–391.

Ougham, H.J. and Howarth, C.J. (1988). Temperature shock proteins in plants. In: *Plants and Temperature*, Symposium of the Society for Experimental Biology, pp. 259–280, eds. S.P. Long and F.J. Woodward. Cambridge: Company of Biologist.

Ougham, H.J., Schünmann, P.H.D., Quigley, A.S. and Howarth, C.J. (1993). Molecular and cellular acclimation of plants to cold. In: *Research in Photosynthesis, Proceedings of the IX International Congress on Photosynthesis, Nagoya, Japan, 1992*, Vol. IV, pp. 121–128, ed. N. Murata. Dordrecht: Kluwer.

Parsons, A.J. and Robson, M.J. (1980). Seasonal changes in the physiology of S.24 perennial ryegrass (*Lolium perenne* L.). I. Response of leaf extension to temperature during the transition from vegetative to reproductive growth. *Ann. Bot.* **46**, 435–445.

Peacock, J.M. (1975). Temperature and leaf growth in *Lolium perenne*. II. The site of temperature perception. *J. Appl. Ecol.* **12**, 115–123.

Pelham, H.R.B. (1985). Activation of heat-shock genes in eukaryotes. *Trends Genet.* **1**, 31–35.

Pollock, C.J. (1990). The responses of plants to temperature change. *J. Agric. Sci.* **115**, 1–5.

Pollock, C.J. and Eagles, C.F. (1988). Low temperature and the growth of plants. In: *Plants and Temperature*, Symposium of the Society for Experimental Biology, pp. 157–180, eds. S.P. Long and F.J. Woodward. Cambridge: Company of Biologists.

Pollock, C.J., Lloyd, E.J., Stoddart, J.L. and Thomas, H. (1984). Growth, photosynthesis and assimilate partitioning in *Lolium temulentum* exposed to chilling temperatures. *Physiol. Plant.* **59**, 257–262.

Pollock, C.J., Tomos, A.D., Thomas, A., Smith, C.J., Lloyd, E.J. and Stoddart, J.L. (1990). Extension growth in a barley mutant with reduced sensitivity to low temperature. *New Phytol.* **115**, 617–623.

Pollock, C.J., Ougham, H.J. and Stoddart, J.L. (1992). The *slender* mutation of barley. In: *Barley: Genetics, Biochemistry, Molecular Biology and Biotechnology*, pp. 265–276, ed. P.R. Shewry. Oxford: CAB International.

Quigley, A.S. (1991). *The Molecular Basis for Varietal Differences in Cold Hardening of Winter Oats (Avena sativa L.).* PhD thesis, University of Wales.

Robson, M.J. and Jewiss, O.R. (1968). A comparison of British and North African varieties of tall fescue (*Festuca arundinacea*). III. Effects of light, temperature and daylength on relative growth rate and its components. *J. Appl. Ecol.* **5**, 191–204.

Santel, H.-J. and Apel, K. (1981). The protochlorophyllide holochrome of barley (*Hordeum vulgare* L.): the effect of light on the NADPH-protochlorophyllide oxidoreductase. *Eur. J. Biochem.* **120**, 95–103.

Schlesinger, M.J., Ashburner, M. and Tissieres, A., eds. (1982). *Heat-Shock: From Bacteria to Man.* New York: Cold Spring Harbor Laboratory.

Schöffl, F., Rossol, I. and Angermüller, S. (1987). Regulation of the transcription of heat-shock genes in nuclei from soybean (*Glycine max*) seedlings. *Plant Cell Environ.* **10**, 113–120.

Schultz, R., Steinmüller, K., Klaas, M., Forreiter, C., Rasmussen, S., Hiller, C. and Apel, K. (1989). Nucleotide sequence of a cDNA coding for the NADPH-protochlorophyllide oxidoreductase (PCR) of barley (*Hordeum vulgare* L.) and its expression in *Escherichia coli. Mol. Gen. Genet.* **217**, 355–361.

Stoddart, J.L. and Lloyd, E.J. (1986). Modification by gibberellin of the growth–temperature relationship in normal and mutant genotypes of several cereals. *Planta* **67**, 364–368.

Stoddart, J.L., Thomas, H., Lloyd, E.J. and Pollock, C.J. (1986). The use of a temperature-profiled position transducer for the study of low-temperature growth in Gramineae. *Planta* **167**, 359–363.

Storti, R.V., Scott, M.P., Rich, A. and Pardue, M.L. (1980). Translational control of protein synthesis in response to heat-shock in *D. melanogaster* cells. *Cell* **22**, 825–834.

Thomas, H. (1983). Analysis of the response of leaf extension to chilling temperatures in *Lolium temulentum* seedlings. *Physiol. Plant.* **57**, 509–513.

Thomas, H. and Norris, I.B. (1977). The growth responses of *Lolium perenne* to the weather during winter and spring at various altitudes in mid-Wales. *J. Appl. Ecol.* **14**, 949–964.

Thomas, H. and Norris, I.B. (1979). Winter growth of contrasting ryegrass varieties at two altitudes in mid-Wales. *J. Appl. Ecol.* **16**, 553–565.

Thomas, H. and Stoddart, J.L. (1984). Kinetics of leaf growth in *Lolium temulentum* at optional and chilling temperatures. *Ann. Bot.* **53**, 341–347.

Thomas, A., Tomos, A.D., Stoddart, J.L., Thomas, H. and Pollock, C.J. (1989). Cell expansion rate, temperature and turgor pressure in growing leaves of *Lolium temulentum* L. *New Phytol.* **112**, 1–5.

Van Volkenburgh, E., Hunt, S. and Davis, W.J. (1983). A simple instrument for measuring cell wall extensibility. *Ann. Bot.* **51**, 669–672.

Velazquez, J.M. and Lindquist, S. (1984). hsp70: nuclear concentration during environmental stress and cytoplasmic storage during recovery. *Cell* **36**, 655–662.

Wilkinson, M.C., Wheatley, P.A., Smith, C.J. and Laidman, D.L. (1990). Higher plant heat-shock protein 70: purification and immuno-chemical analysis. *Phytochemistry* **29**, 3073–3080.

CHAPTER 7

Plants under salt and water stress

J. SHALHEVET

Symbols and abbreviations

Ψ Soil or culture solution total water potential
ψ_o Soil or culture solution solute or osmotic potential
ψ_m Soil matric potential
Ψ^L Leaf total water potential
ψ_s^L Leaf solute or osmotic potential
ψ_p^L Leaf pressure or turgor potential
g_s Stomatal conductance
g_m Mesophyll conductance
A Net assimilation rate
T Transpiration rate
ET   Evapotranspiration rate
WA   Field water application

7.1. Introduction

Plants in nature may be exposed, during their ontogeny, to a variety of environmental stresses, two of which are closely interrelated: water deficiency or water stress, and salt excess or salt stress. The interrelation between the two stress factors is of particular significance for irrigated crops. Salinity has been a part of irrigated agriculture for millennia because, under some conditions, even the best quality water may lead to salt accumulation in the soil to a level detrimental to crops.

Non-saline irrigation water resources are dwindling in many regions, resulting in increasing interest in utilizing saline or brackish water for crop production. Even the use of seawater has been contemplated and tried

(Epstein *et al.*, 1980; Pasternak and San Pietro, 1985). There is, at times, an almost mystical belief that only human inadequacy stands between the dream and the reality of utilizing seawater for crop irrigation (Boyko, 1966). After almost a century of intense interest in the physiology of salt tolerance, our understanding of the mechanism of plant response to salt stress is still very limited. The situation is somewhat better with regard to water stress, where cell dehydration is clearly a primary trigger of the growth-damaging processes.

The interaction of salts with plant physiological processes is obviously complex. There are many salt species, many mechanisms, and many organs, tissues and cells involved. Reasons given in the literature for observed suppressed growth and damage to tissues include:

1. Reduced water uptake and 'physiological drought' (Magistad, 1945).
2. Injury to cell membranes (tonoplast or plasmalemma) (Cramer *et al.*, 1985).
3. Ca^{2+}-Na^+ interaction (Cramer *et al.*, 1990).
4. Na^+-K^+ selectivity, transport and leakage (Storey and Wyn Jones, 1978).
5. The process of osmotic adjustment through solute accumulation in the symplast (Bernstein, 1961).
6. Salt accumulation in the apoplast (cell wall) resulting in cell dehydration (Oertli, 1968b).
7. Damage to developed tissue, resulting in a decrease in photosynthetic surface and in a lack of sufficient metabolites for growing tissue (Rapp *et al.*, 1983).
8. The cost of osmotic adjustment, compartmentation and exclusion (Yeo, 1983).
9. Hormonal balance in the plant (cytokinins and ABA) (Davis *et al.*, 1986).
10. Nutrient deficiencies, especially N^+ and K^+ (Kafkafi, 1984).

These processes are not mutually exclusive, and may act consecutively or concomitantly. Some operate over a short term and others over a long term (Munns and Termaat, 1986). Some play a more important role on certain crops, some on others.

Recently, the major interest has been focused on the breeding of crops for salt tolerance. The interest in breeding, which was dormant during the early decades of salinity research, was reawakened by the publications of Epstein and Jefferies (1964). There have been numerous reviews during the past decade related to breeding for salt tolerance (Epstein, 1976; Epstein and Rains, 1987; Shannon and Noble, 1990; Miles, 1991).

To the best of our knowledge, there are a limited number of commercially available varieties specifically selected for salt tolerance (Shannon and Noble, 1990). Nevertheless, tolerant varieties of some crops are available, e.g. barley (Epstein *et al.*, 1980), wheat (Kingsbury *et al.*, 1984), lettuce (Shannon, 1984), melon (Pasternak and de Malach, 1987) and rice (Yeo and Flowers, 1982).

The use of tissue culture to speed up the breeding process has proved disappointing. Increased salt tolerance of cultured cells has rarely led to increased salt tolerance in regenerated plants (Miles, 1991). Because salt tolerance is a whole-plant response, individual cells can seldom serve as a model (Epstein and Rains, 1987; Miles, 1991).

Future developments in genetic engineering may provide useful tools for speeding up the breeding process if the genes responsible for tolerance traits are identified. So far, almost nothing is known about the genes affecting salt tolerance.

7.2. Additivity of salt and water stress

7.2.1. Response to integrated soil water stress

Under field conditions the two main components of the total soil water potential (Ψ) which are related to plant growth – the matric (ψ_m) and the solute (ψ_o) potentials – are additive in lowering the free energy of the water. It was assumed, therefore, that they would also be additive in their effect on plant growth through a reduction in the availability of water.

Indeed, the early results obtained by Ayers *et al.* (1943), Wadleigh and Ayers (1945) and Wadleigh *et al.* (1946) showed yield reduction of beans and guayule to be a function of an integrated total soil water stress. The latter was the sum of the mean ψ_o and ψ_m, integrated over time. A typical relationship is given in Fig. 7.1. Water, stress treatments were created by irrigating at given ψ_m for all salinity treatments. This resulted in various irrigation intervals, depending on the salinity of the irrigation water: the more saline the water, the longer was the interval. The resulting relationship is curvilinear, with decreasing detrimental effect of Ψ as the total stress level

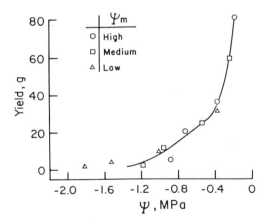

FIGURE 7.1 *The relation of bean yield to the integrated water potential, Ψ, obtained by four levels of osmotic potential, ψ_o, using sodium chloride and calcium chloride and three levels of matric potential: high ($\bigcirc$), medium ($\square$) and low ($\triangle$). (Adapted from Ayers* et al., *1943.)*

increases, irrespective of the source of stress, matric or solute. Thus, a given decrement in ψ_m resulted in the same yield reduction as a similar decrement in ψ_o.

Already at this early stage, the above-mentioned investigators suspected that there must have been factors other than water availability affecting yield. They noticed no wilting at low ψ_o, whereas there was wilting at equivalent low ψ_m. Also, salt stress resulted in a decreased content of reducing sugars in the stem, but no such effect was found with water stress (Wadleigh and Ayers, 1945). These investigators also noted that, at low salt stress, increasing water stress resulted in tissue dehydration; at high salt stress there was no dehydration, commensurate with the no-wilting observation.

More recently, Sepaskhah and Boersma (1979) studied the interaction of salt stress (addition of sodium chloride) and water stress (addition of polyethylene glycol, PEG). The osmotica were equilibrated with the soil solution across a cellulose acetate semipermeable membrane using wheat (*Triticum aestivum*) as a test crop. Replotting the results of Sepaskhah and Boersma produced a similar response function to that shown in Fig. 7.1 (see Fig. 7.2). In Fig. 7.2 the data points representing the different ψ_o levels at high ψ_m (bold curve) and those representing the various ψ_m levels at high ψ_o (main curve) have been connected. These pairs of observations led the investigators to conclude that decreasing ψ_m is more detrimental to growth than an equivalent decrease in ψ_o. Yet, overall, it seems that the two water potential components had an equivalent effect on yield, similar

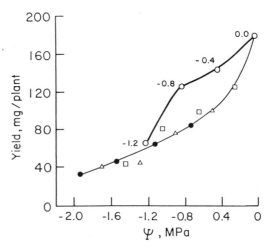

FIGURE 7.2 *The relation of wheat dry matter yield to the integrated water potential, Ψ, obtained by four levels of osmotic potential using sodium chloride (ψ_o, marked in figure) and four levels of matric potential using PEG ($\psi_m = -0.03$ (◯), -0.25 (☐), -0.50 (△), -0.75 (●) MPa). (Adapted from Sepaskhah and Boersma, 1979.)*

to the results of Wadleigh and Ayers (1945). Storey and Wyn Jones (1978), using salt and PEG directly on the growing root systems of barley (*Hordeum vulgare* cv. Arimar) at ψ_m and ψ_o of -1.1 MPa, found sodium chloride to be more detrimental to yield (39% yield reduction) than PEG (11% reduction) after 24 days of growth. Kawasaki *et al.* (1983) found no differences in dry matter yield of beans, sorghum and corn between isosmotic concentrations of sodium chloride and PEG. The contents of potassium, calcium and magnesium in the tops and roots were severely depressed by salt but not by PEG. In a study of sour orange citrus seedlings (*Citrus aurantium*) growing under different salt and PEG concentrations, Zekri and Parsons (1990) found both root and shoot growth to be less adversely affected by salt (39 and 47% yield reduction, respectively) than by PEG (47 and 56% reduction, respectively).

7.2.2. Salt stress preconditioning for water stress

Jensen (1981), using potassium nitrate and PEG to create solute and water stress of -0.27 and -0.47 MPa, found no osmotic adjustment in bean plants (*Phaseolus vulgaris*) treated with PEG, and a consequent tissue dehydration, drop in turgor potential (ψ_p^L) and wilting. Net photosynthesis and stomatal conductance were also depressed. Under salt stress, there was complete osmotic adjustment and turgor maintenance. Plants preadjusted to salt stress and then transferred to PEG maintained their ψ_p^L and no wilting was observed. Richardson and McCree (1985) also found that pretreatment of sorghum (*Sorghum bicolor*) with -0.56 MPa of sodium chloride and calcium chloride resulted in reduced damage due to water stress. The leaf water potential, Ψ^L, at which the leaf blade of corn (*Zea mays*) ceased to elongate was much lower (-1.2 MPa) when plants were pretreated with a salt solution of -0.44 MPa than when not pretreated (-0.4 MPa) (Stark and Jarrell, 1980).

All of these findings point to the increased ability of the preconditioned plants to survive better, rather than to produce better, under water stress. Jensen (1982) did not find any differences in dry matter yield or wilting percentages in response to water stress between barley plants (*Hordeum distichum*) pretreated with saline water and not pretreated. Although damage from water stress following presalinization may be reduced by osmotic adjustment, the converse does not apply. Water stress does not condition plants to salt stress. On the contrary, by adding saline water to water-stressed plants, there is a recovery of plant growth and turgor.

Shalhevet and Hsiao (unpublished data) followed the change in assimilation, transpiration and stomatal conductance as cotton plants under a water stress of ψ_m -0.62 and -1.15 MPa were shifted to a salt stress of similar magnitude by irrigation with saline solution (Table 7.1). There was an immediate rise (within 1 h) in transpiration and net assimilation rate as well as in stomatal and mesophyll conductance. The lower the value of Ψ, the greater was the change. When ψ_m was -1.15 MPa, assimilation and transpiration nearly doubled when ψ_o of -0.92 MPa was applied. Stomatal conductance tripled, but mesophyll conductance increased by only 12%. On

TABLE 7.1. *Net assimilation rate (A), transpiration rate (T) and stomatal (g_s) and mesophyll (g_m) conductance of cotton leaves under water and salt stress at given soil, Ψ, and leaf, Ψ^L, water potentials.*

Treatment	Time after salinization (h)	Ψ (Mpa)	Ψ^L (Mpa)	A (μmol m^{-2} s^{-1})	T (mmol m^{-2} s^{-1})	g_s (mol m^{-2} s^{-1})	g_m (mol m^{-2} s^{-1})
Control		−0.17	−1.00	32.1	7.68	0.717	0.179
Water stress	0	−0.62		17.7	3.15	0.189	0.148
Salt stress	1	−0.57	−1.41	19.5	3.58	0.216	0.160
	23			21.3	3.88	0.393	0.173
Water stress	0	−1.15		6.8	1.21	0.055	0.080
Salt stress	1.4	−0.92	−1.82	13.0	2.22	0.158	0.091
	17			18.6	2.87	0.257	0.128
Control	0	−0.15		25.5	6.21	0.688	0.123
Salt stress	5	−0.53		4.6	0.78	0.043	0.041
	24			12.4	2.14	0.137	0.092

the other hand, applying saline water to well watered plants led to a sharp drop in assimilation and transpiration (Table 7.1).

7.2.3. Osmotic adjustment and turgor maintenance

The obvious difference between water and salt stress is in leaf turgor and the growth processes that are influenced by it. Fig. 7.3 shows this very distinctly and it has been described in numerous reports. In the experiments of Shalhevet and Hsiao (1986), the end result was a growth rate under water stress half as large as under salt stress (Fig. 7.4) in the Ψ range of interest. At Ψ lower than -1.3 MPa, the leaf extension rate became similar for both water and salt stress, while differences in ψ_p^L between the two treatments were still large. Obviously, turgor is not the only factor influencing the differentiation in growth between salt- and water-stressed plants.

The results of Shalhevet and Hsiao (1986) are similar to those of Sepaskhah and Boersma (1979) only when changes in ψ_o under high ψ_m are compared with changes in ψ_m under high ψ_o (see bold curve in Fig. 7.2). When both stress components influence growth, the results of Sepaskhah and Boersma show a relationship to the integrated stress.

The rapid response of plants to changing osmotic potential led Termaat et al. (1985) to suggest that the triggering mechanism resides in the roots. They applied pneumatic pressure of 0.48 MPa to roots in a pressure chamber to offset a possible effect on ψ_p of a solute potential of -0.48 MPa in the external solution. The increased pressure did not result in changes in leaf expansion, photosynthesis or osmotic pressure of the cell sap.

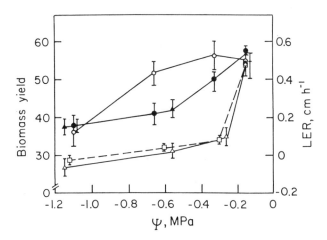

FIGURE 7.3 *The relation of leaf extension rate (LER) and biomass dry matter yield (□) of cotton to the integrated soil water potential, Ψ, obtained by four levels of day (△) and night (▲) water stress and four levels of day (○) and night (●) salt stress (Shalhevet and Hsiao, 1986).*

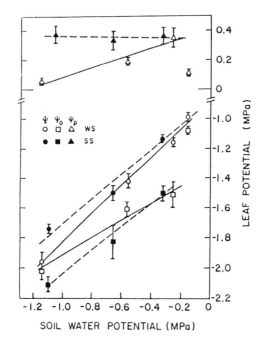

FIGURE 7.4 *The relation of cotton leaf water potential (Ψ^L, $\bigcirc$), osmotic potential (ψ_o^L, $\square$) and pressure potential (ψ_p^L, $\triangle$) to the integrated soil water potential (Ψ) obtained by water deficit (open symbols) and salt excess (filled symbols) (Shalhevet and Hsiao, 1986).*

Non-halophytes seldom control solute uptake, especially at high solute concentrations (Munns and Termaat, 1986). Thus, as solute concentration and time of exposure to salinity increase, salt concentration in the leaves may also increase, resulting in osmotic 'overadjustment' (Boyer, 1965; Yeo and Flowers, 1982).

Termaat *et al.* (1985) suggested that the root messenger triggering response is a growth regulator. Passioura and Gardner (1990) postulated the same mechanism in response to water stress. They applied pressure to the roots of wilting plants to offset the reduction in ψ_p^L, but could not record an improvement in leaf expansion rate. Parra and Romero (1980) also found a greater reduction in the yield of beans (55%) under mild water stress ($\psi_m = -0.04$ to $-0.06\,\text{MPa}$) than under a much greater salt stress (40% yield reduction due to $\psi_o = -0.43\,\text{MPa}$). The reasons for these large differences are not apparent.

7.2.4. *Combined salt and water stress during soil drying*

The additivity of salt and water stress is of special interest under field irrigation conditions using saline water. As soil dries out between irrigations,

ψ_o and ψ_m decrease simultaneously. It would be expected, therefore, that relative crop yield will be affected more when the interval between irrigations is long than when it is short. In a field experiment with sweetcorn, no interaction in relative yield was found between the salinity level of the irrigation water and irrigation interval (Shalhevet et al., 1983, 1986). The principal reason for the lack of interaction was the slower development of water stress under high salt stress because of lower transpiration rates.

Sorghum plants grown in containers exhibited the same phenomenon (Richardson and McCree, 1985). Plants watered with saline water were better able to continue leaf expansion and carbon gain than those watered with non-saline water. This was a consequence of the smaller leaf area and lower transpiration per unit leaf area of salinized plants, which resulted in reduced water loss. Richardson and McCree (1985) suggested that the adjustment of salinized plants should mitigate the effect of using saline water for irrigation, as was indeed found by Shalhevet et al. (1983, 1986).

Similar results were obtained by Stark and Jarrell (1980). The leaf elongation rate of corn plants watered with non-saline water dropped to zero at a higher predawn Ψ^L ($-1.25\,MPa$) when saline water was used. Using similar techniques, Jensen (1982) arrived at the same result with barley as the test crop. Kirkham (1978) showed a barley yield gain of 170% when the crop was irrigated at intervals of 12 days with water of $10.5\,g\,NaCl\,l^{-1}$ compared with non-saline water. By comparison, there was a 74% yield reduction with the same level of salinity when the plants were irrigated daily. These differences reflected the slower development of water stress under saline conditions, counterbalancing the detrimental effect of solute stress. The absolute yields at the longer irrigation interval were, of course, much lower than under daily irrigation.

The greater yield depression at high Ψ under water stress than salt stress was attributed to changes in soil hydraulic conductivity (Parra and Romero, 1980). Although this cannot be completely discounted, it is unlikely. Soil hydraulic conductivity becomes limiting to water movement toward a dense root system only at soil water contents lower than those that prevailed in the experiments described above (Weatherley, 1982). Furthermore, hydraulic conductivity is not involved when ψ_m is reduced using PEG.

It appears that at high water potentials the ability of the plant to adjust osmotically at a faster rate, and at a lower metabolic cost, under salt stress plays an important role in yield response. Whenever lower ψ_m is a significant component of Ψ, its effect becomes dominant and the plant responds to the integrated soil water potential.

7.3. Physiological responses to salt and water stress

7.3.1. Specific and osmotic effects

Reviews of plant response to saline conditions traditionally consider separately osmotic and specific ion effects. The latter is subdivided into toxicity and nutritional imbalances (e.g. Epstein, 1985; Lauchli and Epstein, 1990). In view of the available information regarding the so-called 'osmotic effect', this concept should be re-examined.

Bernstein (1961, 1963) demonstrated that plant roots adjust rapidly to changes in the external salt concentration. This osmotic adjustment of the roots took place within 12 h, during the dark period, and that of the tops during the next light period. The osmotic adjustment of the tops was described earlier (Eaton, 1941). We have already shown (Section 7.2) that the major difference in plant response to salt and water stresses is the occurrence of osmotic adjustment and turgor maintenance under salt stress (Shalhevet and Hsiao, 1986, and others). This finding questions the hypothesis that reduced water uptake by roots is the major cause of damage due to salinity, and that water and salt stresses are essentially similar in their effect on growth in that both cause tissue desiccation. This finding also speciously strengthens the argument that salt effect is specific and toxic in nature. We suggest that, under realistic conditions, the specific ion effect and the osmotic effect may be essentially the same phenomenon. We therefore prefer the term solute effect to the term osmotic effect.

A great deal of confusion was created in the literature by the unfortunate use of a single salt as a model for salt effect. Most glycophytes were evolved in environments of balanced ionic composition (Epstein and Rains, 1987). Although these plants may tolerate large variations in the proportions of the various ions that contribute to salinity, they do not favour extreme conditions of imbalance. Thus, the use of single salts may create toxic effects, which are not necessarily related to salinity as it occurs in nature.

The salt most commonly used in experiments (NaCl) is the most prevalent in the environment. The use of Na^+ without the concomitant increase in Ca^{2+} has been shown to cause specific damage to cell membranes (Cramer et al., 1985) and to leakage of K^+ into the growth medium (Storey and Wyn Jones, 1978). A small (10 mM) addition of Ca^{2+} to the growth medium resulted in the correction of this specific damage (Kent and Lauchli, 1985). Under some conditions K^+ may cause more severe damage than Na^+; this has been demonstrated in sorghum (Weimberg et al., 1984) and in barley (Cramer et al., 1990).

Greenway and Munns (1980) concluded their thorough review of the mechanism of salt tolerance in non-halophytes by warning against studying plant response to salinity using increasing concentrations of sodium chloride in the growth medium. They felt that this practice 'not only causes problems in interpretation, but also is irrelevant in an ecological sense'. Maas and Grieve (1987) state that 'Unfortunately, there is an increasing tendency of many investigators to salinize media with a single salt, usually NaCl. The results of these studies are usually purported to describe plant response to saline conditions. This inference ignores the fundamental distinction between saline and sodic conditions. Many agronomic plant species are seriously injured by high $Na^+ : Ca^{2+}$ ratios.' We concur with this view, as do Epstein and Rains in their recent review (1987). Munns and Termaat (1986) argue in favour of a 'specific effect' component in addition to an osmotic component in plant response to NaCl salinity. This may be so, but specific ion effects of single salt solutions are of no practical interest.

7.3.2. A generalized solute effect

The obvious differences in the response symptoms to solute and water stress notwithstanding, the similar reduction in growth in response to these two components is often apparent (see Section 7.2). Evidently, overall osmotic adjustment is a necessary but not sufficient condition for promoting growth. Despite osmotic adjustment, growth reduction is the first symptom of increasing salinity. Osmotic adjustment occurs during the check in growth (Bernstein, 1961). It would be quite remarkable if the water and solute stress had a similar quantitative effect on yield, but resulting from completely dissimilar and independent mechanisms. Bernstein (1963) felt that 'despite turgor maintenance through osmotic adjustment, there still remains a very good argument in favor of an osmotic mechanism of response, if one can be proposed.'

An osmotic mechanism of response was proposed by Oertli (1968a,b), based on the proposition that salts may accumulate in the apoplast – especially the cell wall – to levels higher than in the symplast, resulting in cell dehydration. Theoretical considerations as well as experimental evidence supported this hypothesis. Margin and leaf burn, as well as lesions at transpiration sites on the leaf, are external symptoms that lend support to the existence of such a mechanism. Even boron toxicity was attributed to excessive localized concentration of B^+ resulting in cell dehydration and death.

Fenn et al. (1970) sampled leaves of avocado (Persea americana), a crop considered particularly sensitive to Na^+ and Cl^-, which was treated with 20 mEq Cl^- l^{-1}. They demonstrated that chloride distribution within the leaf, as well as the concentration in expressed xylem sap, implicated extracellular salt accumulation in salinity response and leaf burn. Chloride accumulated in the apoplast to a ψ_o level of -1.2 MPa (300 mEq l^-) or more.

Munns and Passioura (1984) (see also Flowers and Yeo, 1986) lent support to Oertli's hypothesis by showing that the osmotic pressure of xylem sap of old leaves of intact barley plants increased suddenly after 20–30 days of treatment, indicating a level of cell saturation of approximately $\psi_o^L = -0.05$ MPa. When cells are solute saturated, salts accumulate very rapidly in the much smaller volume of the cell wall. When the ψ_o^L of the xylem reached -0.2 to -0.3 MPa, the leaves died. The expansion rate of young leaves declined linearly with time, unrelated to the change in ψ_o^L of older leaves. This hypothesis explains the death of old leaves. The reduced photosynthetic leaf area deprives young leaves of assimilates for continued normal growth, resulting in slower growth, in spite of osmotic adjustment. In addition, sometimes young leaves may not always adjust osmotically as do old leaves, as was shown by Hoffman et al. (1980) for predawn measurement of pepper leaves growing in saline solutions. The lower night turgor may be responsible for reduced leaf expansion. Midday measurements showed equal ψ_p^L for control and salinized treatments. Westgate and Boyer (1984) obtained similar results by applying water stress to maize. The reduced growth of young leaves was unrelated to the local tissue salt concentration. Different growth rates were found in expanding tissues of barley having similar sodium chloride concentrations (Munns and Passioura, 1984).

Kingsbury *et al.* (1984) compared the response to saline medium (20% sea water, salinity = $10.6\,dS\,m^{-1}$) of two selected lines of wheat (*Triticum aestivum*) differing in tolerance to salinity. The sensitive line showed a significantly lower relative growth rate. Yet ψ_p^L as well as Na^+, Ca^{2+}, K^+ and Cl^- uptake were similar for the two varieties. Thus, there was a total solute effect independent of specific ion toxicity or 'osmotic effect' in the conventional sense of reduced tissue hydration.

In an experiment by Termaat and Munns (1986), sodium chloride was found to be more detrimental to growth of barley and wheat than an isosmotic concentration of macronutrients. They did not feel that a toxic concentration of sodium chloride or the tissue water relation was responsible for the differences and concluded that a signal from the root, possibly abscisic acid (ABA; Zhao *et al.*, 1991), was the most likely cause of the damage. The triggering for this signal could be a change in the water potential, Ψ, of the growth medium, irrespective of whether the change originated from changes of ψ_m or ψ_o.

We plotted the yield data of Termaat and Munns against tissue solute potential, ψ_o^L (Fig. 7.5), and found a good common linear relationship for barley and wheat seedlings.

Recently an experiment was conducted with *roses* to study the effect of salinity under changing or constant levels of Na^+ and Ca^{2+} on plant development and flower yield and quality (Yfrach, 1992, unpublished MSc thesis). Under isosmotic concentrations, the 'balanced' nutrient solution (constant $Na^+:Ca^{2+}$ series) had a lesser effect on yield than did a solution with high $Na^+:Ca^{2+}$ (changing $Na^+:Ca^{2+}$ series). However, when plotting yield against ψ_o^L, a uniform relationship emerged (Fig. 7.6). Evidently, at the same

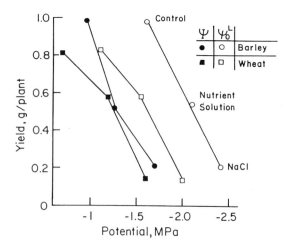

FIGURE 7.5 *The relation of dry matter yield of barley (○, ●) and wheat (□, ■) to the osmotic potential (open symbols) of the expressed cell sap and leaf water potential (filled symbols) obtained by either − 0.56 MPa of macronutrients or − 0.56 MPa of sodium chloride. (Adapted from Termaat and Munns, 1986.)*

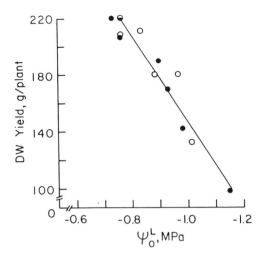

FIGURE 7.6 *The relation of dry matter yield of roses to the osmotic potential, ψ_o^L, of the expressed cell sap for constant $Na^+ : Ca^{2+}$ ($\bigcirc$) and increasing $Na^+ : Ca^{2+}$ ($\bullet$). (Adapted from Yfrach, unpublished data.)*

external ψ_o, high $Na^+ : Ca^{2+}$ ratio results in lower tissue ψ_o^L than when the ratio is low. The high salt concentration treatment of the balanced solution ($\psi_o = -0.27\,MPa$) resulted in osmotic adjustment of about 0.25 MPa, whereas the same treatment having the high $Na^+ : Ca^{2+}$ ratio resulted in osmotic overadjustment of 0.42 MPa, reflecting an unregulated salt uptake (Munns and Termaat, 1986). This unregulated uptake may be due to disturbances of membrane selectivity when Ca^{2+} is displaced by Na^+ ions present at high concentrations (Cramer et al., 1985).

The reduction in ψ_o^L is a result of changes in the amount of electrolytes reaching the cell when the concentration of the outside solution increases, as well as from cell volume changes. Jensen (1982) reported that 42% of the reduction in ψ_o^L in barley was due to increases in solute content and 58% to dehydration. Evidently, the mechanism of osmotic adjustment involves both processes, which change simultaneously to arrive at a final ψ_p^L compatible with the changes in root medium Ψ. The adjustment occurs in response to the lowest Ψ of the root zone against which roots can still absorb water, as concluded by Shalhevet and Bernstein (1968) from a split-root experiment. Water uptake was drastically reduced from the portion of the root zone that was salinized up to a level of $-0.7\,MPa$, yet it did not cease, showing the existence of a potential gradient from roots to leaves.

It seems that osmotic adjustment in plants in response to salt stress is a mechanism for survival rather than a mechanism for maintaining growth rates (Munns, 1988). Without adjustment tissues will dehydrate rapidly and die. Death as a result of salt accumulation is a much slower process.

7.3.3. *Mechanism of solute effect*

All the preceding results point to a solute effect related more to the total tissue solute concentration than to toxicity of a specific salt. The exact mechanism of this effect may be open to speculation. It could involve the very process of osmotic adjustment and the cost involved in ion accumulation, exclusion or compartmentation, as was pointed out by Shannon (1984). The need for increased production of metabolites for long-term osmotic adjustment may deprive plants of growth substances, as was suggested by Matsuda and Riazi (1981), who treated barley seedlings with sodium chloride and PEG at various levels of Ψ and found growth to be related to ψ_o^L, while ψ_p^L remained constant for both osmotica.

The approach proposed by Oertli (1968a) stresses the importance of looking at the various components of tissues, organs and of the whole plant. Unquestionably, compartmentation is an important mechanism through which plants can avoid damage. Salts may be prevented from reaching a sensitive tissue (young leaves) by sequestering in a less sensitive tissue of the plant (roots and old leaves), or by accumulation in the vacuole to prevent entry into the sensitive cytoplasm. The plant may accumulate the less damaging metabolites in preference to electrolytes. Yeo and Flowers (1984) showed how ions vary in concentration between the epidermis and endodermis of rice roots: Na^+ decreased and K^+ and PO_4^- increased in concentration, while Cl^- remained essentially constant.

Salt stress normally results in adjustment through accumulation of electrolytes, but in many cases both electrolytes and metabolites are involved (Shalhevet and Hsiao, 1986), complementing each other in maintaining a desired level of adjustment of ψ_o^L.

There is still confusion regarding causes and effects in plant responses to salt and water stress. For lack of proper understanding of the overall mechanism or mechanisms involved, discussions revolve around symptoms. Munns (1988) justly questioned the validity of the significance imparted to the osmotic adjustment process in relation to turgor maintenance and growth. The lack of turgor maintenance usually results in reduced growth. The maintenance of turgor, however, does not necessarily ensure maintenance of growth. Growth may be related more closely to ψ_o^L than to ψ_p^L.

Regardless of the mechanism, water uptake by plants still obeys the rules of flow of solutions in media containing semipermeable membranes. The flow is controlled by the potential gradients and the resistances in the flow pathway (Dalton *et al.*, 1975). Flow will be reduced as the potential difference decreases and resistances increase, independent of the causes of these changes.

On the basis of the preceding discussion one may present the following simplified scheme of plant response to the initiation of moderate (sublethal) salt stress, applied by a balanced salt solution:

1. Within the first few hours there is a reduction in expansive growth, photosynthesis and transpiration. The drop in ψ_o is sensed by the roots

and transmitted to the tops through a hydraulic, hormonal or inorganic messenger.

2. Within days, the uptake of inorganic solutes causes a drop in ψ_o^L (osmotic adjustment) of the fully developed leaves and in partial recovery of photosynthesis and transpiration. The increase in solute concentration in the protoplasm interacts with cell physiological processes resulting in arrested growth. Inorganic solutes are sequestered in the vacuole while the cytoplasm increases in organic solutes. Expanding leaves may not fully adjust, resulting in reduced turgor during the night and in a consequent reduction in leaf elongation.

3. Because most glycophytes do not effectively regulate ion fluxes into the xylem, inorganic solutes continue to accumulate in the symplast (vacuole) until a saturation level is reached. Salts begin to overflow and accumulate in the apoplast (cell wall), very rapidly reaching osmotic concentrations greater than inside the cells. The result is osmotic dehydration and death, first of mature leaves and eventually also of young leaves.

4. As the leaf area reduces, the total amount of assimilates available for growth falls, further suppressing tissue growth. At the same time, root growth continues draining assimilates from the tops. The process of osmotic adjustment also takes place through the production of organic solutes, further depriving the plant of necessary assimilates for growth.

5. The rate of the processes and their severity depend on the external salt concentration (osmotic potential) and the make-up of the external solution. The greater the concentration and the more imbalanced the solution, the shorter will each step be. Single salt solutions may disrupt membrane integrity and very high, lethal, concentrations may result in death within days.

7.4. Crop water production function under salt and water stress

7.4.1. Principles

The relative magnitude of the effects of salt and water stress on yield is not yet clear and experimental results are conflicting (Section 7.2). Nevertheless, there is ample evidence to show that growth and evapotranspiration (ET) are reduced to the same extent by both components.

The function relating yield to transpiration, T, is the production function, the slope of which is water use efficiency. The production function as defined above is linear, crop-specific and independent of ambient conditions, as long as T is normalized on potential T (de Wit, 1958; Shalhevet and Bielorai, 1977).

Under field conditions, it is seldom possible to measure T directly. Consequently, the above definition of production function is impractical. The common practical definition is yield as a function of field water application. The water quantities applied in the field include, in addition to T, water losses in evaporation and deep seepage (and runoff, in the case of gravity irrigation).

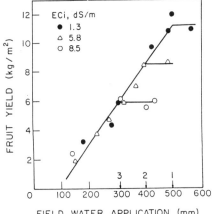

FIGURE 7.7 *The relation of marketable tomato fruit yield to field water application for the period 3 June to 6 August for three levels of irrigation water salinity: 1.3 (●), 5.8 (△) and 8.5 (○) dS m⁻¹. Arrows indicate the water requirements for maximum yield for the three salinity levels. (J. Peretz, J. Shalhevet and A. Meiri, 1985, unpublished data.)*

The evaporation component may shift the intersect on the water quantity axis to the right (Fig. 7.7), but will not change the slope. The deep seepage component, which may result from inefficiencies in water application or from the need for salt leaching, may change the shape of the curve from linear to exponential.

The production function using saline water should be similar to that of non-saline water as long as leaching is not required during the irrigation seasons. Two features may change, however. The maximum yield possible (the yield plateau) is controlled by soil salinity, resulting from the irrigation water salinity. The results depicted in Fig. 7.7 show this clearly. Secondly, the intercept on the water quantity axis may shift to the right because the canopy is slower to develop than under non-saline water irrigation, and complete cover may never be achieved. Shalhevet *et al.* (1969) found a 50% evaporation loss of the total ET of peanuts irrigated with saline water. Plotting the yield values of the plateau in Fig. 7.5 against mean salt salinity will yield a typical salt tolerance response function with its negative slope and threshold salinity (see figures in Feigin *et al.*, 1991 and in Shalhevet, 1992).

7.4.2. Experimental results

Shalhevet (1984) summarized data of nine experiments in which both salinity and water stress were variables. The data support the above description of the production function. Additional results have been published more recently, showing the same trend in tomatoes (Vinten *et al.*, 1986), maize (Frenkel

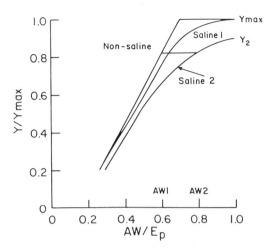

FIGURE 7.8   *The relation of relative yield to the ratio of field water application and pan evaporation, AW/E$_p$, for three levels of irrigation water salinity. (Adapted from Letey* et al., *1985.)*

et al., 1990), barley (Al Khafaf *et al.*, 1990) and cotton (Meiri *et al.*, 1992). In all of these experiments, unintentional deep seepage was minimized by a strict control over water application.

The type of limiting factor response function shown in Fig. 7.7 is not unique to salinity and may be obtained with other factors that limit plant growth. Bernstein (1977) indicated that the inhibitory effect of growth retardants and salinity is simply additive in reducing growth. Shimshi (see Shalhevet *et al.*, 1979) described the production function of maize at various levels of nitrogen fertilizer application. The function relating yield to water application at five levels of nitrogen application was linear with a yield plateau at each level, very similar to Fig. 7.7.

7.4.3. Model computations

Letey *et al.* (1985; Letey and Knapp, 1990) combined the water production function under non-saline irrigation (a function describing the steady-state salt accumulation in the soil under saline water irrigation and leaching) and the salt tolerance function for a given crop, to construct a model for estimating production functions for saline water irrigation. Vinten *et al.* (1991) employed a similar model, except for using a non-steady-state approach to the computation of salt accumulation in the soil. A typical outcome of the model computation is given in Fig. 7.8. The deviation from linearity reflects the leaching fraction $(AW_2 - AW_1)/AW_2$. The distance $Y_m - Y_2$ is the yield loss due to salinity at maximum leaching (the leaching requirements). For a discussion of the limitation of estimating the leaching requirements, see Shalhevet (1984).

Russo (1987) and Russo and Bakker (1987) obtained results for lettuce, maize and cotton which, by and large, conform to the model of Letey *et al.* (1985). The production functions obtained for non-saline water were also non-linear, with a yield response to the ratio of water application to class A pan evaporation of up to 2.0. Such values are exceptionally high and may point to excessive non-ET water loss.

REFERENCES

Al Khafaf, S., Adnan, A. and Al-Asadi, N.M. (1990). Dynamics of root and shoot growth of barley under various levels of salinity and water stress. *Agric. Water Man.* **18**, 63–75.

Ayers, A.O., Wadleigh, C.H. and Magistad, O.L. (1943). The interrelationship of salt concentration and soil moisture content with the growth of beans. *J. Am. Soc. Agron.* **35**, 796–810.

Bernstein, L. (1961). Osmotic adjustment of plants to saline media. I. Steady state. *Am. J. Bot.* **48**, 909–918.

Bernstein, L. (1963). Osmotic adjustment of plants to saline media. II. Dynamic phase. *Am. J. Bot.* **50**, 360–370.

Bernstein, L. (1977). Physiological basis of salt tolerance in plants. In: *Genetic Diversity in Plants*, Proceedings of an International Symposium, 1976, pp. 283–290.

Boyer, J.S. (1965). Effect of osmotic water stress on metabolic rates of cotton plants with open stomata. *Plant Physiol.* **40**, 229–234.

Boyko, H., ed. (1966). *Salinity and Aridity – New Approaches to Old Problems.* The Hague: Junk.

Cramer, G.R., Lauchli, A. and Polito, V.S. (1985). Displacement of Ca^{2+} by Na^+ from the plasmalemma of roots. *Plant Physiol.* **79**, 207–211.

Cramer, G.R., Epstein, E. and Lauchli, A. (1990). Effect of sodium, potassium and calcium on salt stressed barley. I. Growth analysis. *Physiol. Plant.* **80**, 83–88.

Dalton, F.N., Raats, P.A.C. and Gardner, W.R. (1975). Simultaneous uptake of water and solutes by plant roots. *Agron. J.* **67**, 334–339.

Davis, W.J., Metcalfe, J., Lodge, T.A. and da Costa, A.R. (1986). Plant growth substances and the regulation of growth under drought. *Aust. J. Plant Physiol.* **13**, 105–125.

de Wit, C.T. (1958). Transpiration and crop yields. Institute of Biological and Chemical Research on Field Crops and Herbage. *Verse-Landbouwk. Onderz.* **64**, 15–36.

Eaton, F.M. (1941). Water uptake and growth as influenced by inequalities in the concentration of the substrate. *Plant Physiol.* **16**, 545–564.

Epstein, E. (1976). Genetic potential for solving problems of soil mineral stress adaptation of crops to salinity. In: *Plant Adaptation to Mineral Stress in Problem Soils*, Proceedings of Workshop, pp. 73–82, ed. M.J. Wright. Special Publication, Cornell University Agricultural Experimental Station.

Epstein, E. (1985). Salt tolerant crops: origins, development and prospects of the concept. *Plant Soil* **89**, 1–12.

Epstein, E. and Jefferies, R.L. (1964). The genetic basis of selective ion transport in plants. *Annu. Rev. Plant Physiol.* **15**, 169–184.

Epstein, E. and Rains, W. (1987). Advances in salt tolerance. *Plant Soil* **99**, 17–29.

Epstein, E., Rush, D.W., Kingsbury, R.W., Kelly, D.B., Cunningham, G.A. and Wrona, A.F. (1980). Saline culture of crops: a genetic approach. *Science* **210**, 399–404.

Feigin, A., Ravina, I. and Shalhevet, J. (1991). *Irrigation with Treated Sewage Effluent: Management for Environmental Protection*, Advanced Series in Agricultural Science 17. Berlin: Springer Verlag.

Fenn, L.B., Oertli, J.J. and Bingham, F.T. (1970). Specific chloride injury in *Persea americana*. *Soil Sci. Soc. Am. Proc.* **34**, 617–620.

Flowers, T.J. and Yeo, A.R. (1986). Ion relation of plants under drought and salinity. *Aust. J. Plant Physiol.* **13**, 75–91.

Frenkel, H., Mantell, A., Vinten, A. and Meiri, A. (1990). Double-line source sprinkler system for determining the separate and interactive effects of water and salinity on forage corn. *Irrig. Sci.* **11**, 227–231.

Greenway, H. and Munns, R. (1980). Mechanism of salt tolerance in non-halophytes. *Annu. Rev. Plant Physiol.* **31**, 149–190.

Hoffman, G.J., Shalhevet, J. and Meiri, A. (1980). Leaf age and salinity influence water relations of pepper leaves. *Physiol. Plant.* **48**, 463–469.

Jensen, C.R. (1981). Influence of water and salt stress on water relationships and carbon dioxide exchange of top and roots in beans. *New Phytol.* **87**, 285–295.

Jensen, C.R. (1982). Effect of soil water osmotic potential on growth and water relationships in barley during soil water depletion. *Irrig. Sci.* **3**, 111–121.

Kafkafi, U. (1984). Plant nutrition under saline conditions. In: *Soil Salinity Under Irrigation*, Ecological Studies 51, pp. 319–331, eds. I. Shainberg and J. Shalhevet. Berlin: Springer Verlag.

Kawasaki, T., Akiba, T. and Moritsugu, M. (1983). Effect of high concentration of NaCl and polyethylene glycol on growth and ion absorption in plants. *Plant Soil* **75**, 75–85.

Kent, L.M. and Lauchli, A. (1985). Germination and seedling growth of cotton: salinity–calcium interactions. *Plant Cell Environ.* **8**, 155–159.

Kingsbury, R.W., Epstein, E. and Pearcy, R.W. (1984). Physiological responses to salinity in selected lines of wheat. *Plant Physiol.* **74**, 417–425.

Kirkham, M.B. (1978). Salt water irrigation frequency for barley. *Ann. Arid Zone* **17**, 12–18.

Lauchli, A. and Epstein, E. (1990). Plant responses to saline and sodic conditions. In: *Agricultural Salinity Assessment and Management*, ASCE Manuals and Reports on Engineering Practice No. 71, pp. 113–137, ed. K.K. Tanji. New York: American Society of Civil Engineers.

Letey, J. and Knapp, K. (1990). Crop–water production functions under saline conditions. In: *Agricultural Salinity Assessment and Management*, ASCE Manuals and Reports on Engineering Practice No. 71, pp. 305–326, ed. K.K. Tanji.

Letey, J., Dinar, A. and Knapp, K.C. (1985). Crop water production function model for saline irrigation water. *Soil Sci. Soc. Am. J.* **49**, 1005–1009.

Maas, E.V. and Grieve, C.M. (1987). Sodium-induced calcium deficiency in salt-stressed corn. *Plant Cell Environ.* **10**, 559–564.

Magistad, O.C. (1945). Plant growth relation on saline and alkali soils. *Bot. Rev.* **11**, 181–230.

Matsuda, K. and Riazi, A. (1981). Stress-induced osmotic adjustment in growing regions of barley leaves. *Plant Physiol.* **68**, 571–576.

Meiri, A., Frenkel, H. and Mantell, A. (1992). Cotton response to water and salinity under sprinkler and drip irrigation. *Agron. J.* **84**, 44–50.

Miles, D. (1991). Increasing salt tolerance of plants through cell culture requires greater understanding of tolerance mechanism. *Aust. J. Plant Physiol.* **18**, 1–15.

Munns, R. (1988). Why measure osmotic adjustment? *Aust. J. Plant Physiol.* **15**, 717–726.

Munns, R. and Passioura, J.B. (1984). Effect of prolonged exposure to NaCl on the osmotic pressure of leaf xylem sap from intact transpiring barley plants. *Aust. J. Plant Physiol.* **11**, 497–507.

Munns, R. and Termaat, A. (1986). Whole-plant response to salinity. *Aust. J. Plant Physiol.* **13**, 143–160.

Oertli, J.J. (1968a). Extracellular salt accumulation, a possible mechanism of salt injury in plants. *Agrochemia* **12**, 461–469.

Oertli, J.J. (1968b). Effect of external salt concentration on water relation of plants: 6. *Soil Sci.* **105**, 302–310.

Parra, M.A. and Romero, G.C. (1980). On the dependence of salt tolerance of beans (*Phaseolus vulgaris* L.) on soil water matric potentials. *Plant Soil* **56**, 3–16.

Passioura, J.B. and Gardner, P.A. (1990). Control of leaf expansion in wheat seedlings growing in drying soil. *Aust. J. Plant Physiol.* **17**, 149–157.

Pasternak, D. and de Malach, Y. (1987). Saline water irrigation in the Negev Desert. In: *Conference on Agriculture and Food Production in the Middle East*, Athens, Greece. Beer Sheva: Publication Section, Ben-Gurion University of the Negev.

Pasternak, D. and San Pietro, A., eds. (1985). *Biosalinity in Action: Bioproduction with Saline Water*. Dordrecht: Martinus Nijhoff.

Rapp, J.C., Ball, C.M. and Terry, N. (1983). A comparative study of the effects of NaCl salinity on respiration, photosynthesis and leaf extension. *Plant Cell Environ.* **6**, 675–682.

Richardson, S.G. and McCree, K.J. (1985). Carbon balance and water relations of sorghum exposed to salt and water stress. *Plant Physiol.* **79**, 1015–1020.

Russo, D. (1987). Lettuce yield: irrigation water quality and quantity relationship in a gypsiferous desert soil. *Agron. J.* **79**, 8–14.

Russo, D. and Bakker, D. (1987). Crop-water production functions for sweet corn and cotton irrigated with saline water. *Soil Sci. Soc. Am. J.* **51**, 1554–1562.

Sepaskhah, A.R. and Boersma, L. (1979). Shoot and root growth of wheat seedlings exposed to several levels of matric potential and NaCl-induced osmotic potential of soil water. *Agron. J.* **71**, 746–752.

Shalhevet, J. (1984). Management of irrigation with brackish water. In. *Soil Salinity under Irrigation*, Ecological Studies 51, pp. 298–318, eds. I. Shainberg and J. Shalhevet. Berlin: Springer Verlag.

Shalhevet, J. (1992). Using water of marginal quality for crop production: major issues. In: *Water Use Efficiency in Agriculture*, pp. 17–52, eds. J. Shalhevet, C. Liu and Y. Xu. Rehovot: Priel Publications.

Shalhevet, J. and Bernstein, L. (1968). Effect of vertically heterogeneous soil salinity on plant growth and water uptake. *Soil Sci.* **106**, 85–93.

Shalhevet, J. and Bielorai, H. (1977). Crop water requirements in relation to climate and soil. *Soil Sci.* **125**, 240–247.

Shalhevet, J. and Hsiao, T.C. (1986). Salinity and drought, a comparison of their effects on osmotic adjustment, assimilation, transpiration and growth. *Irrig. Sci.* **7**, 249–264.

Shalhevet, J., Reiniger, P. and Shimshi, D. (1969). Peanut response to uniform and non-uniform soil salinity. *Agron. J.* **61**, 384–387.

Shalhevet, J., Mantell, A., Bielorai, H. and Shimshi, D. (1979). *Irrigation of Field and Orchard Crops under Semi-arid Conditions*, 2nd edn. Publication 1, Bet Dagan, Israel: International Irrigation Information Center.

Shalhevet, J., Heuer, B. and Meiri, A. (1983). Irrigation interval as factor in the salt tolerance of eggplant. *Irrig. Sci.* **4**, 83–93.

Shalhevet, J., Vinten, A. and Meiri, A. (1986). Irrigation interval as a factor in sweet corn response to salinity. *Agron. J.* **78**, 539–545.

Shannon, M.C. (1984). Breeding, selection and genetics of salt tolerance. In: *Salinity Tolerance in Plants*, pp. 231–254, eds. R.C. Staple and G.H. Toenniesson. New York: Wiley.

Shannon, M.C. and Noble, C.L. (1990). Genetic approaches for developing economic salt-tolerant crops. In: *Agricultural Salinity Assessment and Management*, ASCE Manuals and Reports on Engineering Practice No. 71, pp. 161–185, ed. K.K. Tanji. New York: American Society of Civil Engineers.

Stark, J.C. and Jarrell, W.M. (1980). Salinity-induced modifications in the response of maize to water deficit. *Agron. J.* **72**, 745–748.

Storey, R. and Wyn Jones, R.G. (1978). Salt stress and comparative physiology in the Gramineae. I. Ion relations of two salt- and water-stressed barley cultivars, California Mariont and Arimer. *Aust. J. Plant Physiol.* **5**, 801–816.

Termaat, A. and Munns, R. (1986). Use of concentrated macronutrient solutions to separate osmotic from NaCl-specific effect on plant growth. *Aust. J. Plant Physiol.* **13**, 509–522.

Termaat, A., Passioura, J.B. and Munns, R. (1985). Shoot turgor does not limit shoot growth of NaCl-affected wheat and barley. *Plant Physiol.* **77**, 869–872.

Vinten, A.J.A., Shalhevet. J., Meiri, A. and Peretz, J. (1986). Water and leaching requirements of industrial tomatoes irrigated with brackish water. *Irrig. Sci.* **7**, 13–25.

Vinten, A.J.A., Frenkel, H., Shalhevet, J. and Elston, D.A. (1991). Calibration and validation of a modified steady-state, model of crop response to saline water irrigation under conditions of transient root zone salinity. *J. Contam. Hydrol.* **7**, 123–144.

Wadleigh, C.H. and Ayers, A.D. (1945). Growth and biochemical composition of bean plants as conditioned by soil moisture tension and salt concentration. *Plant Physiol.* **20**, 106–132.

Wadleigh, C.H., Gauch, H.G. and Magistad, O.C. (1946). Growth and rubber accumulation in guayule as conditioned by soil salinity and irrigation regime. *Tech. Bull US Dept. Agric.* **925**, 1–34.

Weatherley, P.E. (1982). Water uptake and flow in roots. In: *Physiological Plant Ecology.* II. *Water Relations and Carbon Assimilation,* pp. 79–105, eds. O.L. Lange, P.S. Nobel, C.B. Osmond and H. Zigler. Berlin: Springer Verlag.

Weimberg, R., Lerner, H.R. and Poljakoff-Mayber, A. (1984). Changes in growth and water-soluble solute concentration in *Sorghum bicolor* stressed with sodium and potassium salts. *Physiol. Plant.* **62**, 472–480.

Westgate, M.E. and Boyer, J.S. (1984). Transpiration and growth induced water potential in maize. *Plant Physiol.* **74**, 882–889.

Yeo, A.R. (1983). Salinity resistance: physiologies and prices. *Physiol. Plant.* **58**, 214–222.

Yeo, A.R. and Flowers, T.J. (1982). Accumulation and localization of sodium ions within the shoots of rice varieties differing in salinity resistance. *Physiol. Plant.* **56**, 343–348.

Yeo, A.R. and Flowers, T.J. (1984). Mechanisms of salinity resistance in rice and their role as physiological criteria in plant breeding. In: *Salinity Tolerance in Plants,* pp. 151–170, eds. R.C. Staples and G.H. Toenniessen. New York: Wiley.

Zekri, M. and Parsons, L.R. (1990). Comparative effects of NaCl and poly-ethylene glycol on root distribution, growth and stomatal conductance of sour orange seedlings. *Plant Soil* **129**, 137–143.

Zhao, K., Munns, R. and King, R.W. (1991). Abscisic acid levels in NaCl-treated barley, cotton and saltbush. *Aust. J. Plant Physiol.* **18**, 17–24.

CHAPTER 8

Physiological basis of stress imposed by ozone pollution

T.A. MANSFIELD and M. PEARSON

8.1. Introduction

Ozone is the most intensively investigated component of the atmosphere belonging to a group of compounds that are 'secondary pollutants'. These are not released directly as a result of human activities, but they are produced when 'primary pollutants' take part in reactions in the atmosphere. These reactions usually involve sunlight, and the pollutants produced are often given the collective name 'photochemical smog'.

Most of the ozone naturally present in the atmosphere is at high altitudes (above 12 km), i.e. it is within the stratosphere. Very high above the Earth's surface the high-energy photons in sunlight break down oxygen into a mono-atomic form, and it is when these atoms recombine at somewhat lower altitudes that ozone is formed. Most of this remains within the stratosphere, but some of it is transported downwards via the troposphere into the atmospheric boundary layer. Here the highly reactive ozone disappears as it comes into contact with surfaces, and vegetation plays an important part in its removal. Estimates of the concentrations of ozone in unpolluted air vary, and an average of 20 ppb is often quoted, with short-term increases up to 50 ppb.

The fact that ozone is a normal component of the atmosphere around plants means that we should not regard it merely as an unnatural stress-inducing agent. Plants (and animals too) possess some cellular defence mechanisms that enable them to tolerate the presence of low concentrations of ozone. It is when episodes of ozone pollution exceed the operational capacity of these defences that we can expect physiological problems to occur. The degree of enrichment of the troposphere with ozone as a result of human activities has increased steadily over the last two decades, and there is no doubt that the concentrations now occurring in many parts of the world amount to an environmental stress of considerable importance.

8.2. Anthropogenic ozone in the troposphere

The atmospheric pollutants that are mainly responsible for causing increases in the concentration of ozone in the troposphere are nitrogen oxides and hydrocarbons. During most combustion processes there is heat-induced combination of atmospheric nitrogen and oxygen to form nitric oxide, NO. This is subsequently oxidized to NO_2, which is subject to photolysis when it absorbs radiation between 280 and 430 nm in wavelength:

$$NO_2 + \text{radiation} \rightarrow NO + O_{(^3P)} \quad \text{(rate coefficient } K_1) \quad (8.1)$$

In the presence of an energy-absorbing third molecule, M, this reaction then produces ozone:

$$O_{(^3P)} + O_2 + M \rightarrow O_3 + M \quad (8.2)$$

Nitric oxide that is either formed in reaction 8.1 or newly produced in combustion processes reacts with ozone thus:

$$NO + O_3 \rightarrow NO_2 + O_2 \quad \text{(rate coefficient } K_2) \quad (8.3)$$

In air polluted with NO_x the outcome of these reactions for the production and removal of ozone is

$$\text{ozone concentration} = \frac{K_1[NO_2]}{K_2[NO]} \quad (8.4)$$

The residence time of NO_2 in the atmospheric boundary layer is relatively short because it is rapidly removed by deposition processes, particularly by uptake into plant leaves. Because of such removal of NO_2 and the continued production of NO during combustions, it is clear from reaction 8.4 that this series of reactions involving NO_x cannot cause an increase in tropospheric ozone.

There are, however, other processes that play a part. The oxidation of NO to NO_2 can be achieved by peroxy radicals $(RO_2 \cdot)$, which are present in the atmosphere as a result of the photochemical breakdown of hydrocarbons. $RO_2 \cdot$ reacts with NO thus:

$$RO_2 \cdot + NO \rightarrow RO \cdot + NO_2 \quad (8.5)$$

When reactions 8.1 and 8.2 occur alongside 8.5, the overall outcome is

$$RO_2 \cdot + O_2 + \text{radiation} \rightarrow RO \cdot + O_3 \quad (8.6)$$

The emissions of unburnt hydrocarbons alongside NO during combustion processes are therefore largely responsible for the increases in tropospheric ozone that have occurred in recent years. A more detailed account of the transformation of primary to secondary pollutants has been given by Fowler (1992).

There is great spatial and temporal variation in the concentrations of ozone of anthropogenic origin. At sites of high elevation (above 800 m) there is often little diurnal variation, but at low elevations the diurnal cycle can be very marked, with the highest concentrations occurring for several hours after midday and a steep decline at night. On all sites, however, the occurrence of 'high' concentrations (e.g. 100 ppb) is episodic, and this is a feature that is particularly relevant to our evaluation of ozone as an agent of stress for plants. There are appreciable annual variations with higher concentrations in summer, but the precise pattern varies from year to year because of the strong influence of meteorological conditions. High irradiances and high temperatures generally favour ozone production via reactions 8.1, 8.2 and 8.5.

8.3. Damage to cells and tissues caused by ozone

Ozone is a highly reactive molecule and it is unlikely that it can penetrate unchanged far into the interior of cells. Although it is likely to be destroyed by contact with plant surfaces the major factor regulating the deposition velocity appears to be stomatal conductance (Colbeck and Harrison, 1985; Kerstiens and Lendzian, 1989). Calculations of the concentration of ozone in the leaf intercellular air space, using methods analogous to those for carbon dioxide, suggest that it is close to zero (Laisk *et al.*, 1989). Findings of this nature indicate a considerable concentration gradient and suggest that ozone is very reactive within the leaf. When it diffuses into a leaf via the stomata it first makes contact with moist cell walls and then with the plasma membranes. In the aqueous phase in the cell walls free radicals such as $HO\cdot$ and $HO_2\cdot$ will be produced; and if ozone succeeds in passing through the cell walls to reach the plasma membranes it is likely to react with constituents such as unsaturated fatty acids and proteins. Changes in membrane permeability (indicated by 'leakiness' of cells) have often been observed in ozone-treated tissues, and have been interpreted as evidence of direct attack on plasma membranes.

Until recently the discussion of effects of ozone on plants was mostly directed to the action of the aqueous phase free radicals, or to the damage to macromolecules caused by any ozone that succeeds in penetrating to the plasma membranes or beyond. It has, however, been recognized for some time that the reaction in the gaseous phase between ozone and unsaturated hydrocarbons such as ethylene may be important, and positive evidence of this came from the work of Mehlhorn and Wellburn (1987). They fumigated seedlings of *Pisum sativum* with a range of concentrations of ozone from 50 to 150 ppb for 7 h daily for 3 weeks after germination. Surprisingly, even the highest concentration did not cause any marked injury to the plants. On the other hand, there was severe visible injury to the leaves when 3-week-old seedlings that had been grown in clean air were treated with the same concentrations of ozone for just one period of 7 h. Further studies suggested that these dramatic differences could be attributed to the amounts of ethylene being produced by the plants. Those that had been fumigated for 3 weeks with 150 ppb ozone were found to be emitting ethylene at a rate of only

$0.2\,\text{nmol g(dry wt)}^{-1}\text{h}^{-1}$, whereas the untreated controls were producing more than 10 times as much (2.5 nmol). Plants grown in clean air for 3 weeks, then exposed to 150 ppb ozone for 7 h, were found to be emitting ethylene at a rate of $5.7\,\text{nmol g(dry wt)}^{-1}\text{h}^{-1}$.

Ethylene is one of several volatile unsaturated hydrocarbons that are emitted by plants. These compounds can react with ozone to form both organic free radicals and stable but very reactive products. The reactions are complex in nature and cannot be presented as a precise equation, but they take the general form:

$$RCH = CHR + O_3 \rightarrow RO_2\cdot + ROOH + \text{other products}$$

It is the production of ROOH (hydroperoxides) that has been given most attention. The quantities of hydrocarbons emitted by plants (especially trees) can significantly affect urban ozone levels, and they are frequently not considered in evaluations of the benefits of reducing emissions of anthropogenic hydrocarbons (Chameides *et al.*, 1988). Hewitt *et al.* (1990) made a detailed study of plants that are sources of isoprene, an alkene that constitutes a large fraction of global emissions of hydrocarbons by plants. They used *Macuna deeringiana* (velvet bean), which emits isoprene strongly from older but not from younger leaves. With the aid of high-performance liquid chromatography with a hydroperoxide specific detection system, they found a 135-fold greater content of ROOH in older than in younger leaves after exposure to 120 ppb ozone.

Mehlhorn *et al.* (1991) followed up their earlier studies and showed that ozone toxicity was related to rates of ethylene biosynthesis in several different

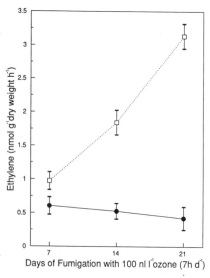

FIGURE 8.1 *Rate of ethylene formation by pea seedlings after different periods of exposure to 100 ppb ozone (●) or clean air (□). (After Mehlhorn et al., 1991.)*

species. They found that application of aminoethoxyvinylglycine (AVG), an inhibitor of ethylene biosynthesis, greatly reduced ozone sensitivity, and they obtained strong evidence confirming that ozone is more toxic to plants when they experience brief rather than prolonged exposure to the pollutant. The reduction in ethylene production in plants given a daily treatment with ozone was convincingly confirmed in this study (Fig. 8.1).

Recent observations have some important implications for the way in which the phytotoxicity of ozone should be studied experimentally. High concentrations of ozone usually occur in short episodes in the outside environment, and hence plants are unlikely to develop resistance mechanisms involving reduced ethylene emissions of the kind suggested in Fig. 8.1.

8.4. Ozone pollution and stomatal behaviour

The production of toxic derivatives by reactions between biogenic hydrocarbons emitted by plants, and ozone in the atmosphere, is likely to occur near the surfaces of leaves. The cells most exposed to attack are those surrounding the substomatal cavity, and the stomatal guard cells themselves are in a particularly vulnerable position. There have been many observations of effects of ozone on stomata, and in general moderate to high concentrations have been found to cause decreases in stomatal conductance accompanied by reductions in photosynthesis. It has usually not been clear whether the effect of the pollutant is directly upon the stomata, or whether photosynthetic carbon dioxide uptake is reduced so that substomatal carbon dioxide concentration rises, stomatal closure occurring in consequence. Stomata are known to respond to changes in intercellular carbon dioxide concentration (Mott, 1988), and therefore any agent that reduces photosynthesis is likely to cause them to close partially. One of the most detailed analyses to be carried out was that by Farage *et al.* (1991), who fumigated leaves of wheat with 200–400 ppb ozone for 4–16 h. They concluded that, in the early stages of treatment (within 8 h), ozone exerted some direct effects on the stomata, but later the primary action of the pollutant was on carboxylation efficiency within the mesophyll. Experiments on *Vicia faba* by Aben *et al.* (1990) enabled a clear distinction to be drawn between effects on stomata and on photosynthesis. They treated plants with 60 ppb ozone for 8 h daily over a 2-week period, and found major differences in response at different times of year. Plants fumigated in this way in open-top chambers between May and July did not display any marked visible symptoms of injury, but there was a decrease in stomatal conductance without an accompanying drop in photosynthesis. In contrast, plants exposed to the same amount of ozone between August and October became chlorotic and the rate of photosynthesis at saturating light and carbon dioxide was reduced. The authors concluded that the stomatal complex was more sensitive to ozone than was photosynthesis in the mesophyll, and they also suggested that under conditions when photosynthesis is affected the stomatal closure caused by decreased photosynthesis can overrule the direct effect of ozone on the stomata. Other recent work by Matyssek *et al.* (1991) on ozone-treated *Betula pendula* has suggested that carbon gain was less limited by reduced stomatal conductance than by the decline in capacity for carbon dioxide fixation in the mesophyll.

These data, together with those from other studies in the past, suggest that effects of ozone on the stomata and gas exchange are complex. The timing of exposure to the pollutant, and the environmental conditions, may alter the nature of the response. However, when careful analysis is conducted it is sometimes possible to distinguish a direct effect on stomata, which precedes action on the underlying mesophyll.

In some other studies, no effects of ozone on stomata, or even increases in conductance and/or inhibition of closure, have been reported (Barnes *et al.* 1990a; Reich and Lassoie, 1984; Leonardi and Langebartels, 1990). In beech (*Fagus sylvatica*), Taylor and Dobson (1989) found that ozone caused stomatal closure in leaves of the first flush, but there was an opposite effect on those of the second flush.

8.4.1. *Effects of ozone on stomatal behaviour of beech*

During the summer of 1991 we conducted a long-term fumigation of 3-year-old beech trees with ozone. The treatments were intermittent, and were of a magnitude that depended on ambient temperatures and solar radiation. The trees were contained in daylit hemispherical chambers (Fig. 8.2) that have been described in detail by Lucas *et al.* (1987).

In view of the evidence presented above that plants may become 'acclimated' to regular treatment with ozone, careful attention was given to the precise

FIGURE 8.2 *Daylit fumigation chambers (Solardomes) used at Lancaster University, UK for fumigating plants with ozone and other pollutants.*

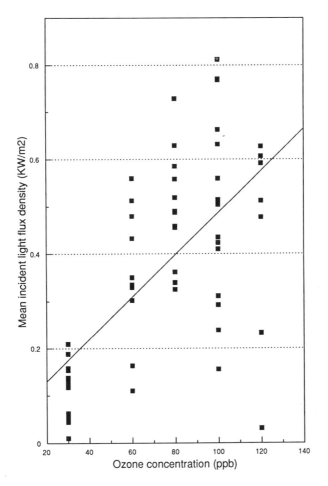

FIGURE 8.3 *Relationship between ozone concentration and average incident light flux density during an experiment on* Fagus *sylvatica.*

nature of the treatments. A basal level of 30 ppb ozone was always present, and higher concentrations were imposed on this on some days for periods of 6 h. These elevated concentrations were 60, 80, 100 or 120 ppb. Early on each day, the current and forecast weather conditions were considered, and what appeared to be the most appropriate ozone concentration was selected. The highest concentrations were applied during warm, sunny weather. When the weather was persistently cloudy, progressive steps downward were made daily until the basal level was reached. The relationship between the applied concentration and incident light flux is shown in Fig. 8.3, and the actual concentrations over a 51-day period are shown in Fig. 8.4. On 2 days in each week no supplementary ozone was applied.

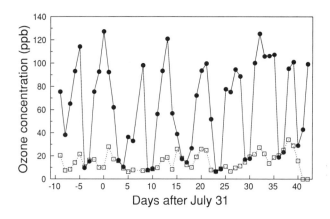

FIGURE 8.4 *Episodic ozone treatments (●) used in long-term experiments on Fagus sylvatica. Control levels (□) are also shown.*

Some of the trees were randomly selected to be subjected to soil moisture stress at a predetermined time in the experiment. Measurements of stomatal resistance made with a diffusion porometer began on 1 August (Fig. 8.4). On this occasion the target ozone concentration was 100 ppb, and subsequent measurements were made on days when the same target concentration was applied. On 8 September water was withheld from some of the trees, and measurements were made 2, 5 and 9 days after this.

Prior to the imposition of water stress, ozone caused a highly significant rise in stomatal resistance, the magnitude of which increased steadily with time (Fig. 8.5). Determinations with the porometer were made twice daily at 08.15 and 11.00 h, and the effect of ozone proved to be greater at the later time, even though the stomatal resistance was not significantly different between the two times in the unpolluted controls. At 08.15 h the resistance was 35% higher, and at 11.00 h it was 60% higher, as a result of the ozone treatment. Each day's 6 h ozone episode began at 10.00 h, and thus it appeared that there was a substantial response of the stomata within 1 h. The previous day's treatment had, however, ended at 16.00 h and so the change detected at 08.15 provided evidence of an after-effect of the pollution treatment, which persisted at least overnight. The imposition of water stress caused noticeable effects after 2 days, by which time the stomatal resistance in the unwatered trees had increased. Well watered trees continued to show the same response as before, but in the unwatered treatment the response was different (Fig. 8.6). Here control trees possessed significantly higher resistances ($P < 0.024$). These data suggest that when water stress is experienced the stomata of polluted trees do not close to the same extent as in the unpolluted controls. Stomatal closure in response to water shortage is one of the fundamental mechanisms of stress regulation in land plants, and some disruption of this protective response has been revealed by this

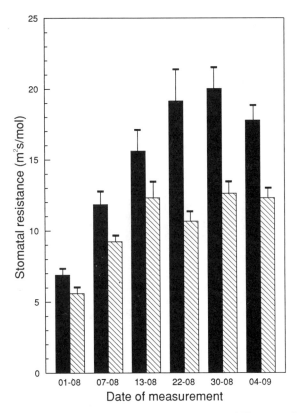

FIGURE 8.5 *Stomatal resistance in well watered trees of* Fagus sylvatica *in the presence (■) and absence (▨) of ozone.*

experiment. In contrast, in the well watered trees, ozone pollution caused a considerable increase in stomatal resistance, which must indicate reduced carbon dioxide intake for photosynthesis (although whether stomatal closure is a cause or a consequence of its reduced intake in the mesophyll remains to be determined for this species).

The long-term implications of such alterations in stomatal behaviour are important. It appears that episodes of ozone pollution may reduce photosynthesis, and consequently affect productivity, when the trees have an adequate supply of water. On the other hand, during periods of water stress, there is a failure to show the normal degree of stomatal closure and this may disrupt the control of water status. A combination of these effects in years when the trees are subject to drought may cause death of younger distal shoots and could account for the observed decline in vigour and dieback found recently on trees in some parts of Europe. We regard these preliminary experiments on young trees as an important indication of how ozone *may* affect mature trees in the field, but experiments under field conditions will now have to be performed to test the hypothesis.

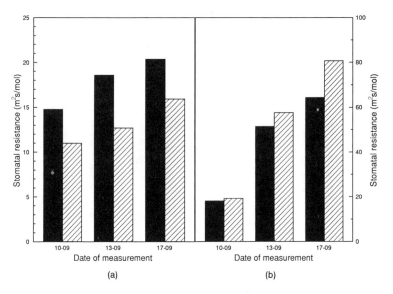

FIGURE 8.6 *Effects of ozone pollution on stomatal resistance in (a) well-watered (■) and (b) droughted (■) trees of* Fagus sylvatica. ⊠, *controls.*

8.5. Effects of ozone and sensitivity to frost

As mentioned earlier, the plasma membrane is exposed to the deposition either of ozone itself or of its reaction products, and many of the macromolecular components are likely to be susceptible to damage. It has been speculated that changes in fatty acid, lipid, sterol or protein composition might have deleterious consequences, and one that has attracted particular attention is tolerance of cold stress. Because the plasma membrane is a primary site of freezing injury (Steponkus, 1984), effects on its composition may have important influences on cold tolerance.

A large body of evidence has accumulated which suggests that much of the recent forest decline in Europe and the USA may be attributable to interactions between two or more different stresses (McLaughlin, 1985). In particular, a great deal of consideration has been given to the possible role ozone may play in predisposing trees to winter injury. Circumstantial evidence came from observations in the field of a clear gradient of injury, which increases in severity with higher elevation (Blank, 1985; Friedland *et al.*, 1984). Experiments undertaken by Brown *et al.* (1987) on 3-year-old clonal saplings of Norway spruce (*Picea abies* (L.) Karst.) provided some of the first experimental evidence of increased sensitivity to frost resulting from prior exposure to ozone. They found that older needles of clones which had received approximately 100 ppb ozone (47 days) developed visible injury in the form of brown necrosis and needle loss after a frost in November. Similar experiments using detached shoots from plants previously exposed to precise pollution regimens were performed by Barnes and Davison (1988) and by Lucas *et al.* (1988). The duration of the pollution regimen (70 and 65 days) was similar in both experiments, but Barnes and Davison fumigated clonal *Picea abies* with ozone at a concentration of approximately 120 ppb, whereas Lucas *et al.* (1988) employed *Picea sitchensis* and a range of concentrations: 5 (the control), 70, 120 and 170 ppb. In both studies, shoots were exposed to a range of freezing temperatures in programmable freezing cabinets and the degree of visible injury was assessed. Both studies produced clear evidence that early frosts might be damaging to both these species of spruce after exposure to high ozone concentrations in the summer. In an attempt to remove some of the associated errors involved in subjective assessments of visible injury, Murray *et al.* (1989) introduced a refined method of measuring electrolyte leakage from damaged tissues. This involved immersing a section of tissue which had been subjected to a freezing treatment in a volume of deionized water, and following the change in its conductivity over time. The rate of increase, adjusted for shoot size by comparison to the maximum conductivity obtained after autoclaving, gave a first-order rate constant, K. These values were shown to vary directly with the extent of tissue damage and related well to subjective assessments of physical injury.

In an experiment performed in our laboratory in 1988, trees of red spruce (*Picea rubens*) were exposed to 70 ppb ozone, applied daily from 09.00 to 16.00 h from June to September. Frost tolerance tests were performed in September and October using a range of subzero temperatures, the degree of damage being assessed using the electrolyte leakage method. The freezing temperatures used to assess the degree of tolerance were -3, -7, -11 and $-15\,°C$ in September and, were -7, -13, -19 and $-25\,°C$ in October. The ozone pollution to which the plants had been exposed earlier was found to increase the rate of electrolyte leakage significantly: $P < 0.001$ and $P < 0.001$ for September and October, respectively (Fig. 8.7). It was also apparent from these data that the absolute magnitude of the K values decreased from September to October, providing a clear indication of the natural progression of frost tolerance. A K value of 0.004 is an indication that the threshold for irreversible damage has been reached (Murray *et al.*, 1989). Although this value was not quite attained in the earlier frost test, it was nevertheless clear that significantly more membrane damage had occurred as a result of ozone pollution.

In many tree species the development of cold hardening is associated with increases in the unsaturation of the fatty acid component of membrane lipids. A high proportion of polyunsaturated fatty acids allows membranes to maintain their fluidity and to remain semipermeable at low temperatures (Levitt, 1980). A failure to develop the normal degree of lipid unsaturation during frost hardening might be expected to result in membrane dysfunction, and hence cellular damage when low temperatures are experienced (Wolfenden and Wellburn, 1991). Examination of the seasonal changes in phospholipids and galactolipids indicated that the maximum degree of frost hardiness was associated with increases in these lipids (Alberdi *et al.*, 1990). Seasonal changes have also been seen in the ratio of saturated to unsaturated fatty acid components of the chloroplastic lipid monogalactosyl diglyceride (MGDG) (Öquist, 1982). Long-term exposure (3 years) of Norway spruce to ozone by

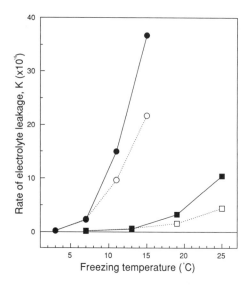

FIGURE 8.7 *Electrolyte leakage from* Picea rubens *grown in clean air or treated daily for 7 h with 70 ppb ozone.* ● *and* ■, *ozone treated;* ○ *and* □, *controls. Frost tolerance tests were performed in September (●, ○) and October (■, □).*

Wolfenden and Wellburn (1991) revealed significant changes in the ratio of some of the the fatty acid components of MGDG, and suggested that ozone may interfere with the biosynthesis of octadecatetraenoic acid during frost hardening, this being the second largest fatty acid component of MGDG. It appears likely that ozone, acting directly or through free radical production, is able to affect critical aspects of the cold hardening process.

The very reactive nature of ozone means that it has the potential to affect many cellular and possibly subcellular processes. The episodic nature of the occurrence of this pollutant not only increases the apparent toxicity, but the highest concentrations are normally correlated with meteorological conditions prevalent during periods of greatest photosynthetic gain. Reactions with biogenic hydrocarbons are likely to produce a cocktail of reactive products whose potential impacts are still being explored, but which are almost certainly negative. Isoprene, the biosynthesis of which appears to be regulated by ATP generation in the light (Loreto and Sharkey, 1990), accounts for up to 40% of the total biogenic hydrocarbon load in the atmosphere. As with elevated ozone concentrations, the emission of isoprene is likely to be greatest during periods of high solar radiation.

Ozone as a single agent of stress poses a number of problems for plants, but interactions with abiotic factors such as drought and cold may cause a multitude of effects which push the plant closer to the limit of its tolerance. Even when such interactions do not immediately have serious deleterious effects, if they occur for several years in succession there may be cumulative action leading to severe injury or premature death. Predictions of future tropospheric ozone concentration as a component of global climate change are a cause for great concern. The increases are expected to be of the order of 10–50% over the next 3–4 decades (Hough and Derwent, 1990; Thompson, 1990).

It is often assumed that global increases in atmospheric carbon dioxide concentration will lead to increases in biomass production, but the possibility that these may be wholly or partly negated by rising ozone concentrations is usually neglected. Recent experimental evidence suggests that the benefits of carbon dioxide enrichment are much reduced in the presence of ozone (e.g. Van der Eerden *et al.*, 1993). Although some encouraging progress has been made in the past 10 years, we still require a much deeper understanding of the mode of action of ozone on plants. This would provide a sounder basis for predicting effects and also for selecting (or genetically modifying) plants to incorporate the characteristics necessary for tolerance of an ozone-polluted environment. The results of an investigation by Barnes *et al.* (1990b) drew attention to the importance of our lack of understanding of tolerance mechanisms. They tested the ozone sensitivity of ten cultivars of spring wheat that were bred and introduced for cultivation in Greece between 1932 and 1980. It was found, contrary to expectations, that more modern cultivars displayed greater sensitivity (Fig. 8.8). Without a better understanding of the basic mechanisms underlying the injuries caused by ozone, it is not possible to advise plant breeders how to avoid this inadvertent incorporation of an undesirable character into modern cultivars.

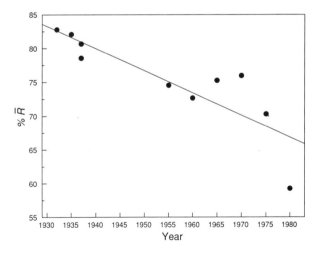

FIGURE 8.8 *Relative growth rates of ten wheat cultivars in ozone (90 ppb, 7 h daily) expressed as a percentage of those in clean air. Abscissa shows the year of introduction of the cultivar. (After Barnes* et al., *1990b.)*

ACKNOWLEDGEMENTS

We are grateful to the EC STEP, programme and the UK Department of the Environment for financial support.

REFERENCES

Aben, J.M.M., Janssen-Jurkovicova, M. and Adema, E.H. (1990). Effects of low-level ozone exposure under ambient conditions on photosynthesis and stomatal control of *Vicia faba* L. *Plant Cell Environ.* **13**, 463–469.

Alberdi, M., Fernandez, R.C., Romero, M. and Rios, D. (1990). Lipid changes in cold hardened leaves of *Nothofagus dombeyi. Phytochemistry* **29**, 2467–2471.

Barnes, J.D. and Davison, A.W. (1988). The influence of ozone on the winter hardiness of Norway spruce (*Picea abies* (L.) Karst). *New Phytol.* **108**, 159–166.

Barnes, J.D., Eamus, D., Davison, A.W., Ro-Poulson, H. and Mortensen, L. (1990a). Persistent effects of ozone on needle water loss and wettability in Norway spruce. *Environ. Pollut.* **63**, 345–363.

Barnes, J.D., Verlissariou, D., Davison, A.W. and Holevas, C.D. (1990b). Comparative ozone sensitivity of old and modern Greek cultivars of spring wheat. *New Phytol.* **116**, 707–714.

Blank, L.W. (1985). A new type of forest decline in Germany. *Nature* **314**, 311–314.

Brown, K.A., Roberts, T.M. and Blank, L.W. (1987). Interactions between ozone and cold sensitivity in Norway spruce: a factor contributing to the forest decline in central Europe. *New Phytol.* **105**, 149–155.

Chameides, W.L., Lindsay, R.W., Richardson, J. and Kiang, C.S. (1988). The role of biogenic hydrocarbons in urban photochemical smog: Atlanta as a case study. *Science* **241**, 1473–1475.

Colbeck, I. and Harrison, R.M. (1985). Dry deposition of ozone: some measurements of deposition velocity of vertical profiles to 100 metres. *Atmos. Environ.* **19**, 1807–1818.

Farage, P.K., Long, S.P., Lechner, E.G. and Baker, N.R. (1991). The sequence of change within the photosynthetic apparatus of wheat following short-term exposure to ozone. *Plant Physiol.* **95**, 529–535.

Fowler, D. (1992). Air pollution transport, deposition and exposure to ecosystems. In: *Air Pollution and Biodiversity*, eds. J.R. Barker and D.T. Tingey. New York: Van Nostrand Reinhold.

Friedland, A.J., Gregory, R.A., Karenlampi, L. and Johnson, A.H. (1984). Winter damage to foliage as a factor in red spruce decline. *Can. J. For. Res.* **14**, 963–965.

Hewitt, C.N., Kok, G.L. and Fall, R. (1990). Hydroperoxides identified in ozone exposed plants: a mechanism for air pollution damage to alkene emitters. *Nature* **344**, 56–58.

Hough, A.M. and Derwent, R.G. (1990). Changes in the global concentration of tropospheric ozone due to human activities. *Nature* **344**, 645–648.

Kerstiens, G. and Lendzian, K.J. (1989). Interactions between ozone and plant cuticles. I. Ozone deposition and permeability. *New Phytol.* **112**, 13–19.

Laisk, A., Kull, O. and Moldau, H. (1989). Ozone concentration in leaf intercellular air space is close to zero. *Plant Physiol.* **90**, 1163–1167.

Leonardi, S. and Langebartels, C. (1990). Fall exposure of beech saplings (*Fagus sylvatica* L.) to ozone and simulated acidic mists: effects on gas exchange and leachability. *Water Air Soil Pollut.* **54**, 143–153.

Levitt, J. (1980). *Responses of Plants to Environmental Stresses*, Vol. 1, *Chilling, Freezing and High Temperature Stress*, ed. T.T. Kozlowski. New York: Academic Press.

Loreto, F. and Sharkey, T.D. (1990). A gas exchange study of photosynthesis and isoprene emission in *Quercus rubra* L. *Planta* **182**, 523–531.

Lucas, P.W., Cottam, D.A. and Mansfield, T.A. (1987). A large-scale fumigation system for investigating interactions between air pollution and cold stress on plants. *Environ. Pollut.* **43**, 15–28.

Lucas, P.W., Cottam, D.A., Sheppard, L.J. and Francis, B.J. (1988). Growth responses and delayed winter hardening in Sitka spruce following summer exposure to ozone. *New Phytol.* **108**, 495–504.

Matyssek, R., Gunthardt-Goerg, M.S., Keller, T. and Scheidegger, C. (1991). Impairment of gas exchange and structure in birch leaves (*Betula pendula*) caused by low ozone concentrations. *Trees* **5**, 5–13.

McLaughlin, S.B. (1985). Effects of air pollution on forests: a critical review. *J. Air Pollut. Control Ass.* **35**, 512–534.

Mehlhorn, H. and Wellburn, A.R. (1987). Stress ethylene formation determines plant sensitivity to ozone. *Nature* **327**, 417–418.

Mehlhorn, H., O'Shea, J.M. and Wellburn, A.R. (1991). Atmospheric ozone interacts with stress ethylene formation by plants to cause visible plant injury. *J. Exp. Bot.* **42**, 17–24.

Mott, K.A. (1988). Do stomata respond to CO_2 concentrations other than intercellular? *Plant Physiol.* **86**, 200–203.

Murray, M.B., Cape, J.N. and Fowler, D. (1989). Quantification of frost damage in plant tissues by rates of electrolyte leakage. *New Phytol.* **113**, 307–311.

Öquist, G. (1982). Seasonally induced changes in acyllipids and fatty acids of chloroplast thylakoids of *Pinus silvestris*. *Plant Physiol.* **69**, 869–875.

Reich, P.B. and Lassoie, J.P. (1984). Effects of low level O_3 exposure on leaf diffusive conductance and water-use efficiency in hybrid poplar leaves. *Plant Cell Environ.* **7**, 661–668.

Steponkus, P.L. (1984). Role of plasma membrane in freezing injury and cold acclimation. *Annu. Rev. Plant Physiol.* **35**, 543–584.

Taylor, G. and Dobson, M.C. (1989). Photosynthetic characteristics, stomatal responses and water relations of *Fagus sylvatica*: impact of air quality at a rural site in southern Britain. *New Phytol.* **113**, 265–273.

Thompson, A.M. (1990). Effects of atmospheric chemical and climate change on tropospheric ozone. *Ozone Sci. Eng.* **12**, 177–194.

Van der Eerden, L., Tonneijck, A., Jarosz, W., Bestebroer, S. and Dueck, T. (1992). Influence of nitrogenous air pollutants on carbon dioxide and ozone effects on vegetation. *Proceedings of NATO Advanced Research Workshop, Interacting Stresses on Plants in a Changing Climate,* eds. M.B. Jackson and C.R. Black. Berlin: Springer Verlag.

Wolfenden, J. and Wellburn, A.R. (1991). Effects of summer ozone on membrane lipid composition during subsequent frost hardening in Norway spruce [*Picea abies* (L.) Karst]. *New Phytol.* **118**, 323–329.

CHAPTER 9

Plant adaptation to environmental stress: metal pollutant tolerance

P. J. PETERSON

9.1. Introduction

A great diversity of geological materials have been available for soil formation around the world. Different climatic conditions and biological processes over millions of years have produced a vast array of soil types, and evolutionary developments of associated plant assemblies have taken place during this time (Wild, 1978). Also within the past tens of thousands of years some vegetational types have been particularly modified by extreme climatic events such as the extension of the northern ice sheets. More recently, emissions from mining and smelter industries, from chemical factories, and widespread use of metal-containing fungicides and wood preservatives and of galvanized fences and other structures have contaminated local soils. Long-range transport of gaseous and particulate emissions from large industrial complexes have further contaminated soils and vegetation on a regional basis.

These different geochemical environments and climatic conditions have therefore markedly influenced the evolutionary development of plants and plant assemblages. The specialized seleniferous, nickeliferous and uraniferous floras, for example, have developed over geological time, whereas recent pollutant emissions have given rise to apparent metal-tolerant vegetation within relatively recent years (Peterson, 1971, 1979, 1983). In addition to contaminated soils, other soils may be deficient in one or more of the essential plant nutrients; such examples include the highly acid and infertile soils underlain by sands and siliceous materials that have also given rise to vegetational discontinuities such as the 'ecological islands' of California and elsewhere, the Shale barrens of the Appalachian Mountains and the Chimanimani Mountain Flora of Zimbabwe.

In addition to accumulating the essential macro- and micro-elements necessary for growth and reproduction, a number of plant species may selectively accumulate the other elements present in the soil to varying degrees (Fig. 9.1). This necessarily implies that some species are tolerant of these high concentrations. Other species, which do not grow on such soils but occur on surrounding soils, must by inference be non-tolerant of the geochemical conditions.

Tolerance is usually considered to be metal specific and an inherited characteristic (Baker, 1987). With endemic species found on the very old geochemical anomalies, metal tolerance has been described as an absolute phenomenon (Wild, 1978), whereas with the recently evolved metal-tolerant varieties, variously called races, subspecies, ecotypes or physiotypes, a gradation between low and high degrees of tolerance can be measured (Peterson, 1983). Whether the development of tolerance imposes a cost in terms of energy and hence a disadvantage for the plant in evolutionary terms needs further attention (Wilson, 1988).

Element accumulation and taxonomic status of species have been described earlier (Peterson, 1983). Little work has been done, however, to see whether the evolution of metal tolerance is accompanied by chromosome changes. Cobon and Murray (1983), working with metal-tolerant *Silene maritima* and non-tolerant coastal populations, found that there was no indication of chromosome change accompanying the evolution of tolerance. Indeed, they concluded that the genes controlling metal tolerance were probably distributed throughout the chromosome complement.

The genetic basis for metal tolerance of plants colonizing metalliferous mine wastes has been well documented for copper, lead, zinc, cadmium and

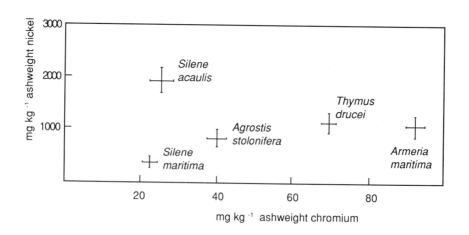

FIGURE 9.1 *Relationship between the concentrations of chromium and nickel in samples of the tops of plants collected from Unst, Shetlands serpentine site. The arithmetic means are shown with crosses, the limits of which show the standard deviations. (After Shewry and Peterson, 1976.)*

arsenic, particularly for several grass species and the dicotyledonous herb *Silene vulgaris* (Antonovics *et al.*, 1971; Porter and Peterson, 1977; Paliouris and Hutchinson, 1991). The genetic control of copper tolerance in *Mimulus guttatus* has been shown to be determined primarily by a single major gene (Macnair, 1983) indicating that copper tolerance has evolved in a manner analogous to other adaptations such as mimicry, industrial melanism and pesticide resistance. Earlier work (Macnair, 1977) on the same plant suggested that two major genes are involved in copper tolerance. Miranda *et al.* (1990), also working with copper-tolerant *M. guttatus*, have isolated two genes that encode a protein with class I metallothionein domains. This represents the first description of a metallothionein gene in a flowering plant.

Clearly, species differ in their sensitivity to metal pollutants in mine wastes. Some apparently do not possess tolerant genotypes, thus limiting their potential for evolutionary development on potentially toxic soils (Bradshaw and Hardwick, 1989). There is, however, little information available on the patterns of genetic variations in metal-tolerant plants. On the basis of isoenzyme variations (particularly esterase) in copper-tolerant *Agrostis stolonifera* (Wu *et al.*, 1975) and arsenic-tolerant *A. stolonifera* and *A. tenuis* (Benson *et al.*, 1981) it was concluded that, despite the very strong selection pressures from the geochemical environment, the tolerant plants had remained genetically diverse. The extreme concentrations of arsenic in the soils and associated plant growth over a hundred or so years did not lead to an obvious reduction in the number of clones in the population. In both of these cases, tolerance had not been lost over continuous vegetative culture for several years or in seed material. Cadmium tolerance in *Holcus lanatus* plants collected from a lead–zinc–cadmium smelter complex was, however, shown to be inducible and could be lost when transferred to uncontaminated soils (Baker, 1984; Baker *et al.*, 1990). Studies on the genetic basis of metal-tolerant endemic species to extreme nickel-rich soils, for example, is a relatively neglected area, yet promising for fundamental research. Extreme accumulations of zinc and lead to above $10\,000\ \mu g\ g^{-1}$ (dry wt) have been reported in *Viola calaminaria* and *Thlaspi calaminare* and would seem to be worthy of detailed biochemical and physiological study (Reeves and Brooks, 1983).

This paper outlines a number of possible tolerance mechanisms that may enable plants to adapt to metalliferous soils of geological origin as well as to soils contaminated by recent mine and smelter wastes and atmospheric metal depositions.

9.2. Synthesis of low molecular weight substances

In addition to the accumulation of metals such as cadmium and zinc and metalloids arsenic and selenium by plants, a range of low molecular weight organometals have been isolated from plants accumulating elements to high concentration. In several instances, the substances are known to be formed enzymatically; in others, chemical reactions could explain their synthesis, although they may be biosynthesized. The substances included in this review include selenoamino acids, nickel organic acids, chromium trioxalate, gold cyanide, fluoroacetate, zinc organic acids and copper–proline.

9.2.1. Selenoamino acids

Selenium-rich soils with their characteristic floras have been recognized for many years from various regions where they are often toxic to livestock and on occasions to indigenous people (Peterson, 1971). *Astragalus* species belonging to the Leguminosae and *Stanleya pinnata* (Cruciferae) are typical species from the USA that accumulate selenium to extreme concentrations of up to 10% of the dry weight of seeds. Selenium is metabolized to the non-protein amino acids, selenomethyl selenocysteine and selenocystathionine, and can be found largely as the γ-glutamyl derivatives in seeds (Peterson and Butler, 1967, 1971; Nigam and McConnell, 1976). The Australian accumulator *Neptunia amplexicaulis* also accumulates selenocystathionine and both isomeric γ-glutamyl derivatives in its seeds. However, it does not accumulate cystathionine to such high levels. Biosynthesis of selenocystathionine and cystathionine by partially purified enzymes from *N. amplexicaulis* has been reported (Ramadan, 1980).

Non-accumulators and selenium-sensitive plants, on the other hand, synthesize predominantly selenocysteine and selenomethionine, which are incorporated into proteins via aminoacyl-tRNA synthetase enzymes. Selenium accumulators contain very little selenium in their proteins, which implies that tolerance is related to enzymatic selection. Confirmation of this has come from a study of protein synthesis in selenium-accumulating and non-accumulating *Astragalus* species. The cysteinyl-tRNA synthetase from the accumulator *A. bisulcatus* did not link selenocysteine to its tRNA, but the same enzyme from non-accumulating *Astragalus* species was able to form cysteinyl-tRNA as well as selenocysteinyl-tRNA and incorporate seleno-cysteine as a substrate, yet the amino acid is not found naturally in protein of *A. bisulcatus* (Peterson and Butler, 1967; Burnell, 1981).

Selenium accumulators also synthesize volatile selenium compounds. Principally these are dimethyl diselenide and dimethyl selenide, depending on plant species. Significant amounts of selenium of up to 0.1% Seh^{-1} can be lost from such plants under extreme conditions (Zieve and Peterson, 1984). Whether this process can be considered to represent one aspect of tolerance remains to be determined.

9.2.2. Nickel organic acids

During the past 20 years, around 120 taxa, mostly endemic to nickel-rich soils derived from ultramafic outcrops, have been reported as accumulating nickel to high concentrations (Brooks *et al.*, 1979; Reeves *et al.*, 1983; Kruckeberg *et al.*, 1993). Such plants have been termed 'hyperaccumulators' when containing more than 1000 μg Ni g^{-1} (dry wt). This term does not imply any physiological basis for nickel accumulation but is used to exemplify such extreme accumulation. Nickel concentrations in plants normally vary around $0.05\,\mu g$ Ni g^{-1} (dry wt).

Many of these hyperaccumulators have been identified on the basis of herbarium specimens, although a number have also been verified from field-collected material. They have been found particularly in Italy, New Caledonia, Australia, Zimbabwe, Canada and North America, and belong to the Cruciferae, Brassicaceae, Caryophyllaceae, Violaceae and Cunoniaceae, to mention a few of the major families. An assessment of nickel accumulation and taxonomic status has been made earlier (Peterson, 1983).

In many of the nickel hyperaccumulators from New Caledonia, for example, the citrato-nickel (II) complex was the predominant form of nickel. Indeed, a strong correlation was reported between the concentration of nickel and citric acid in the leaves of 17 accumulators (Lee *et al.*, 1977, 1978). However, high concentrations of the hydrated Ni^{2+} were also recorded in some species, i.e. *Sebertia acuminata*.

Nickel complexes with malic and malonic acids have also been reported in other species, including *Alyssum bertolonii* from Italy, in *Pearsonia metallifera* from Zimbabwe, and other species (Pelosi *et al.*, 1976; Brooks *et al.*, 1981). The occurrence of high concentrations of organic acids in the hyperaccumulators of nickel seems well established, but their adaptive significance is uncertain. Are the nickel-organic acids formed metabolically or passively in the cell vacuole? It is also important to recall that nickel toxicity has been associated with an induced iron deficiency in many plants and that iron chlorosis conditions give rise to increased concentrations of organic acids, especially citrate and malate (Proctor and Woodell, 1971; Wallace, 1981). The exact role of organic acids in plant adaptation to nickel-rich soils still remains to be elucidated. Presumably the extreme concentrations of Ni^{2+} in *Sebertia acuminata* implies that this species has developed a different method of nickel tolerance.

9.2.3. Chromium trioxalate

Few chromium accumulators have been described among higher plants, and little research has been undertaken on the chemical form(s) of chromium present within the plants (Smith *et al.*, 1989). In one study, the trioxalatochromate III ion has been described from the leaves and roots of *Leptospermum scoparium* collected from serpentine soils in New Zealand (Peterson, 1979). Chromium was, however, transported as chromate in the xylem sap (Lyon *et al.*, 1969) and was presumably converted to the complex within the leaf tissues. Whether the process is enzyme mediated was not clear. Presumably the chromium–organic acid complex was less toxic than the high concentrations of chromate ions.

9.2.4. Gold cyanide, platinum and silver

On the basis of analytical chemistry, a number of plants have been shown to possess the ability to naturally accumulate gold up to $1000\,ng$ Au g^{-1}

(dry wt) from the soil, particularly in auriferous areas, and other species can accumulate the platinum group metals from geochemically anomalous areas such as the Merensky Reef (Peterson and Minski, 1985).

Gold has been reported to occur as the cyanide complex [Au (CN)$_2^-$] in the cyanogenic gold accumulator *Phacelia sericea* (Girling *et al.*, 1979) and is readily transported as the soluble ion throughout the plant (Girling and Peterson, 1978). Inorganic gold, on the other hand, was largely restricted to the roots of experimental plants and was adsorbed to cell walls, mitochondrial membranes and ribosomes, where inhibiting effects have been noted. Presumably, gold cyanide formation is a mechanism that enables the plant to avoid toxic effects and thus accumulate the element.

Accumulation of platinum, palladium and iridium, in particular, has been shown to be species specific. They can reach relatively elevated concentrations in particular plants growing over anomalous areas where they can presumably tolerate high concentrations (Peterson and Minski, 1985). In several experimental plant species, on the other hand, growth was inhibited by either ionic platinum or palladium but the mechanisms of these actions were not studied. Whether tolerance mechanisms are involved is unknown.

Silver concentrations in plants can exceed 10 000 ng Ag g^{-1} (dry wt), particularly on silver-rich soils and mine spoils (Peterson and Minski, 1985; Jones *et al.*, 1986) and is again species specific. The presence of silver-tolerant plants on lead–silver mine wastes, for example, has not been reported, although plants are tolerant of the lead. Silver humates and fulvates are absorbed and transported more readily than ionic silver in experimental plants (Jones and Peterson, 1986) and in a comparable way to gold humates and fulvates (Jones and Peterson, 1989). Whether the presence of such compounds reduces the inhibitory effects of ionic silver at the cellular level is unknown.

9.2.5. *Fluoroacetate*

Fluoride accumulator plants reported from various geographical regions of the world belong to diverse families (Oelrichs and McEwan, 1961; Meyer and O'Hagan, 1992). Fluoroacetate now identified from over 30 plant species was first isolated from the species *Dichapetalum cymosum* growing in low-fluoride soils. Concentrations of up to 0.23 mg fluoroacetate g^{-1} (fresh wt) have been reported. It is highly toxic to animals (and, to a lesser extent, other plant species) when activated to fluoroacetyl coenzyme A (fluoroacetyl-CoA) because it leads to inhibition of the citric acid cycle. Fluoroacetyl-CoA competes with acetyl-CoA for the enzyme citrate synthase, where it reacts with oxaloacetate and is involved in the synthesis of the toxic 2R,3R-fluorocitrate instead of citrate.

Although fluorocitrate in the young leaves of *D. cymosum* produces lethal effects on animals consuming it, the plant is able to tolerate high concentrations without adverse effects. At least one of the enzymes of the citric acid cycle of *D. cymosum* must therefore be different from those of other plants. Recent work by Meyer *et al.* (1992) has shown that an enzyme in a crude mitochondrial extract of *D. cymosum* can exclusively hydrolyse fluoroacetyl-CoA to fluoroacetate without hydrolysing acetyl-CoA. Fluoro-citrate is therefore not formed. Indeed, earlier work has shown that aconitase from *D. cymosum* is as sensitive to fluorocitrate as aconitase from other organisms. The presence of this fluoroacetyl-CoA hydrolase enzyme explains how this plant is tolerant to fluoride uptake and resultant fluoroacetate biosynthesis.

Meyer and O'Hagan (1992) also discuss the metabolism of fluoroacetate into fluoro-fatty acids in *Dichapetalum toxicarium*, although tolerance is unclear. Similarly, the relevance of volatilization of organofluorine compounds by *Acacia georginae* homogenates (Peters and Shorthouse, 1971) to tolerance is unclear.

9.2.6. Zinc organic acids

A number of zinc-tolerant plants have been shown to accumulate higher concentrations of malate, and to a lesser extent citrate, than non-tolerant clones, leading to the view that zinc malate, for example, is involved as a tolerance mechanism (Ernst, 1975; Mathys, 1977). Indeed, malate increased in concentration in tolerant plants with increasing zinc in the culture solution (Thurman and Rankin, 1982). However, the amount of organic acids in the zinc-tolerant plants was sufficient to complex only approximately one-third of the zinc, and the stability constants for zinc citrate and zinc malate are of a similar order of magnitude to those of the copper complexes, yet the plants are not copper tolerant (Thurman and Rankin, 1982). These results would not seem to equate with a tolerance mechanism.

9.2.7. Copper–proline

Copper–amino acid complexes have been isolated from the roots of copper-tolerant *Armeria maritima* and the endemic *Becium homblei* from Zimbabwe (Reilly, 1972; Farago and Mullen, 1979). Of particular interest was the occurrence of high concentrations of proline and a copper–proline complex, but their importance in the development of copper tolerance is not known. Recent research on copper-tolerant *A. maritima* by Farago and Mehra (1992) again stresses the occurrence of copper–proline in tolerant plants, whereas this was not so for the non-tolerant ecotypes. About 20% of the total copper in the roots and 12% in the leaves was water soluble as the copper–proline complex. Indeed, proline accounts for approximately 40% of the soluble amino acid fraction. The remaining 80% or so of the total copper in the plants was reported to be bound to cell walls.

9.3. Synthesis of high molecular weight compounds

Considerable discussions have taken place over the mechanisms of metal tolerance in zinc-, lead-, copper- and cadmium-tolerant populations of certain grass species, such as *A. tenuis*, *A. stolonifera*, *A. capillaris*, *Festuca rubra*, *Holcus lanatus* and the dicotyledonous herb *S. vulgaris* (Symeonidis *et al.*,

1985; Verkleij and Prast, 1989; Baker *et al.*, 1990). Early studies reported the biosynthesis of sulphur-rich metallothionein-like proteins by various crop plants (also blue-green algae and fungi), particularly when grown in solution cultures containing elevated concentrations of cadmium or copper (Bartolf *et al.*, 1980; Wagner and Trotter, 1982), as well as in copper-tolerant *Agrostis gigantea* (Rauser and Curvetto, 1980), copper-tolerant *Silene cucubalus* (Lolkema *et al.*, 1984) and copper-tolerant *M. guttatus* (Miranda *et al.*, 1990).

More recent studies refer to the induction of small sulphur-rich peptides (phytochelatins) following cadmium and copper stress in a range of crop plants and in *S. vulgaris* (Grill *et al.*, 1985; Robinson and Jackson, 1986; Rauser, 1987; Verkleij and Prast, 1989; Gupta and Goldsbrough, 1991). These relatively small peptides are composed of three amino acids (L-glutamic acid, L-cysteine and L-glycine) bound as the general formula (γ-glutamic acid–cysteine)$_n$–glycine, where $n = 2$–8. They are thus referred to as poly(γ-glutamylcysteinyl)glycines. These peptides bind metal ions, and in the presence of such ions the amounts of the poly(γ-glutamylcysteinyl)glycines increase. The presence of peptide bonds through the γ-carboxyl group suggests that these peptides are not encoded by structural genes, but by the products of biosynthetic pathways (Robinson and Jackson, 1986). Cells that are resistant to supraoptimal concentrations of particular metal ions produce considerable amounts of the poly(γ-glutamylcysteinyl)glycines. The synthesis of these phytochelatins can be induced by cadmium, copper, mercury, lead and zinc (Grill *et al.*, 1985). Indeed, Gupta and Goldsbrough (1991) have shown that more than 90% of the cadmium in the most tolerant selected tomato cell lines occurred as cadmium–phytochelatin complexes. Much research remains to be done, however, to characterize the metal complexes, their stability and especially their role within cells.

The amino acid composition of these phytochelatins can be compared with the glutamic acid–cysteine-rich glutathiones. In view of the similarity of glutathione (glutamylcysteinylglycine) to the phytochelatins, the tripeptide may be involved somehow in the biosynthesis of these metal-binding peptides. Indeed, an enzyme from *S. cucubalus* cell suspension cultures named γ-glutamyl-cysteine dipeptidyl transpeptidase (phytochelatin synthase) catalyses the reaction:

$$(glutamylcysteine)–glycine + (glutamylcysteine)_n–glycine$$
$$\Rightarrow (glutamylcysteine)_{n+1}–glycine + glycine$$

Thus it catalyses the transfer of the γ-glutamylcysteine dipeptide moiety of glutathione to an acceptor glutathione molecule or a growing chain of (glutamylcysteine)$_n$–glycine oligomers, thus synthesizing the phytochelatin (Grill *et al.*, 1989). In culture medium free of heavy metal ions, no phytochelatin induction was observed, thus lending credence to the essential role of this peptide in the regulation of the cellular balance of 'free' and complexed metal ions. On the other hand, Schat and Kalff (1992), working with copper-tolerant and non-tolerant *S. vulgaris*, concluded that phytochelatins were not decisively involved in differential copper tolerance. Rather, phytochelatin levels were a consequence of tolerance, an independent measure of stress.

It is of interest that cadmium-bearing electron-dense granules have been reported inside root parenchyma cells of *Agrostis gigantea* and *Zea mays* exposed to cadmium ions. They were found specifically in the nuclei of differentiating root cells, rather than in the mature root cells (Rauser and Ackerley, 1987), suggesting a role in a dynamic pattern of cadmium localization in root cells. Whether the cadmium-containing electron-dense granules represent aggregates of phytochelatin, for example, is unknown but worthy of further study. Whether these granules relate to the cadmium sulphide crystallite particles coated with phytochelatin peptides in cadmium-stressed tomato plants, recently reported by Reese *et al.* (1992), remains to be determined.

Rauser (1987), Verkleij and Prast (1989) and Baker *et al.* (1990) have shown that, in general, in a range of plant species, cadmium in the culture medium markedly induces glutathione biosynthesis up to 20 times higher than control plants. Elevated levels of glutathione were not produced in non-tolerant plants, although several exceptions were noted, perhaps implying alternative mechanisms of tolerance, which deserve further attention.

It is of interest that a cadmium-binding phosphoglycoprotein (cadmium mycophosphatin) has been isolated from the mushroom *Agaricus macrosporus*, which accumulates cadmium to relatively high levels (Meisch and Schmitt, 1986). The poly(γ-glutamylcysteinyl) glycines are structurally similar to the peptides cadystin B and cadystin A, which occur in some yeasts (Kondo *et al.*, 1985).

9.4. Enzyme adaptations

Until recently, the best known example of the presence of an enzyme with an altered substrate specifically towards an essential organic compound and its contaminant analogue was that of the cysteinyl-tRNA synthetase described earlier. In essence, this enzyme from the selenium-accumulator *Astragalus bisulcatus* distinguished between selenocysteine and cysteine. Selenocysteine was not incorporated into its proteins, but cysteine was. Selenium-sensitive *Astragalus* species that lacked this altered specificity of the cysteinyl-tRNA synthetase incorporated both amino acids into proteins, leading to lethal effects (Burnell and Shrift, 1979).

Now we need to add to this list the occurrence of a fluoroacetyl-CoA hydrolase enzyme from *D. cymosum*. This hydrolase could not use acetyl-CoA as substrate (Meyer *et al.*, 1992). Thus the proposal by Eloff and von Sydow (1971) that fluorocitrate synthesis does not take place in *D. cymosum* plants because citrate synthetase has different affinities for fluoroacetyl-CoA and acetyl-CoA must now be rejected.

As both of these plant species are well known accumulators, probably palaeo-accumulators, it can be expected that as more of these species are examined biochemically and enzymatically further examples of altered substrate specificity will be reported.

Many attempts to show substantial differences in the sensitivity of enzymes to various metals in the metal-tolerant grasses and other plants from mine wastes have not been successful (Mathys, 1975; Brookes *et al.*, 1981; Thurman and Rankin, 1982). However, Wainwright and Woolhouse (1975) and Cox and Thurman (1978), working with cell-wall acid phosphatases from copper- and zinc-tolerant plants, were able to show that enzymes from tolerant and non-tolerant clones possessed different K_i values.

9.5. Cellular adaptations

In addition to the possible tolerance mechanisms mentioned earlier, other differences between tolerant and non-tolerant plant varieties have been reported, but their significance is unclear. For example, differences in the cell, vacuole, nucleus and nucleolar size and RNA and protein contents of root meristems in zinc-tolerant compared with non-tolerant cultivars of *Festuca rubra* have been reported (Powell *et al.*, 1986a; Davies *et al.*, 1991a). How these changes relate to tolerance is unclear. Zinc has also been shown to extend the cell cycle in root meristems, particularly of non-tolerant cultivars, which again illustrates differences between cultivars although their physiological significance is not known (Powell *et al.*, 1986b). The effect of zinc on cellular processes and the sensitivity of the root meristems to zinc may make them suitable as early indicators of metal toxicity (Davies *et al.*, 1991b).

Early reports mention the restricted transport of metals from roots to shoots and that this phenomenon relates in some way to tolerance (Peterson, 1969; Antonovics *et al.*, 1971; Baker, 1978). Others, however, mention that copper is translocated more rapidly to the shoots of copper-tolerant *S. cucubalus* than in non-tolerant plants (Lolkema and Vooijs, 1986). Some reports stress specific metal binding sites in constituents of root cell walls as being involved in tolerance (Peterson, 1969; Turner, 1970; Farago *et al.*, 1980), although others doubt this hypothesis (Lolkema and Vooijs, 1986).

Similar uptake and translocation of zinc and cadmium in zinc-tolerant plants and cadmium in cadmium-tolerant plants and possible co-tolerance (see Baker *et al.*, 1990) has led to the suggestion that similar tolerance mechanisms apply to both elements (Verkleij and Prast, 1989). Indeed, evidence suggests that both elements are accumulated in the vacuole (Rauser and Ackerley, 1987; Wagner, 1988). Brookes *et al.* (1981) examined compartmental flux analysis using radioactive zinc supplied to tolerant and non-tolerant clones of *Deschampsia caespitosa*. Only the zinc-tolerant clones could actively pump zinc from the cytoplasm of root cells across the tonoplast into the vacuoles at high zinc concentrations. Working with copper tolerance in *S. cucubalus*, De Vos *et al.* (1991) showed that the tolerant plants had an increased resistance to copper-induced damage of the plasmalemma of root cells. Thus the level of copper tolerance was related to the ability of the plants to prevent the copper-induced alteration of the permeability barrier. The prevention of potassium loss via the plasmalemma is possibly an important aspect of copper tolerance. Strange and Macnair (1991) working with copper tolerance in *Mimulus guttatus* suggest that the primary copper tolerance mechanism probably resides in the cell membrane. This conclusion is based on the greater potassium effect induced by copper stress in the non-tolerant plants. It is argued by Strange and Macnair (1991) that the results provide further evidence against the involvement of phytochelatins in the *primary* tolerance mechanism.

In the case of arsenic, on the other hand, fewer studies have been undertaken to ascertain the tolerance mechanism. Concentrations of up to 1% (dry wt) have been recorded in individual plants of *A. tenuis* and *A. stolonifera* growing on arsenic-rich mine spoil where the predominant ionic species was arsenate (Porter and Peterson, 1975, 1977; Benson *et al.*, 1981). Arsenic was accumulated to higher levels in the shoots of tolerant plants than in roots, and to higher levels than in non-tolerant plants, implying cellular adaptation. Tolerance was also shown to be an inherited characteristic, recently confirmed by Watkins and Macnair (1991). These findings have been confirmed by Meharg and Macnair (1990, 1991a, 1991b) working with arsenic-tolerant *Holcus lanatus* and *D. caespitosa*, who have shown that the arsenic-tolerant genotypes of these species have a reduced arsenate influx in the roots, as determined by isotherm studies. The high-affinity uptake system was suppressed or absent in tolerant plants. Yet the fate of arsenic in these plants has yet to be established.

Copper-mediated oxidative stress brought about by a lowering of the protective glutathione concentration in cells of the copper-tolerant *S. cucubalus* during production of phytochelatin has been suggested as another important process in metal tolerance that needs detailed examination (De Vos *et al.*, 1992). The oxidation of cellular thiols, resulting in the production of free radicals and subsequent lipid peroxidation of microsomes isolated from roots, is potentially harmful to cellular processes (De Vos *et al.*, 1989).

9.6. Conclusions

The mechanisms involved in plant tolerance, outlined on the preceding pages, range from chelation with organic ligands and/or biotransformation of inorganic ions to organic substances, through to changes in the rates of physiological and biochemical processes, such as differences in enzyme specificity, ion transport, etc., no doubt the result of evolutionary adaptation. Volatilization of accumulated elements from leaves, leaf exudations, loss of epicuticular waxes containing metals, and leaf fall are other processes perhaps involved in tolerance. The strategies adopted by plants to avoid, restrict or alleviate potentially toxic conditions are thus many and varied, with no underlying general principles; the processes are directed by the plant species and the chemistry of the ions accumulated.

Detailed biochemical and physiological studies at both the whole-plant and cellular levels have proceeded slowly over a number of decades; much still remains to be done. Insufficient attention has been given to an understanding of the action of metal toxicity in higher plants. It seems that metal toxicity leads to the oxidative breakdown of the lipid fraction of the biomembrane (Van Assche *et al.*, 1990), yet data to substantiate such effects are lacking (Clijsters *et al.*, 1990). It has been proposed that future research on metal phytotoxicity should involve the study of inhibition and induction of enzymes at the different cell membranes, especially the plasma membrane *in vivo* (Van Assche and Clijsters, 1990).

The most recent research by Meyer *et al.* (1992), working with cell-free extracts of *D. cymosum*, referred to earlier, has shown specific enzymatic hydrolysis of fluoroacetyl-CoA to fluoroacetate, thus removing it from competition with acetyl-CoA in the citric acid cycle. This provides a biochemical explanation for the fluoride tolerance of this species first recognized over 40 years ago.

One problem that has hampered physiological studies of tolerance has been our inadequate knowledge of the ionic forms present in the soil solution that are absorbed by plants. More studies involving chemical speciation in the xylem and phloem sap, as carried out by, for example, Lyon *et al.* (1969) are required. The chemical forms of elements in the soil solution and their relationship to element transport and accumulation, such as those undertaken on silver and gold humates and fulvates compared with the ionic forms, are certainly necessary (Jones and Peterson, 1986, 1989). The application of advanced analytical methodology, such as X-ray microprobe analysis, will open up new advances for research and should be vigorously pursued. Tissue culture experiments, aimed at an understanding of cellular processes and their relationship to biochemical tolerance, are also important for comparison with the whole-plant studies usually undertaken.

Another area almost completely neglected is the relevance of mycorrhizal fungi in or on the roots of plants growing on metal-rich soils. Studies with mycorrhizal *Betula* systems look promising (Brown and Wilkins, 1985). Similarly, the roles of root nodules and associated microorganisms, particularly in endemic and metal-tolerant leguminosae, have been largely neglected.

The ecological significance of tolerance at the population level has not been adequately tackled except for experiments with metal-tolerant grasses. In such cases, selection for inherited copper tolerance in *A. stolonifera*, for example, has been reported to occur readily and within a generation from the tolerant or partly tolerant individuals that exist in 'normal' populations at very low frequency, i.e. about two per 1000 individuals (Wu and Bradshaw, 1972; Wu *et al.*, 1975). On the other hand, different conclusions have been drawn from experiments on cadmium tolerance in the grass *Holcus lanatus*. Baker (1984) and Baker *et al.* (1990) report that cadmium tolerance could be lost in transplanting plants to uncontaminated soils, thus questioning the genecological basis of cadmium tolerance and its heritability.

In conclusion, this article has covered a range of mechanisms widely perceived as enabling plants to survive their own accumulation and production of potentially toxic substances. Much research still remains to be done on the plants referred to in this article and on the other accumulators not mentioned, including terrestrial fungi, lichens, mosses and aquatic plants from various taxonomic classes. Only then will we be in a position to satisfactorily understand these processes and their relevance to plant tolerance.

ACKNOWLEDGEMENT

The environmental toxicological implications of these studies have been supported in part by the European Science Foundation.

REFERENCES

Antonovics, J., Bradshaw, A.D. and Turner, R.G. (1971). Heavy metal tolerance in plants. *Adv. Ecol. Res.* **7**, 1–85.

Baker, A.J.M. (1978). Ecophysiological aspects of zinc tolerance in *Silene maritima* With. *New Phytol.* **80**, 635–642.

Baker, A.J.M. (1984). Environmentally-induced cadmium tolerance in the grass *Holcus lanatus* L. *Chemosphere* **13**, 585–589.

Baker, A.J.M. (1987). Metal tolerance. *New Phytol.* **106** (Suppl.), 93–111.

Baker, A.J.M., Ewart, K., Hendry, G.A.F., Thorpe, P.C. and Walker, P.L. (1990). The evolutionary basis of cadmium tolerance in higher plants. In: Proceedings of the 4th International Conference on Environmental Contamination, pp. 23–29, ed. J. Barcelo. Edinburgh: CEP Consultants.

Bartolf, M., Brennan, E. and Price, C.A. (1980). Partial characterization of a cadmium-binding protein from the roots of cadmium-treated tomato. *Plant Physiol.* **66**, 438–441.

Benson, L.M., Porter, E.K. and Peterson, P.J. (1981). Arsenic accumulation, tolerance and genotypic variation in plants on arsenical mine wastes in south-west England. *J. Plant Nutr.* **3**, 655–666.

Bradshaw, A.D. and Hardwick, K. (1989). Evolution and stress: genotypic and phenotypic components. *Biol. J. Linn. Soc.* **37**, 137–155.

Brookes, A., Collins, J.C. and Thurman, D.A. (1981). The mechanism of zinc tolerance in grasses. *J. Plant Nutr.* **3**, 695–705.

Brooks, R.R., Morrison, R.S., Reeves, R.D., Dudley, T.R. and Akman, Y. (1979). Hyperaccumulation of nickel by *Alyssum* Linnaeus (Cruciferae). *Proc. R. Soc.* **B203**, 387–403.

Brooks, R.R., Shaw, S. and Marfil, A.A. (1981). The chemical form and physiological function of nickel in some Iberian *Alyssum* species. *Physiol. Plant.* **51**, 167–170.

Brown, M.T. and Wilkins, D.A. (1985). Zinc tolerance of mycorrhizal *Betula*. *New Phytol.* **99**, 101–106.

Burnell, J.N. (1981). Selenium metabolism in *Neptunia amplexicaulis*. *Plant Physiol.* **67**, 316–324.

Burnell, J.N. and Shrift, A. (1979). Cysteinyl-tRNA synthase from *Astragalus* species, *Plant Physiol.* **63**, 1045–1097.

Clijsters, H., Van Assche, F., Vangronsveld, J. and Gora, L. (1990). Plant cell membranes as a common target for pollutants and environmental stress? In: *Proceedings of the 4th International Conference on Environmental Contamination*, pp. 41–44, ed. J. Barcelo. Edinburgh: CEP Consultants.

Cobon, A.M. and Murray, B.G. (1983). Evidence for the absence of chromosome differentiation in populations of *Silene maritima* With. growing on heavy-metal contaminated sites. *New Phytol.* **94**, 643–646.

Cox, R.M. and Thurman, D.A. (1978). Inhibition by zinc of soluble and cell wall phosphatases of roots of zinc-tolerant and non-tolerant clones of *Anthoxanthum ordoratum*. *New Phytol.* **80**, 17–22.

Davies, K.L., Davies, M.S. and Francis, D. (1991a). Zinc-induced vacuolation in root meristematic cells of *Festuca rubra* L. *Plant Cell Environ.* **14**, 399–406.

Davies, M.S., Francis, D. and Thomas, J.D. (1991b). Rapidity of cellular changes induced by zinc in a zinc-tolerant and non-tolerant cultivar of *Festuca rubra* L. *New Phytol.* **117**, 103–108.

De Vos, C.H.R., Schat, H., Vooijs, R. and Ernst, W.H.O. (1989). Copper-induced damage to the permeability barrier in roots of *Silene cucubalus*. *J. Plant Physiol.* **135**, 165–169.

De Vos, C.H.R., Schat, H., De Waal, M.A.M., Vooijs, R. and Ernst, W.H.O. (1991). Increased resistance to copper-induced damage of the root cell plasmalemma in copper-tolerant *Silene cucubalus*. *Physiol. Plant.* **82**, 523–528.

De Vos, C.H.R., Vonk, M.J., Vooijs, R. and Schat, H. (1992). Glutathione depletion due to copper-induced phytochelatin synthesis causes oxidative stress in *Silene cucubalus*. *Plant Physiol.* **98**, 853–858.

Eloff, J.N. and Von Sydow, B. (1971). Experiments on the fluoroacetate metabolism of *Dichapetalum cymosum* (Gifblaar). *Phytochemistry* **10**, 1409–1415.

Ernst, W. (1975). Physiology of heavy metal resistance in plants. In: *Heavy Metals in the Environment*, Symposium Proceedings, Toronto, University of Toronto, Vol. 2, pp. 121–136.

Farago, M.E. and Mehra, A. (1992). Uptake of elements by the copper-tolerant plant *Armeria maritima*. In: *Metal Compounds in Environment and Life: Interrelation between Chemistry and Biology*, Vol. 4, pp. 163–169, eds. E. Merian and W. Haerdi. Northwood: Science and Technology Letters.

Farago, M.E. and Mullen, W.A. (1979). Plants which accumulate metals. Part IV. A possible copper–proline complex from the roots of *Armeria maritima*. *Inorg. Chim. Acta Lett.* **32**, 93–94.

Farago, M.E., Mullen, W.A., Cole, M.M. and Smith, R.F. (1980). A study of *Armeria maritima* (Mill.) Willdenow growing in a copper-impregnated bog. *Environ. Pollut.* **A21**, 225–244.

Girling, C.A. and Peterson, P.J. (1978). Uptake, transport and localization of gold in plants. *Trace Subs. Environ. Health* **12**, 105–118.

Girling, C.A., Peterson, P.J. and Warren, H.V. (1979). Plants as indicators of gold mineralization at Watson Bar, British Columbia, Canada. *Econ. Geol.* **74**, 902–907.

Grill, E., Winnacker, E.-L. and Zenk, M.H. (1985). Phytochelatins: the principal heavy-metal complexing peptides of higher plants. *Science* **230**, 674–676.

Grill, E., Löffler, S., Winnacker, E.-L. and Zenk, M.H. (1989). Phytochelatins, the heavy-metal binding peptides of plants, are synthesized from glutathione by a specific γ-glutamyl cysteine dipeptidyl transpeptidase (phytochelatin synthase). *Proc. Natl. Acad. Sci. USA* **86**, 6838–6842.

Gupta, S.C. and Goldsbrough, P.B. (1991). Phytochelatin accumulation and cadmium tolerance in selected tomato cell lines. *Plant Physiol.* **97**, 306–312.

Jones, K.C. and Peterson, P.J. (1986). The influence of humic and fulvic acids on silver uptake by perennial ryegrass and its relevance to the cycling of silver in soils. *Plant Soil* **95**, 3–8.

Jones, K.C. and Peterson, P.J. (1989). Gold uptake by perennial ryegrass: the influence of humates on the cycling of gold in soils. *Biogeochemistry* **7**, 3–10.

Jones, K.C., Davies, B.E. and Peterson, P.J. (1986). Silver in Welsh soils: physical and chemical distribution studies. *Geoderma* **37**, 157–174.

Kondo, N., Isobe, M., Imai, K. and Goto, T. (1985). Synthesis of metallothionein-like peptides cadystin A and B occurring in fission yeast and their isomers. *Agric. Biol. Chem.* **49**, 71–83.

Kruckeberg, A.R., Peterson, P.J. and Samiullah, Y. (1993). Hyperaccumulation of nickel by *Arenaria rubella* (Caryophyllaceae) from Washington State *Madrono.* **40**, 25–30.

Lee, J., Reeves, R.D., Brooks, R.R. and Jaffré, T. (1977). Isolation and identification of a citrato complex of nickel from nickel-accumulating plants. *Phytochemistry* **16**, 1503–1505.

Lee, J., Reeves, R.D., Brooks, R.R. and Jaffré, T. (1978). The relationship between nickel and citric acid in some nickel-accumulating plants. *Phytochemistry* **17**, 1033–1035.

Lolkema, P.C. and Vooijs, R. (1986). Copper tolerance in *Silene cucubalus*. *Planta* **167**, 30–36.

Lolkema, P.C., Donker, M.H., Schouten, A.J. and Ernst, W.H.O. (1984). The possible role of metallothioneins in copper tolerance of *Silene cucubalus*. *Planta* **162**, 174–179.

Lyon, G.L., Peterson, P.J. and Brooks, R.R. (1969). Chromium-51 transport in the xylem sap of *Leptospermum scoparium* (Manuka). *NZ J. Sci.* **12**, 541–545.

Macnair, M.R. (1977). Major genes for copper tolerance in *Mimulus guttatus*. *Nature* **268**, 428–430.

Macnair, M.R. (1983). The genetic control of copper tolerance in the yellow monkey flower, *Mimulus guttatus*. *Heredity* **50**, 283–293.

Mathys, W. (1975). Enzymes of heavy-metal resistant and non-resistant populations of *Silene cucubalus* and their interaction with some heavy metals *in vitro* and *in vivo*. *Physiol. Plant.* **33**, 161–165.

Mathys, W. (1977). The role of malate, oxalate and mustard oil glucosides in the evolution of zinc-resistance in herbage plants. *Physiol. Plant.* **40**, 130–136.

Meharg, A.A. and Macnair, M.R. (1990). An altered phosphate uptake system in arsenate-tolerant *Holcus lanatus* L. *New Phytol.* **116**, 29–35.

Meharg, A.A. and Macnair, M.R. (1991a). Uptake, accumulation and translocation of arsenate in arsenate-tolerant and non-tolerant *Holcus lanatus* L. *New Phytol.* **117**, 225–231.

Meharg, A.A. and Nacnair, M.R. (1991b). The mechanisms of arsenate tolerance in *Deschampsia caespitosa* (L.) Beauv. and *Agrostis capillaris* L. *New Phytol.* **119**, 291–297.

Meisch, H.-U. and Schmitt, J.A. (1986). Characterization studies on cadmium: mycophosphatin from the mushroom *Agaricus macrosporus*. *Environ. Health Perspect.* **65**, 29–32.

Meyer, M. and O'Hagan, D. (1992). Rare fluorinated natural products. *Chem. Br.* **28**, 785–788.

Meyer, J.J.M., Grobbelaar, N., Vleggaar, R. and Louw, A.I. (1992). Fluoro-acetyl-coenzyme A hydrolase-like activity in *Dichapetalum cymosum. J. Plant Physiol.* **139**, 369–372.

Miranda, J.R. de, Thomas, M.A., Thurman, D.A. and Tomsett, A.B. (1990). Metallothionein genes from the flowering plant *Mimulus guttatus. FEBS Lett.* **260**, 277–280.

Nigam, S.N. and McConnell, W.B. (1976). Isolation and identification of two isomeric glutamylselenocystathionines from the seeds of *Astragalus pectinatus. Biochim. Biophys. Acta* **437**, 116–121.

Oelrichs, P.B. and McEwan, T. (1961). Isolation of the toxic principle in *Acacia georginae. Nature* **190**, 808–809.

Paliouris, G. and Hutchinson, T.C. (1991). Arsenic, cobalt and nickel tolerances in two populations of *Silene vulgaris* (Moench) Garcke from Ontario, Canada. *New Phytol.* **117**, 449–459.

Pelosi, P., Fiorentini, R. and Galoppini, C. (1976). On the nature of nickel compounds in *Alyssum bertolonii* Desv. Part 2. *Agric. Biol. Chem.* **40**, 1641–1649.

Peters, R.A. and Shorthouse, M (1971). Identification of a volatile constituent formed by homogenates of *Acacia georginae* exposed to fluoride. *Nature* **231**, 123–124.

Peterson, P.J. (1969). The distribution of zinc-65 in *Agrostis tenuis* Sibth. and *A. stolonifera* L. tissues. *J. Exp. Bot.* **20**, 863–875.

Peterson, P.J. (1971). Unusual accumulation of elements by plants and animals. *Sci. Prog.* **59**, 505–526.

Peterson, P.J. (1979). Geochemistry and ecology. *Phil. Trans. R. Soc. B* **288**, 169–177.

Peterson, P.J. (1983). Unusual element accumulations as a taxonomic character. In: *Anatomy of the Dicotyledons*, Vol. 2, pp. 167–173, eds. C.R. Metcalfe and L. Chalk. Oxford: Clarendon Press.

Peterson, P.J. and Butler, G.W. (1967). Significance of selenocystathionine in an Australian selenium-accumulating plant *Neptunia amplexicaulis. Nature* **213**, 599–600.

Peterson, P.J. and Butler, G.W. (1971). The occurrence of selenocystathionine in *Morinda reticulata* Benth., a toxic seleniferous plant. *Aust. J. Biol. Sci.* **24**, 175–177.

Peterson, P.J. and Minski, M.J. (1985). Precious metals and living organisms. *Interdisc. Sci. Rev.* **10**, 159–169.

Porter, E.K. and Peterson, P.J. (1975). Arsenic accumulation by plants on mine waste (United Kingdom). *Sci. Tot. Environ.* **4**, 365–371.

Porter, E.K. and Peterson, P.J. (1977). Arsenic-tolerance in grasses growing in mine waste. *Environ. Pollut.* **14**, 255–267.

Powell, M.J., Davies, M.S. and Francis, D. (1986a). Effects of zinc on cell, nuclear and nucleolar size, and on RNA and protein content in the root meristem of a zinc-tolerant and a non-tolerant cultivar of *Festuca rubra. New Phytol.* **104**, 671–679.

Powell, M.J., Davies, M.S. and Francis, D. (1986b). The influence of zinc on the cell cycle in the root meristem of a zinc-tolerant and a non-tolerant cultivar of *Festuca rubra* L. *New Phytol.* **102**, 419–428.

Proctor, J. and Woodell, S.R.J. (1971). The plant ecology of serpentine. I. Serpentine vegetation of England and Scotland. *J. Ecol.* **59**, 375–395.

Ramadan, S.S. (1980). *The Metabolic Pathway of Cystathionine and its Selenium Analogue in Plants and Micro-organisms.* PhD thesis, University of London.

Rauser, W.E. (1987). Changes in glutathione content of mine seedlings exposed to cadmium. *Plant Sci.* **51**, 171–175.

Rauser, W.E. and Ackerley, C.A. (1987). Localization of cadmium in granules within differentiating and mature root cells. *Can. J. Bot.* **65**, 643–646.

Rauser, W.E. and Curvetto, N.R. (1980). Metallothionein occurs in roots of *Agrostis* tolerant to excess copper. *Nature* **287**, 563–564.

Reese, R.N., White, C.A. and Winge, D.R. (1992). Cadmium-sulfide crystallites in Cd-(γ-EC)$_n$G peptide complexes from tomato. *Plant Physiol.* **98**, 225–229.

Reeves, R.D. and Brooks, R.R. (1983). Hyperaccumulation of lead and zinc by two metallophytes from mining areas of central Europe. *Environ. Pollut.* **31A**, 277–285.

Reeves, R.D., MacFarlane, R.M. and Brooks, R.R. (1983). Accumulation of nickel and zinc by western North American genera containing serpentine-tolerant species. *Am. J. Bot.* **70**, 1297–1303.

Reilly, C. (1972). Amino acids and amino acid copper complexes in water-soluble extracts of copper-tolerant and non-tolerant *Becium homblei* Z. *Pflanzenphysiologie* **66**, 294–296.

Robinson, N.J. and Jackson, P.J. (1986). 'Metallothionein-like' metal complexes in angiosperms; their structure and function. *Physiol. Plant.* **67**, 499–506.

Schat, H. and Kalff, M.M.A. (1992). Are phytochelatins involved in differential metal tolerance or do they merely reflect metal-imposed strain? *Plant Physiol.* **99**, 1475–1480.

Shewry, P.R. and Peterson, P.J. (1976). Distribution of chromium and nickel in plants and soil from serpentine and other sites. *J. Ecol.* **64**, 195–212.

Smith, S., Peterson, P.J. and Kwan, K.H.M. (1989). Chromium accumulation, transport and toxicity in plants. *Toxicol. Environ. Chem.* **24**, 241–251.

Strange, J. and Macnair, M.R. (1991). Evidence for a role for the cell membrane in copper tolerance of *Mimulus guttatus*, Fischer ex DC. *New Phytol.* **119**, 383–388.

Symeonidis, L., McNeilly, T. and Bradshaw, A.D. (1985): Interpopulation variations in tolerance to cadmium, copper, lead, nickel and zinc in mine populations of *Agrostis capillaris* L. *New Phytol.* **101**, 317–324.

Thurman, D.A. and Rankin, J.L. (1982). The role of organic acids in zinc tolerance in *Deschampsia caespitosa*. *New Phytol.* **91**, 629–635.

Turner, R.G. (1970). The subcellular distribution of zinc and copper within the roots of metal-tolerant clones of *Agrostis tenuis* Sibth. *New Phytol.* **69**, 725–731.

Van Assche, F. and Clijsters, H. (1990). Effects of metals on enzyme activity in plants. *Plant Cell Environ.* **13**, 195–206.

Van Assche, F., Vangronsveld, J. and Clijsters, H. (1990). Physiological aspects of metal toxicity in plants. In: *Proceedings of the 4th International Conference on Environmental Contamination*, pp. 246–250, ed. J. Barcelo. Edinburgh: CEP Consultants.

Verkleij, J.A.C. and Prast, J.E. (1989). Cadmium tolerance and co-tolerance in *Silene vulgaris* (Moench.) Garcke [= *S. cucubalus* (L.) Wib.]. *New Phytol.* **111**, 637–645.

Wagner, G.J. (1988). Responses of plant cells to cadmium exposure. *J. Cell Biochem. Suppl.* **12D**, 335.

Wagner, G.J. and Trotter, M.M. (1982). Inducible cadmium binding complexes of cabbage and tobacco. *Plant Physiol.* **69**, 804–809.

Wainwright, S.J. and Woolhouse, H.W. (1975). Physiological mechanisms of heavy metal tolerance in plants. In: *The Ecology of Resource Degradation and Renewal*, pp. 231–257, eds. M.J. Chadwick and G.T. Goodman. Oxford: Blackwell.

Wallace, A. (1981): Some physiological aspects of iron deficiency in plants. *J. Plant Nutr.* **3**, 637–642.

Watkins, A.J. and Macnair, M.R. (1991). Genetics of arsenic tolerance in *Agrostis capillaris* L. *Heredity* **66**, 47–54.

Wild, H. (1978). The vegetation of heavy metals and other toxic soils. In: *Biogeography and Ecology of Southern Africa*, pp. 1301–1332, ed. M.J.A. Weger. The Hague: Junk.

Wilson, J.B. (1988). The cost of heavy-metal tolerance: an example. *Evolution* **42**, 408–413.

Wu, L. and Bradshaw, A.D. (1972). Aerial pollution and the rapid evolution of copper tolerance. *Nature* **238**, 167–169.

Wu, L., Bradshaw, A.D. and Thurman, D.A. (1975). The potential for evolution of heavy metal tolerance in plants. III. The rapid evolution of copper tolerance in *Agrostis stolonifera. Heredity* **34**, 165–187.

Zieve, R. and Peterson, P.J. (1984). Volatilization of selenium from plants and soil. *Sci. Tot. Environ.* **19**, 197–202.

CHAPTER 10

Structural and functional responses to fire and nutrient stress: case studies from the sandplains of South-West Australia

J. S. PATE

10.1. Introduction: the edaphic and climatic environments of sandplain ecosystems

The deep sands typical of much of the heathlands and open woodlands of the kwongan of South-West Australia represent some of the most oligotrophic rooting substrates found anywhere in the world (Beard, 1984; Bettenay, 1984). Organic contents are extremely low and almost entirely confined to the top 10–15 cm layer, where the bulk of the macronutrients nitrogen, phosphorus and sulphur are also located. Due to deposition of oceanic salt, sodium and chlorine may be abundant relative to calcium, potassium and magnesium. Litter accumulating between fires is particularly depauperate in essential nutrients because presenescence retrieval of such elements from foliage occurs with extraordinary efficiency (see Pate and Dell, 1984). Ash from ground fires is accordingly of limited nutritional value as opposed to that from hot canopy fires in which green foliage is burnt when still fully loaded with nutrients. However, the element nitrogen mostly escapes in gaseous form after fire.

Data gathered so far strongly suggest that productivity of sandplain vegetation is driven primarily by availability of phosphorus, except in the aftermath of fire when nitrogen becomes limiting. As expected of a mediterranean climate, plants experience heat and moisture stress from mid-summer (December) to early autumn (March/April). Wetter but colder conditions prevail from late autumn through winter, when low temperatures may limit growth. A warm, still wet spring and early summer period follows, during which extension growth and reproduction of most species are accomplished. As described by Dodd *et al.* (1984), soils show progressive drying from the surface to at least 1 m depth from September to April, a slow recharge of the profile from April say to June, followed by only 1 or 2 months of full saturation when water percolates freely to the water table.

Thus, kwongan vegetation copes continuously with low availability of nutrients, seasonally with wide fluctuations in temperature and soil moisture status, and periodically with drought and fire. It consequently comes as no surprise to find among the component species a bewildering array of structural or functional adaptations relating to the above constraints. This chapter selects a number of specializations relating specifically to fire or nutrient stress. All have been examined recently by the author and his colleagues in sandplain flora near Perth, Western Australia, some of the examples being widely expressed and of general significance to plant growth and survival, others less commonly observed but equally fascinating in terms of the processes that may have shaped their evolution.

10.2. Responses to fire

Vegetation of mediterranean type climates, such as in South-West Australia, has a long history of association with fire, to the extent that certain species now appear to be markedly dependent on periodic burning for their reproduction and seedling establishment (Frost, 1984; Bell et al., 1984; Keeley, 1986) or, if surviving fire, clearly benefit from the reduced competition and improved availability status of the post-fire environment.

10.2.1. Fire ephemerals

Plants of this response category germinate exclusively after hot burns to form conspicuous elements of the early post-fire flora. They either complete their growth cycle within the growing season following a summer fire (monocarpic fire ephemerals) or commence flowering in the second season and reproduce prolifically for the next 2 or 3 years before dying (polycarpic fire ephemerals). Study of a range of fire ephemerals of the sandplains of South-West Australia (Pate et al., 1985) has shown them to possess extremely high relative growth rates in comparison with other growth forms, and an ability to enrich their biomass with generally higher levels of nitrogen, phosphorus and potassium than similarly aged seedlings of non-ephemeral character. Fire ephemerals also show marked inherent plasticity in mature size, enabling them to respond differentially in terms of biomass across sites at which ash resources are distributed unevenly. However, harvest indices for dry matter are uniformly high regardless of plant size, e.g. the smallest and largest successfully reproducing individuals of the freely tillering monocarpic grass *Stipa elegantissima* exhibit a 1000-fold difference in final dry weight, but the harvest indices for dry matter of all individuals lie within the narrow range 0.21–0.27 (see Pate et al., 1985).

Mean seed production per plant per season has been shown to vary from 39 to 1480 in the 18 monocarpic species studied and from 1400 to 33 000 per plant in the second plus third seasons of the two dioecious polycarpic species *Gyrostemon ramulosus* and *Tersonia brevipes* (Gyrostemonaceae) (Pate et al., 1985). Mature male plants of the sexually dimorphic *T. brevipes* are less than one-fifth of the size of females, and by this means a disproportionately large share of the ash bed nutrient resource is channelled into female biomass and seed production.

Monocarpic fire ephemerals mobilize up to 91% of the phosphorus and 70% of the nitrogen of their vegetative parts to seeds, indicating how effectively potentially limiting nutrients are passed to the seed bank of a species. However, seeds of most species are small, not particularly well stocked with mineral reserves, but in many cases well endowed with oil and starch. This indicates selection for high propagule number and longevity, while relying on the high absorptive capacities of the rapidly expanding fibrous root system of the seedling and the assurance of an immediate nutrient supply from the ash bed to compensate for poor starting capital of mineral reserves in the seed.

Judging from the extremely high initial growth rates of fire ephemerals, their rates of net photosynthesis and partitioning of assimilates to new leaf area must be extremely high compared with other species. They are also particularly well adapted to the cool autumn and winter conditions under which the formative stage(s) of their growth cycles are completed.

10.2.2. Obligate seeders and resprouters

Two especially common fire response traits encountered among mediterranean type ecosystems are those commonly referred to as the 'obligate seeder' and 'resprouter' strategies. Plants of the former group are killed outright by fire and recruit thereafter exclusively from canopy- or soil-stored banks of seed, whereas plants of the resprouter category survive fire and rapidly replace their whole shoot or foliar canopy by sprouting from heat-resistant buds on root crowns, below-ground lignotubers or trunks. Seeders and resprouters germinate prolifically following fire, and juvenile stages of woody taxa from the two types exhibit highly divergent patterns of growth and dry matter partitioning during the first few years of growth (Pate *et al.*, 1990b). Thus, by its third or fourth season, the typical young seeder has achieved an almost three-fold greater total plant dry weight and a more than four-fold higher shoot:root dry weight than cohabiting seedlings of the typical resprouter. As shown by Bowen and Pate (1991) for the carbon budgets of congeneric resprouter and seeder juveniles of Proteaceae, an almost three-fold greater allocation of carbon of net photosynthate occurs to shoots of a seeder than in a counterpart resprouter (Fig. 10.1).

Mean starch contents of shoots and roots of 32 juvenile seeder species studied by Pate *et al.* (1990b) turn out to be low (5 and $8 \, mg \, g^{-1}$, respectively) in comparison with corresponding values for the shoots ($15 \, mg \, g^{-1}$ dry wt) and especially the roots ($55 \, mg \, g^{-1}$) of 34 similarly aged species of resprouters. The resprouter is clearly disadvantaged in terms of growth rate when investing less than one-quarter of its photosynthate into shoot growth (Fig. 10.1), but the relatively large starch pool of its root has obvious long-term significance for initiating new shoot growth after fire.

Anatomical studies of the distribution of root starch have shown interesting differences between seeders and sprouters (see Pate *et al.*, 1990b). The storage potential of the root is determined primarily by the space devoted to

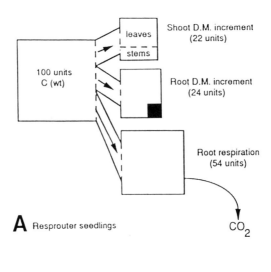

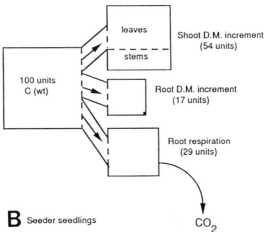

FIGURE 10.1 *Carbon allocation profiles for 18–30-month-old seedlings of three resprouter and three seeder representatives of the Proteaceae. Black rectangles in the compartments marked root dry mass (DM) increment denote the proportional investment into starch. All data are expressed in terms of net gain from the atmosphere by the shoot of 100 units by weight of carbon. (From Bowen and Pate, 1991.)*

parenchyma as opposed to conducting and other non-storage dead tissues, and those areas composed of such parenchyma are generally substantially greater in resprouter than seeder representatives of a genus (e.g. in *Banksia, Hovea* and *Bossiaea*). Moreover, potential storage sites tend to be utilized more fully for starch storage in resprouters than seeders, although resprouters themselves rarely come near to realizing the full storage potential suggested by their anatomy. For example, some store starch only in their cortex, others use their noticeably enlarged ray tissue, and others xylem parenchyma and rays, while only a few exhibit significant starch storage in all of these locations.

Comparative studies of the concentrations of the macronutrients phosphorus, nitrogen, calcium, potassium and magnesium in shoot and root dry matter of seeders and sprouters (Pate et al., 1990b) show higher mineral levels in shoots than in roots of plants of both categories but no evidence of key elements such as nitrogen, phosphorus and potassium being more concentrated in roots of resprouters than seeders. A premium is thus placed in resprouters on providing a ready energy and carbon source for refoliation and production of new feeding roots. Once these new roots form and transpiring surfaces are created, the plant is able to exploit the nutrients released from ash or from the below-ground biomass of species killed in the fire.

Comparable studies of adult obligate seeders and resprouters have also been made on certain understorey legumes (Hansen, 1990; Hansen et al., 1991, 1992). In a comparison of the seeder *Bossiaea aquifolium* (Fabaceae) and a cohabiting resprouter (*B. ornata*), Hansen et al. (1991) have shown that a shoot : root dry weight ratio of 10 is achieved in the seeder by its third or fourth growing compared with only 0.8 for the resprouter. This differential is maintained into adulthood, but when the shoot of the resprouter is burnt it takes some 4 or 5 years before the equilibrium shoot : root ratio (0.8) is reattained. Because of greater investment into photosynthesizing biomass, a 6-year-old seeder shows an 18-fold greater plant size (dry weight) than a recruiting resprouter of similar age.

Even more noticeable is the disparity in reproductive effort between the recruits of seeder and resprouter species (e.g. see Groves et al., 1986; Zammit and Westoby, 1987; Lamont and Barrett, 1988). As shown for *Bossiaea* by Hansen et al. (1991), the seeder species commences reproducing in the second season after germination and produces a cumulative total of 301 seeds per plant by year 6, whereas equivalent recruits of the resprouter species do not fruit until year 4 and produce only 14 seeds per plant over the same 6 year period. Even large resprouting individuals of 15–30 years age reproduce only slightly more successfully (360 seeds per plant) than does the seeder recruit (301 seeds) over the same 6 year period. Associated with these differences in reproductive activity, the proportions (%) of the total plant's seasonal increment of dry matter or of a specific mineral that are invested in reproductive organs (the 'reproductive effort' of the species) are generally higher in the seeder than in the resprouter: e.g. in the above species of *Bossiaea*, values of 29% for dry matter, 38% for nitrogen and 55% for phosphorus in 6-year-old seeder species versus 19, 25 and 38% respectively for large plants of the resprouter by their sixth season after fire (Hansen et al., 1991).

Turning to parallel modifications in a family of monocotyledons (*Restionaceae*), a recent study by Pate et al. (1991a) has demonstrated resprouters to have more deeply buried and thicker rhizomes and noticeably lower culm:rhizome dry weight ratios than seeders. Starch levels in rhizomes are generally much higher in resprouters (mean level $71 \pm 16 \, mg \, g^{-1}$ (dry wt) for the 35 species studied) than seeders ($1.6 \pm 1 \, mg \, g^{-1}$ (dry wt) for the 45 species studied) and a comparable differential is evident between the groups in mean levels of stored sugars (seeders 33.9 ± 5.9 versus resprouters 10.6 ± 1.9). Accentuation of carbohydrate storage in rhizomes of resprouters is accomplished through development of high proportional areas of interstitial storage parenchyma relative to vascular tissue and sclerenchyma. Seeders have very small amounts of such parenchyma. Some resprouters (e.g. *Sporadanthus* spp. and *Hypolaena exsulca*) lack starch reserves in their rhizomes but compensate by accumulating high sucrose levels in their rhizomes (*Sporadanthus* spp.) or by sprouting rosettes of juvenile-type foliage leaves (*H. exsulca*) whose high photosynthetic activity provides immediate benefit to the shoot regeneration process (E. Hickman and J.S. Pate, unpublished data).

The significance of root reserves of carbohydrate in generating new shoot biomass after fire has been recently examined in the resprouters *Stirlingia latifolia* (a species of Proteaceae whose reproduction is stimulated by burning) (Bowen and Pate, 1993) and *Conostephium pendulum* (a long-lived shrubby member of the Epacridaceae) (T. Bell and J.S. Pate, unpublished data). Both species recover by multiple sprouting from their upper root stocks, *S. latifolia* being particularly well endowed with starch (pre-fire levels $90–110 \, mg \, g^{-1}$ root (dry wt)), *C. pendulum* not so ($3–10 \, mg \, g^{-1}$ root (dry wt)). If the first flush of resprouting shoots is removed from burnt adult plants of *C. pendulum* only a few further shoots form, as might be expected from the paucity of their root reserves. By contrast, if new resprouting shoots of *S. latifolia* are removed on a monthly basis, further crops of shoots form unabated for up to a year and until virtually all root starch has been utilized.

Under normal circumstances an intact resprouting plant of *S. latifolia* loses some 50–75% of its initial carbohydrate reserve during the first 2–3 months of resprouting, and starch levels remain low until inflorescences have formed and vegetative shoot biomass has attained a magnitude similar to or greater than that before the fire. Final return to a starch pool size typical of unburnt plants does not eventuate until plants have finished fruiting, some 2 years after the fire.

Mention should be made of the very common strategy among burnt resprouters of developing mesophytic-type juvenile foliage of high photosynthetic potential. This is particularly evident in species whose adult foliage is composed of sclerophyllous cladodes or phyllodes. In many tree species (e.g. *Casuarina fraseriana*, *Eucalyptus marginata* and *Banksia attenuata*), trunks and larger branches not succumbing to a canopy fire sprout densely packed investments of new shoots whose combined leaf surfaces offer a most effective photosynthetic catchment during the first season after fire. However, the vast majority of these shoots die and only a few dominant members remain to form an open canopy structure similar to that before the fire. By this time all supernumerary shoots have been shed and the trunk and branches have reattained their normally smooth contour.

10.2.3. Geophytes

The final response to fire to be discussed is that exhibited by geophytes. South-West Australian species of this type are mostly stem- or root-tuberous and only rarely bulbous or cormous (Pate and Dixon, 1982). Most are herbaceous winter active species, completing their annual cycle of growth and reproduction between autumn and late spring and then dying back to a deeply buried storage organ before the heat and water stress of summer. They are thus likely to escape fire in space and time and subsequently benefit from the better light, moisture and nutrient conditions of the post-fire environment. A number of geophytes, particularly orchids, and certain sundews (*Drosera* spp.) flower only after fire, and their storage organs show a progressive build-up of nutrients such as phosphorus in the years immediately following a burn (e.g. Dixon and Pate, 1978; Pate and Dixon, 1982).

10.2.4. Seeds and fire

Seeds in litter or near the soil surface are highly vulnerable to fire, so sandplain plants exhibit a range of adaptations for lodging seed resources in safe refuges within their habitats. Self-burying seeds, as exemplified by composites and awned grasses, are produced by some kwongan species, but this is rare in comparison with the phenomenon of 'bradyspory', in which seeds are retained in fire-resistant woody fruits in the plant canopy (e.g. in many species of Proteaceae, Myrtaceae and Casuarinaceae). Bradysporous species may release seeds from their fruits only after a hot fire (the phenomenon of 'serotiny') or may shed seeds between fires and not necessarily on the death of the plant or shedding of fruiting branches. Seeds of bradysporous species germinate immediately on release from the cone and may be well camouflaged to avoid the attention of birds and mammals foraging in burnt habitats.

A taxonomically wide range of species have ant-dispersed seeds (Bell *et al.*, 1993). Most of these possess specialized oil-containing appendages (elaiosomes), which are consumed by ants after burial in their underground galleries. Understorey legumes (e.g. *Acacia* spp., *Bossiaea* spp. and *Hovea* spp.) utilize ants in this manner and, although seeds may be buried at a depth of 6–8 cm, they still respond positively to heat or gaseous emanations from the fire. Because seeds are mostly localized in special storage chambers, they tend to germinate in groups and may thus experience intense intraspecific competition during seedling establishment (Monk *et al.*, 1981).

10.3. Adaptations to low levels of nutrient

10.3.1. Rooting morphologies and nutrient and water uptake

Species of the sandplain engage in either a summer pattern of growth and reproduction or one strictly confined to the cool wet season between the autumn and the end of spring. Most large woody species, especially members of the Proteaceae and Myrtaceae, have a warm-season pattern of shoot growth and typically possess 'sinker' roots, which extend downwards for many metres to the water table or at least into soil layers that remain wet throughout the summer. Each plant also develops a set of outwardly radiating lateral roots, most of which are confined to the upper soil layers where essential nutrients such as nitrogen and phosphorus are likely to be concentrated. Each autumn, following onset of the season's rains, these lateral roots develop new sets of specialized nutrient-absorbing rootlets, which remain active in absorption until the upper soil dries out in late spring or early summer. Sinker roots provide water through a full year, thereby allowing photosynthesis and growth to continue well into the summer, when the upper soil profile may be totally desiccated. The main period of shoot growth thus occurs some 4–5 months later than the principal period of nutrient uptake by the root (Bowen and Pate, 1991).

In contrast, plants that grow and reproduce only during the cool season of the year (e.g. most herbaceous geophytes, therophytes and a number of small shrubby species (e.g. Epacridaceae and some Myrtaceae)) generally possess shallow fibrous root systems, usually without distinct sinker roots. If plants of this kind retain above-ground biomass through the year, they inevitably become severely stressed during the summer (Dodd et al., 1984). Their root biomass is especially concentrated in the upper nutrient-rich layers of the soil profile and ephemeral feeding rootlets form seasonally on the upper laterals.

10.3.2. Specializations relating to nutrient uptake

As indicated by Lamont (1984) and summarized in a recent review by Pate (1993a), modifications of root structure facilitating uptake of essential nutrients are widely encountered among herbaceous species other than fire ephemerals and are almost universal among woody shrub and tree species. Interesting correlations hold between trophic behaviour, rooting morphology and mycorrhizal status, some specializations occurring very widely, others in relatively few species (Table 10.1). A number of generalizations emerge.

1. Uptake mediated by cluster (proteoid type) roots, as exemplified by virtually all Proteaceae and certain legumes, is extraordinarily well represented, especially in ecosystems dominated by proteaceous biomass. This type of root forms mats very close to the soil surface and is renewed annually after a short season of activity (Lamont, 1984; Bowen and Pate, 1991).
2. Non-mycorrhizal woody species are mostly confined to polycarpic fire ephemerals and root hemiparasitic species. Herbaceous species lacking mycorrhizae are encountered in carnivorous species of highly oligotrophic habitats and in certain geophytes and herbaceous annuals (including fire ephemerals) that flourish in habitat niches well supplied with nutrients from ash or mineralization of litter.
3. All plants engaging in nitrogen-fixing symbioses (e.g. *Macrozamia*, *Casuarina* and legumes of the Fabaceae and Mimosaceae) bear cluster

TABLE 10.1. *Relationships between trophic attributes, growth form and mycorrhizal status among commonly encountered taxa of the sandplains (kwongan) of South-West Australia.*

Autotrophy	
Herbaceous species	
Non-mycorrhizal	Fire ephemerals and certain geophytes
Vesicular–arbuscular mycorrhizal	Many dicotyledonous and monocotyledonous species
Orchid-type mycorrhizal	Orchidaceous geophytes
Woody species	
Ecto + vesicular–arbuscular mycorrhizal	Range of tree and shrub genera
Cluster roots (non-mycorrhizal)	Proteaceae
Vesicular–arbuscular mycorrhizal	Large number of shrubby taxa
Ericoid mycorrhizal	Epacridaceae (hair roots)
Nitrogen-fixing	
Vesicular–arbuscular mycorrhizal	*Macrozamia*, certain Fabaceae
Ecto + vesicular–arbuscular mycorrhizal	Fabaceae, Mimosaceae, Casuarinaceae
Cluster roots (non-mycorrhizal)	*Daviesia, Viminaria, Acacia* (few species)
Carnivory	
Herbaceous species	
Non-mycorrhizal	Droseraceae, Byblidaceae, Utriculariaceae
Holoparasitism	
Herbaceous species	
Non-mycorrhizal	On roots *Orobanche*
	Subepidermal *Pilostyles* (no root)
	On stems (*Cuscuta, Cassytha*) (no root)
Root hemiparasitism	
Woody species	
Non-mycorrhizal	*Olax, Exocarpos, Leptomeria, Anthobolus, Nuytsia, Santalum*
Aerial hemiparasitism	
Woody species	
Non-mycorrhizal	*Amyema* spp., *Lysiana* spp. (no root)
Heterotrophic epiparasitism	
Herbaceous species	
Shared *Rhizoctonia* mycorrhiza	*Rhizanthella* (no roots) on host *Melaleuca*
Mycoautotrophy	
Herbaceous species	
Ectomycorrhizhal	*Lobelia* spp.

roots or vesicular arbuscular and/or ectotrophic mycorrhizae. By meeting a plant's requirements for phosphorus, these modifications allow a species to become highly competitive in nitrogen-deficient habitats over species not capable of nitrogen fixation, e.g. the flourishing of nitrogen-fixing species after fire (Monk *et al.*, 1981; Hansen and Pate, 1987).

4. Ericoid-type mycorrhizal relationships are apparently confined to members of the Epacridaceae. As with counterparts in the Ericaceae (Read, 1991), they are associated with impoverished acid soils in which nitrogen occurs mostly in relatively intractible form as peptides, proteinaceous residues and other organic complexes. Plants of this kind utilize nitrate poorly if at all (e.g. Stewart *et al.*, 1993; G.R. Stewart and J.S. Pate, unpublished data), and their mycorrhizae may possess enzymatic capacities (e.g. acid proteases) for using the above forms of nitrogen (Leake *et al.*, 1989). The role of ericoid-type mycorrhiza in phosphorus nutrition does not appear to have been evaluated.

5. Vesicular arbuscular mycorrhizal associations alone, or in conjunction with ectotrophic mycorrhizae, are widespread among woody and herbaceous species (e.g. Lamont, 1984; Brundrett and Abbott, 1991). Judging from a recent study on the nitrate-reducing capacities and $^{15}N : ^{14}N$ natural abundance ratings of a range of common kwongan species (Stewart *et al.*, 1992; Pate *et al.*, 1993), these mycorrhizal species use forms of nitrogen mostly other than NO_3^-.

10.3.3. *Exploitation of other macroorganisms as sources of nutrients*

Carnivorous species are particularly prevalent in sandplain ecosystems, including a spectacular array of growth and lifeforms of sundews (*Drosera* spp.). Several recent studies of *Drosera* spp. and other carnivorous genera in native habitat have indicated appreciable dependence on arthropod prey in respect of nitrogen and phosphorus (Dixon *et al.*, 1980; Schulze *et al.*, 1991). Such species are mostly confined to severely impoverished habitats in which heterotrophic gain from animal prey might be particularly advantageous (Karlsson and Pate, 1992).

A second, somewhat unusual, example is that of 'heterotrophic epiparasitism', in which a significant proportion of the nutrients acquired by a non-autotrophic receptor species is derived from another species via a shared mycorrhizal partner (Table 10.1). This is exemplified in the kwongan by the underground orchid *Rhizanthella gardneri* in partnership with the fungus *Rhizoctonia* sp. and the host myrtaceous shrub *Melaleuca uncinata* (Dixon *et al.*, 1990). The rootless orchid bears underground capitula-like inflorescenes, which are pollinated by termites. Infection of the stem tubers of the orchid by the fungus occurs via specialized epidermal trichomes. The deep sands in which the *Melaleuca–Rhizoctonia–Rhizanthella* relationship occurs are exceptionally low in organic matter, so it is likely that most, if not all, of the carbon of the dry matter of the orchid comes from photosynthate of the host shrub.

A supposedly similar situation is encountered in certain annual species of *Lobelia*. e.g. *L. gibbosa* (G. Nielsson, K.W. Dixon and J.S. Pate, unpublished data). Seed germinates deeply below ground in autumn and the emerging root is immediately infected by a basidiomycete-type mycorrhiza. This nourishes the rapidly developing corolloid root mass, which eventually reaches as much as 20 g fresh weight and becomes richly endowed with nutrients, especially phosphorus. Shoots then form at the expense of the root mass, and the resulting autotrophic phase reproduces and sets seed before dying the following summer. Again, carbon acquired by the root biomass is likely to be of epiparasitic origin. The term 'mycoautotrophy' (Table 10.1) has been introduced to refer to the life-cycle of species such as *L. gibbosa* in which the first half of the growth cycle is nourished by a fungus and the second half is photosynthetically autotrophic.

Parasitic plants that derive water, nutrients and photosynthate partly or wholly from their hosts (Pate, 1993b,c) are particularly common in kwongan ecosystems, whether as aerial parasites of shrubs or trees (e.g. the stem holoparasites *Cassytha* and *Pilostyles* and the hemiparasitic mistletoes *Amyema* and *Lysiana*) or as root hemiparasites exploiting a wide range of other taxa (e.g. *Olax, Santalum, Nuytsia* and *Exocarpos*). The above-mentioned holoparasites obtain all of their requirements for water, nutrients and photosynthetic products from the host xylem and phloem. The two mistletoe genera listed possess xylem-tapping haustoria, which establish lumen-to-lumen continuities between their tracheary elements and the exposed xylem vessels of the host (Kuo and Pate, 1990). These enable mass flow of xylem fluid to occur from host to parasite, whereas the contact parenchyma of the parasite are probably involved in selective uptake of nutrients (Pate, 1993b).

Growth and survival of mistletoes depends on maintenance of a negative gradient in leaf water potential with a host. This is achieved by higher rates of water loss and stomatal conductance during the day and higher rates of transpiration during the night. High osmotic potentials and water-holding ability (capacitance) of the mistletoe buffer it well against water loss during drought when neither partner is likely to be transpiring to any extent (Pate, 1993c). As shown in a recent study of the mineral balance of *Amyema linophyllum* (Pate *et al.*, 1991b), more than sufficient nutrients enter from the host to meet the requirements of reproduction by the mistletoe. As further evidence of nutrient sufficiency, foliage of mistletoes increases continuously with age in a full range of nutrients until shed from the plant.

Root hemiparasites of kwongan possess multiple haustorial connections on most surrounding woody species, from which they are most effective in obtaining nutrients. They can thus represent the only large shrub or tree species in heathland communities in which all other species are less than 1.5 m in height. Most shrubby root hemiparasites (e.g. *Olax phyllanthi* (Olacaceae), *Leptomeria* and *Exocarpos*) have particularly shallow root systems and must therefore rely on tap-rooted hosts for water during the summer (see Pate *et al.*, 1990a,c,d).

10.4. Efficient utilization and conservation of nutrients within living biomass

With the notable exception of the above-mentioned fire ephemerals and hemiparasites, most kwongan species exhibit generally lower concentrations of essential nutrients in the dry matter of their shoots, roots and leaves than in plants of other ecosystems, especially those of agriculture and horticulture. This is especially so of certain groups of monocotyledon (e.g. data on Restionaceae and Cyperaceae by Meney *et al.*, 1990), woody dicotyledons of heathlands (Pate *et al.*, 1990b) and many proteaceous trees and shrubs (Kuo *et al.*, 1982). However, low concentrations of key nutrients in adult biomass need not necessarily imply unusual ability to function at low cytoplasmic concentrations of these nutrients, because the species being considered may be highly sclerophyllous and poorly endowed with chlorenchyma and storage parenchyma. It would therefore be more meaningful to compare species in terms of nutrient levels per unit of tissue protein. This has yet to be done, but where young apical growing shoots of woody kwongan species have been analysed their nutrient levels per unit dry matter turn out not to be noticeably different from those recorded for mesophytic species (Pate and Dell, 1984).

It may be argued that the sclerophyllous leaf character, combined with leaf longevities of 2 years or more (Pate *et al.*, 1984), will reduce nutrient losses through leaf shedding to a minimum. Equally important, however, is the extraordinary efficiency with which kwongan species mobilize nutrients from leaves back into their parent shoots prior to leaf abscission (Pate and Dell, 1984). Added to this a number of woody species senesce one age group of leaves coincident with seasonal shoot elongation and expansion of new leaves, thus obviating much of the requirement for transient or long-term storage in non-foliar tissues (compare deciduous species in the northern hemisphere). An equivalently effective transfer is observed in the end-of-season retrieval of key nutrients from above-ground stem and leaves to underground storage organs of geophytes (Pate and Dixon, 1982).

Equally relevant to the survival of certain species is the ability to form highly concentrated reserves of certain elements in seeds. This phenomenon is vividly illustrated in Proteaceae (Kuo *et al.*, 1982; Pate *et al.*, 1986), the dry matter of whose seeds may carry 30–500 times higher concentrations of phosphorus and 8–100 times higher levels of nitrogen, calcium and zinc than adjacent leaves or cones. Seeds are accordingly many times more expensive to fill, say in terms of mobilization from vegetative catchments of limiting elements such as nitrogen and phosphorus, than their dry weight would suggest.

Seeds of Proteaceae store oil rather than starch and their principal nitrogenous reserve consists of protein bodies largely restricted to the parenchyma (mesophyll) of their cotyledons. Protein bodies may contain any one or more of seven distinct types of globoid and crystalloid inclusions, most of which contain phytate. Electron probe microanalyses have demonstrated that the inclusion types vary in composition, particularly in respect of ratios of calcium to phosphorus and of phosphorus to sulphur or magnesium. The mean protein level of seeds of the 32 species studied by Pate *et al.* (1986)

is high (39.5% of embryo dry weight), but this underscores the potential value in terms of stored nitrogen because of the extremely high levels of the nitrogen-rich amino acid arginine in the protein and ethanol-soluble fractions of the seeds. The extremely well balanced nature of the seed reserves of Proteaceae is well demonstrated in a study by Stock *et al.* (1990) showing survival of seedlings for 300 days under continuous nutrient starvation without visible signs of any specific nutrient deficiency. Such resilience would have great significance for seedlings germinating under highly oligotrophic conditions in which a large fraction of the cotyledonary resource is expended in developing a deeply penetrating tap root, thereby providing vital access to underground water reserves over the first summer (Bowen and Pate, 1991). It is not until the next season that the seedling adds appreciably to its nutrient pool through the agency of its cluster roots.

A final example, drawn from the studies of Dixon *et al.* (1983), concerns the seed-like aestivating structures of granite outcrop-inhabiting geophytes (e.g. *Isoetes muelleri, Philydrella pygmaea* and *Stylidium petiolare*). These species live in shallow lenses of soil, which become extremely hot and desiccated during the summer. Their dormant corms resemble seeds in terms of small size (1–2 mg dry wt), low water content (15% of fresh weight) and presence of protein bodies with phytate inclusions rich in phosphorus, potassium, calcium and magnesium. This contrasts with the majority of geophytes whose generally much larger corms, bulbs or tubers are composed mostly of water (Pate and Dixon, 1982).

10.5. Conclusion: adaptations of species to every season and circumstance

Space does not permit description of the diverse range of structural and functional modifications of kwongan species relating to survival of heat and drought of summer, although tangential reference has been made to some of these when considering rooting morphologies and seasonality of growth. Most of the specializations described in the chapter are likely to have multivalent effects on growth, reproduction and overall survival of a species, if only through equipping the individual to survive subsequent stress experiences.

It becomes clear that one should aim to examine the interrelated effects of the whole suite of adaptations that a species exhibits in response to stress agencies operating at various stages of its growth cycle. For example, as shown here for proteaceous trees or shrubs, study would need to be undertaken of all stages of reproduction, especially pollination and the setting and filling of seeds with nutrients, the effects of fire on seed release and seedling establishment and the significance of cotyledons versus cluster roots in establishing the shoot and root systems of the seedling. In like manner, attention should be paid to water relations of shoot and lateral and sinker roots in early and later growth, the processes of partitioning of assimilates between shoot and root, and the effects of good or poor seasons on establishment of seed banks in species that are obligate seeders or on the investment of fire-resistant biomass in resprouter species. As a further example, when considering the flooding-tolerant shrubby legume *Viminaria juncea*, study would need to concentrate on the seasonal production of pneumatophores and the adaptive significance of the aerenchymatous investments of nodules, roots and stems to nitrogen fixation during inundation (Walker *et al.*, 1983), the effects of fire and flooding on seed germination and seedling establishment, and the relevance of cluster roots to nutrient uptake (Lamont, 1984). In addition, *V. juncea* exhibits marked morphotypic differentiation in respect to rate of transition from juvenile trifoliolate foliage to the phyllodineous habit of the adult and a comparable level of variability between its ecotypes in their respective abilities to form cluster roots under varying levels of phosphate or in their growth response to flooding, mesic or drought-simulating conditions (Walker and Pate, 1986). Variations of this kind provide a most interesting platform for examining the genetic basis of evolutionary response to stress.

ACKNOWLEDGEMENTS

The author is greatly indebted to a number of colleagues who have contributed to the studies reported in this chapter. Their names are to be found in the joint publications listed in the bibliography. We are indebted to the generosity of the Australian Research Council for consistently supporting studies of Australian flora. In times of financial constraint, when funding agencies attach increasing emphasis to research of an applied nature and of immediate benefit to the community, it is consoling still to find support for curiosity-based basic studies of survival and functioning of allegedly 'unimportant' native plant species.

REFERENCES

Beard, J.S. (1984). Biogeography of the kwongan. In: *Kwongan – Plant Life of the Sandplain*, pp. 1–26, eds. J.S. Pate and J.S. Beard. Nedlands: University of Western Australia Press.

Bell, D., Hopkins, A.J.M. and Pate, J.S. (1984). Fire in the Western Australian kwongan. In: *Kwongan – Plant Life of the Sandplain*, pp. 178–204, eds. J.S. Pate and J.S. Beard. Nedlands: University of Western Australia Press.

Bell, D.T., Plummer, J.A. and Taylor, S.K. (1993). Seed germination ecology in southwestern Western Australia. *Bot. Rev.* **59**, 24–73.

Bettenay, E. (1984). Origin and nature of the sandplains. In: *Kwongan – Plant Life of the Sandplain*, pp. 51–68, eds. J.S. Pate and J.S. Beard. Nedlands: University of Western Australia Press.

Bowen, B.J. and Pate, J.S. (1991). Adaptations of S.W. Australian members of the Proteaceae: allocation of resources during early growth. In: *Proceedings of the International Protea Association Sixth Biennial Conference*, pp. 289–301. Perth: Promaco Conventions.

Bowen, B.J. and Pate, J.S. (1993). The significance of root starch in post-fire shoot recovery of the resprouter *Stirlingia latifolia* R. Br. (Proteaceae). *Ann. Bot.* in press.

Brundrett, M.C. and Abbott, L.K. (1991). Roots of jarrah forest plants. I. Mycorrhizal associations of shrubs and herbaceous plants. *Aust. J. Bot.* **39**, 445–457.

Dixon, K.W. and Pate, J.S. (1978). Phenology, morphology and reproductive biology of the tuberous sundew, *Drosera erythrorhiza* Lindl. *Aust. J. Bot.* **26**, 441–454.

Dixon, K.W., Pate, J.S. and Bailey, W.J. (1980). Nitrogen nutrition of the tuberous sundew *Drosera erythrorhiza* Lindl. with special reference to catch of arthropod fauna by its glandular leaves. *Aust. J. Bot.* **28**, 283–297.

Dixon, K.W., Kuo, J. and Pate, J.S. (1983). Storage reserves of the seed-like, aestivating organs of geophytes inhabiting granite outcrops in south western Australia. *Aust. J. Bot.* **31**, 85–103.

Dixon, K.W., Pate, J.S. and Kuo, J. (1990). The Western Australian subterranean orchid *Rhizanthella gardneri* Rogers. In: *Orchid Biology, Reviews and Perspectives*, pp. 37–62, ed. J. Arditti. Oregon, Canada: Timber Press.

Dodd, J., Heddle, E.M., Pate, J.S. and Dixon, K.W. (1984). Rooting patterns of sandplain plants. In: *Kwongan – Plant Life of the Sandplain*, pp. 146–177, eds. J.S. Pate and J.S. Beard. Nedlands: University of Western Australia Press.

Frost, P.G.H. (1984). The responses and survival of organisms in fire-prone environments. In: *Ecological Effects of Fire in South African Ecosystems*, pp. 274–309, eds. P. deV. Booysen and N.M. Tainton. Berlin: Springer-Verlag.

Groves, R.H., Hocking, P.J. and McMahon, A. (1986). Distribution of biomass, nitrogen, phosphorus and other nutrients in *Banksia marginata* shoots and *B. ornata* shoots of different ages after fire. *Aust. J. Bot.* **34**, 709–725.

Hansen, A. (1990). *Growth, Reproductive Performance and Resource Allocation of Selected Seeder and Resprouter Understorey Legumes of Jarrah Forest in Western Australia*. PhD thesis, University of Western Australia.

Hansen, A.P. and Pate, J.S. (1987). Comparative growth and symbiotic performance of seedlings of *Acacia pulchella* and *A. alata* in defined pot culture or as natural understorey components of a eucalypt forest ecosystem in S.W. Australia. *J. Exp. Bot.* **38**, 13–25.

Hansen, A., Pate, J.S. and Hansen, A.P. (1991). Growth and reproductive performance of a seeder and a resprouter species of *Bossiaea* as a function of plant age after fire. *Ann. Bot.* **67**, 497–509.

Hansen, A., Pate, J.S. and Hansen, A.P. (1992). Growth, reproductive performance and resource allocation of the herbaceous obligate seeder *Gompholobium marginatum* R.Br. (Fabaceae). *Oecologia* **90**, 158–166.

Karlsson, P.S. and Pate, J.S (1992). Contrasting effects of supplementary feeding of insects or mineral nutrients on the growth and nitrogen and phosphorus economy of pygmy species of *Drosera*. *Oecologia* **92**, 8–13.

Keeley, J.E. (1986). Resilience of mediterranean shrub communities to fires. In: *Resilience in Mediterranean-Type Ecosystems*, pp. 95–112, eds. B. Dell, A.J.M. Hopkins and B.B. Lamont. Dordrecht: Junk.

Kuo, J. and Pate, J.S. (1990). Anatomy and ultrastructure of haustoria in selected West Australian parasitic angiosperms. In: *Proceedings of the XIIth International Congress for Electron Microscopy*, pp. 690–691. San Francisco: San Francisco Press.

Kuo, J., Hocking, P.J. and Pate, J.S. (1982). Nutrient reserves in seeds of selected proteaceous species from south-western Australia. *Aust. J. Bot.* **310**, 231–249.

Lamont, B.B. (1984). Specialized modes of nutrition. In: *Kwongan – Plant Life of the Sandplain*. pp. 236–245, eds. J.S. Pate and J.S. Beard. Nedlands: University of Western Australia Press.

Lamont, B.B. and Barrett, G.J. (1988). Constraints on seed production and storage in a root suckering *Banksia*. *J. Ecol.* **76**, 1069–1082.

Leake, J.R., Shaw, G. and Read, D.J. (1989). The role of ericoid mycorrhiza in the ecology of ericaceous plants. *Agric. Ecos. Environ.* **29**, 237–250.

Meney, K., Pate, J.S. and Dixon, K.W. (1990). Phenology of growth and resource deployment in *Alexgeorgea nitens* (Nees) Johnson & Briggs (Restionaceae), a clonal species from south-western Western Australia. *Aust. J. Bot.* **38**, 543–557.

Monk, D., Pate, J.S. and Loneragan, W.A. (1981). Biology of *Acacia pulchella* R.Br. with special reference to symbiotic nitrogen fixation. *Aust. J. Bot.* **29**, 579–592.

Pate, J.S. (1993a). The mycorrhizal association: just one of many nutrient-acquiring specializations in natural ecosystems. In: *Proceedings of the International Symposium on Management of Mycorrhizas in Agriculture, Horticulture and Forestry*, Perth, W. Australia, 28 September–2 October 1992.

Pate, J.S. (1993b). Mineral relationships of parasites and their hosts. In: *Parasitic Flowering Plants*, eds. M.C. Press and J.D. Graves. London: Chapman & Hall.

Pate, J.S. (1993c). Functional attributes of angiosperm hemiparasites and their hosts and predictions of possible effects of global climate change on such relationships. In: *Symposium on Anticipated Effects of a Changing Global Environment on Mediterranean-type Ecosystems*, Valencia, Spain, 13–17 September 1992.

Pate, J.S. and Dell, B. (1984). Economy of mineral nutrients in sandplain species. In: *Kwongan – Plant Life of the Sandplain*, pp. 227–252, eds. J.S. Pate and J.S. Beard. Nedlands: University of Western Australia Press.

Pate, J.S. and Dixon, K.W. (1982). *Tuberous, Cormous and Bulbous Plants*. Nedlands: University of Western Australia Press.

Pate, J.S., Dixon, K.W. and Orshan, G. (1984). Growth and life form characteristics of kwongan species. In: *Kwongan – Plant Life of the Sandplain*, pp. 84–100, eds. J.S. Pate and J.S. Beard. Nedlands: University of Western Australia Press.

Pate, J.S., Casson, N.E., Rullo, J. and Kuo, J. (1985). Biology of fire ephemerals of the sandplains of the kwongan of south-western Australia. *Aust. J. Plant Physiol.* **12**, 641–655.

Pate, J.S., Rasins, E., Rullo, J. and Kuo, J. (1986). Seed nutrient reserves of Proteaceae with special reference to protein bodies and their inclusions.

Ann. Bot. **57**, 747–770.

Pate, J.S., Davidson, N.J., Kuo, J. and Milburn, J.A. (1990a). Water relations of the root hemiparasite *Olax phyllanthi* (Labill) R.Br. (Olacaceae) and its multiple hosts. *Oecologia* **84**, 186–193.

Pate, J.S., Froend, R.H., Bowen, B.J., Hansen, A. and Kuo, J. (1990b). Seedling growth and storage characteristics of seeder and resprouter species of Mediterranean-type ecosystems of S.W. Australia. *Ann. Bot.* **65**, 585–601.

Pate, J.S., Kuo, J. and Davidson, N.J. (1990c). Morphology and anatomy of the haustorium of the root hemisparasite *Olax phyllanthi* (Olacaceae), with special reference to the haustorial interface. *Ann. Bot.* **65**, 425–436.

Pate, J.S., Pate, S.R., Kuo, J. and Davidson, N.J. (1990d). Growth, resource allocation and haustorial biology of the root hemiparasite *Olax phyllanthi* (Olacaceae). *Ann. Bot.* **65**, 437–449.

Pate, J.S., Meney, K.A. and Dixon, K.W. (1991a). Contrasting growth and morphological characteristics of fire-sensitive (obligate seeder) and fire-resistant (resprouter) species of Restionaceae (S. Hemisphere restiads) from south-western Western Australia. *Aust. J. Bot.* **39**, 505–525.

Pate, J.S., True, K.C. and Kuo, J. (1991b). Partitioning of dry matter and mineral nutrients during a reproductive cycle of the mistletoe *Amyema linophyllum* (Fenzl.) Tieghem parasitizing the swamp *Casuarina obesa* miq. *J. Exp. Bot.* **42**, 439–441.

Pate, J.S., Stewart, G.R. and Unkovich, M.J. (1993). ^{15}N natural abundance of plant and soil components of a *Banksia* woodland ecosystem in relation to nitrate utilization, life form, mycorrhizal status and N_2 fixing abilities of component species. *Plant Cell. Environ.* in press.

Read, D.J. (1991). Mycorrhizas in ecosystems: nature's response to the 'law of the minimum'. In: *Frontiers in Mycology*, Fourth International Mycological Congress, Regensberg, 1990, pp. 121–159, ed. D.L. Hawksworth. CAB International.

Schulze, E.-D., Gebauer, G., Schulze, W. and Pate, J.S. (1991). The utilization of nitrogen from insect capture by different growth forms of *Drosera* from southwest Australia. *Oecologia* **87**, 240–246.

Stewart, G.R., Pate, J.S. and Unkovich, M.J. (1993). Characteristics of inorganic nitrogen assimilation of plants in fire-prone Mediterranean-type vegetation. *Plant Cell Environ.* in press.

Stock, W.D., Pate, J.S.,and Delfs, J.C. (1990). Influence of seed size and quality on seedling development under low nutrient conditions in five Australian and South African members of the Proteaceae. *J. Ecol.* **78**, 1005–1020.

Walker, B.A. and Pate, J.S. (1986). Morphological variation between seedling progenies of *Viminaria juncea* (Schrad. & Wendl.) Hoffmans. (Fabaceae) and its physiological significance. *Aust. J. Plant Physiol.* **13**, 305–319.

Walker, B., Pate, J.S and Kuo, J. (1983). Nitrogen fixation by nodulated roots of *Viminaria juncea* (Schrad. and Wendl.) Hoffmans. (Fabaceae) when submerged in water. *Aust. J. Plant Physiol.* **10**, 409–421.

Zammit, C. and Westoby, M. (1987). Population structure and reproductive status of two *Banksia* shrubs at various times after fire. *Vegetatio* **79**, 11–20.

Metabolic and genetic responses

CHAPTER 11

Chemical signalling and the adaptation of plants to conditions where water availability is restricted

W.J. DAVIES, F. TARDIEU and C.L. TREJO

11.1. Introduction

Effects of soil drying on morphology and physiology have been well documented for a wide range of plants (e.g. Sinnott, 1960). Several authors have noted that many of these effects can be mimicked by the application of abscisic acid (ABA) to well watered plants (e.g. Trewavas and Jones, 1991) (Table 11.1) and we now think that many physiological and developmental responses of plants in drying soil can be attributed to modifications in the plant's own ABA balance. The majority of characters highlighted in Table 11.1 can help to confer some degree of drought resistance on the plant. This realization has led to a worldwide search for ABA-induced proteins (e.g. Bray, 1991) in the hope that these may be involved in drought resistance, and has led to an active breeding programme based on the premise that a high capacity for ABA production should confer some degree of drought tolerance (Quarrie, 1991).

All this research activity has been based on a qualitative understanding of how the ABA balance of the plant may be regulated, and how ABA status may confer drought resistance or sensitivity. The current status of our knowledge in these two fields is reviewed below.

11.2. Drought-induced variation in abscisic acid balance and changes in shoot growth and physiology

Soil drying will increase the ABA content of the plant. Comparatively mild treatments can increase ABA concentrations in the xylem of some plants (e.g. maize and sunflower) by up to 100-fold (Zhang and Davies, 1989). In other plants, e.g. *Phaseolus*, a similar treatment may only produce a two-fold increase in concentration (e.g. Trejo and Davies, 1991). Most writings

TABLE 11.1. *Effects of water deficit and abscisic acid application on a range of biochemical physiological and developmental processes* (From Trewavas and Jones, 1991).

Response	Water deficit	Abscisic acid	Correlation
Biochemical and physiological			
Specific mRNA and protein synthesis	Increase	Increase	+ +
Proline and betaine accumulation	Increase	Increase	+ +
Osmotic adaptation	Yes	Yes	+
Photosynthetic enzyme activity	Decrease	Decrease	+
Desiccation tolerance	Increase	Increase	+
Salinity and cold tolerance	Induces	Induces	+ +
Wax production	Increase	Increase	+
Growth			
General growth inhibition	Yes	Yes	+ + +
Cell division	Decrease	Decrease	+ + +
Leaf initiation	Inhibits	Inhibits	+ +
Cell expansion	Decrease	Decrease	+ + +
Germination	Inhibits	Inhibits	+ +
Root growth	Increase/ decrease	Increase/ decrease	+ +
Root/shoot ratio	Increase	Increase	+ +
Suberization	?	Increase	
Morphology			
Production of trichomes	Increase	Increase	+ +
Production of spines	Increase	Increase	+ + +
Stomatal index	Decrease	Decrease	+ +
Tillering in grasses	Decrease	Decrease/ increase	+
Conversion from aquatic to aerial leaf type	Yes	Yes	+ +
Induction of dormancy, terminal buds or perennation organs	Yes	Yes	+ +
Apical dominance			
Reproductive			
Flowering in annuals	Often advanced	Often advanced	+ +
Flower induction in perennials	Inhibits	Inhibits	+ +
Flower abscission	Increases	Increases	+
Pollen viability	Decreases	Decreases	+
Seed set	Decreases	Decreases	+
Embryo maturation	?	Accelerates	

The strength of correlation is indicated as ranging from weak (+) to strong (+ + +).

now emphasize the sensitivity of many plant processes to ABA and it is something of a mystery why some unwatered plants produce concentrations in the micromolar range. Nevertheless, many studies of ABA-induced gene regulation seem to suggest that concentrations even higher than this are required before many new proteins are seen. Consequently, we might expect that many of these proteins would be switched on only by very severe drought stress, and might be associated only with dehydration tolerance. We will stress here the importance of *regulation* of growth and development by drought, rather than its damaging effects. To understand the molecular basis of more subtle regulation of development, it may be necessary to investigate the effects of very much more subtle changes in ABA concentration.

For a range of plants in drying soil it can be demonstrated that there is a relationship between the concentration of ABA in the xylem of the leaf and the growth rate and conductance of that leaf (Fig. 11.1) (Zhang and Davies, 1990a, b). External application of ABA to the plant will raise the concentration of ABA in the xylem to values comparable with those generated by soil drying and, interestingly, will generate a comparable restriction in leaf growth rate and leaf conductance (Fig. 11.1). Trewavas (1981) and others have levelled much criticism at plant growth regulator research where a role has been claimed for a particular regulator, following experiments where plants were treated with concentrations very much higher than those found endogenously. This is not the case here and it seems reasonable to claim a role for ABA in the regulation of leaf growth and leaf gas exchange of plants in drying soil. Zhang and Davies (1991) have removed ABA from the xylem stream of unwatered maize plants using an immunoaffinity column. This treatment of the sap also removed the antitranspirant activity from the sap, confirming the important role for ABA in this plant. It should be noted that Munns and King (1988) have performed similar experiments with unwatered wheat plants and suggest that, in this case, stomata are controlled by another chemical regulator that still has to be identified. In both of these pieces of work the dominant role of a chemical influence on stomatal behaviour is beyond dispute. Soil drying will affect the supply of water to the shoots and we know that any hydraulic changes will also affect stomatal behaviour and leaf growth. It is important to assess the relative importance of these two types of signal.

11.3. Hydraulic versus chemical signals

Most writing on the effects of soil drying on shoot growth and physiology emphasizes the important effects of reductions in leaf water content and turgor. Although these changes can undoubtedly exert very substantial influences on most aspects of leaf biology, careful examination of much of the literature on plant water relations reveals that soil drying can result in reduced stomatal conductance and leaf growth rate even in plants where leaf water relations are not perturbed (e.g. Tardieu *et al.*, 1991). Observations of this kind have led to the suggestion that plants may sense directly the

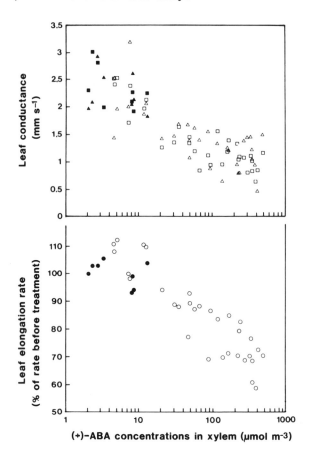

FIGURE 11.1 *Effects of manipulating xylem abscisic acid (ABA) concentrations on the abaxial conductance and elongation rate of maize leaves. A series of ABA solutions was fed to part of maize roots (see text). Leaf conductance measurements were taken 24 h (▲ △) or 48 h (■ □) after ABA feeding. Leaf elongation rate (● ○) is expressed as the percentage of the rate before treatment was given and average rates 48 h before and after ABA feeding were used in the calculation. Each point represents the measurements taken from one plant and means are shown where repeated measurements were made on the plant. Closed points are from plants that were fed with water. (From Zhang and Davies, 1990b.)*

drying of soil around the roots and communicate this information to the shoots as a chemical signal.

Recent unequivocal demonstrations of root-to-shoot signalling have been provided by Gollan *et al.* (1992) and Schurr *et al.* (1992). These workers have used a pressure chamber around the roots of the plant to counteract the effects of soil drying on plant water relations. Roots of plants grown in this way are still in contact with drying soil and it is important to note that limitation in leaf conductance is observed even though the shoot water status of unwatered plants is comparable to that of unpressurized well watered plants.

Another impressive demonstration of a dominant role for roots signals has been provided by Gowing *et al.* (1990) working with split-root plants. Here, roots of single plants are divided between two rooting containers. When the soil in one of these is allowed to dry, leaf growth rate is reduced even though the roots in wet soil supply enough water to keep the shoots turgid. If the roots in contact with drying soil are removed from the plant, leaf growth rate recovers to the rate shown by the well watered plants (Fig. 11.2). There is no possibility that removing the roots can make more water available to the shoots. Therefore, the most likely interpretation of these data is that removal of roots in contact with drying soil removes a source of a leaf growth inhibitor, which is generated in drying roots.

The experiments described above provide clear evidence for root-to-shoot signalling of the effects of soil drying. Other laboratory experiments suggest strongly that ABA synthesized in drying roots and moving in the transpiration stream can have a controlling influence on shoot growth and physiology (Zhang and Davies, 1989, 1990a, b, 1991). More recent experiments have examined the role for root signals in the control of physiology and growth of a maize crop in the field.

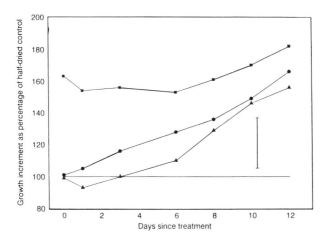

FIGURE 11.2 *Daily increment in leaf area by apple trees with roots divided between two pots. Plants were either watered on both halves of the root system (■) or on one-half of the root system only. Results are for plants that received the above treatments for 24 days and are expressed as percentages of the growth of the plants that remained with half their roots in dry soil (—). Other groups of plants were either rewatered (●) or had the roots in dry soil excised (▲). All points are 3-day pooled means (n = 6) with standard errors shown. (From Gowing et al., 1990.)*

11.4. Chemical signals and the control of stomata in the field

Paired measurements of leaf conductance and ABA concentration in the xylem of the same leaves from field-grown maize plants show a clear relationship between these two variables and suggest that variation in ABA concentration in the range up to $200 \, \mu mol \, m^{-3}$ has a controlling influence on stomata (Tardieu *et al.*, 1992a). Strong supporting evidence for such a role for ABA is provided by a recent paper by Tardieu *et al.* (1993). Here, ABA was fed to maize plants in the field, and the resulting restriction in stomatal conductance was almost exactly that generated by comparable concentrations of endogenous ABA. This means that it is not necessary to invoke the involvement of any other chemical signal in the control of stomatal behaviour by this crop. The evidence from this study is that stomata are acting to regulate the water relations of the shoot, rather than the converse, which has been the traditionally held view.

11.4.1. *What information is conveyed by an abscisic acid signal from maize roots in drying field soil?*

Implicit in the concept of root-to-shoot signalling is the idea that the signal should provide some measure of the degree of soil drying, and that this measure might be used by the plant as a regulator of development. In our study, plants were grown on field plots with compacted or non-compacted soil. These plots were irrigated or remained unwatered. ABA concentrations in the xylem sap before dawn and in the roots increased 25-fold and five-fold respectively, as the soil dried, with a close correlation with soil water status (Tardieu *et al.*, 1992b). In contrast to the laboratory experiments described above, no appreciable increases in xylem ABA concentration and reduction in stomatal conductance were observed with dehydration of part of the root system in the upper soil layers. These responses occurred only when the soil water reserve was close to zero and the transpiration declined (Fig. 11.3).

It is clear that plants in the field do not act in the same way as the split-root plants described above. Other possibilities are that the root signal reflects changes in the amount of water available in the rooting zone or that the signal is a function of some other change in soil properties that accompanies soil drying. Passioura (1988) has proposed that root signals might reflect a reaction by the root to increasing soil strength as the soil dries. In our experimental system, xylem ABA concentration measured during the day was very much higher in plants on compacted soil than in plants on non-compacted soil (Fig. 11.3), but this was not the case before dawn. Because any mechanical message is unlikely to disappear at night, we have argued that ABA accumulation in plants on compacted soil reflects a faster decrease in root water potential and water flux in soil where many roots are clumped in the larger cracks.

The data in Fig. 11.3 suggest that the integrated soil water status is an appropriate variable to determine the ABA message. It is equally clear, however, that the water flux through the plant will also be a substantial contributing influence on the signal. We see this in the effect of soil com-

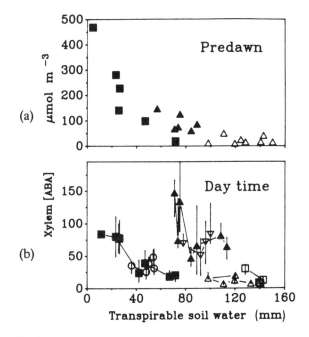

FIGURE 11.3 *ABA concentration in the xylem sap of field-grown maize plants plotted against the transpirable soil water (a) predawn and (b) during the day. Values in Fig. 11.3b correspond to the median, bars: quartiles (non-symmetrical with respect to the mean value). Symbols show different cultural treatments. (From Tardieu et al., 1992b.)*

paction on ABA concentration, but also in the very substantial differences between xylem ABA concentrations recorded before dawn and during the day. The latter result must be a function of the increased transpiration flux during the daylight hours. Our measurements of diurnal variation in ABA concentration (Tardieu *et al.*, 1992b) also show reductions in concentration during the afternoon hours when transpiration fluxes are highest.

The sum of this information seems to show that the root signal will provide the shoot with a measure of the degree of dehydration of the root system, but that during the day, to a great extent, the signal reflects the flux of water through the root system. This raises a possible problem with the proposed control system in that high transpiration fluxes on hot dry days will act to dilute the chemical signal under exactly the conditions where some control of leaf function might be required. Many plants growing in drying soil will show increasing restriction in stomatal conductance in the latter part of the day. If we are to argue that ABA in the xylem has a controlling influence on stomatal conductance, we must reconcile declining leaf conductance in the afternoon hours with decreasing ABA concentrations over the same period. One possible explanation for this apparent contradiction is that stomata respond to the flux of ABA into the leaf rather than to the concentration of the hormone. Tardieu *et al.* (1993) have tested this possibility and failed to find evidence that ABA flux can act as a controlling variable. We have therefore tested the possibility that variation in sensitivity of shoot processes to ABA may be an important part of the control system.

11.4.2. *Variation in apparent sensitivity of shoot processes to a root signal*

There are now many reports that stomatal sensitivity to ABA can be highly variable. It is apparent, for example, that at temperatures between 10 and 15 °C a 100 μM solution of ABA has little or no effect on stomata of *Phaseolus* (Cornic and Ghashghaie, 1991) (Fig. 11.4). When leaf temperature is increased to 25 °C, the same concentration of the hormone will significantly restrict stomatal conductance. Burschka *et al.* (1983) have reported that stomata of Mediterranean plants show a diurnal variation in sensitivity to applied ABA,

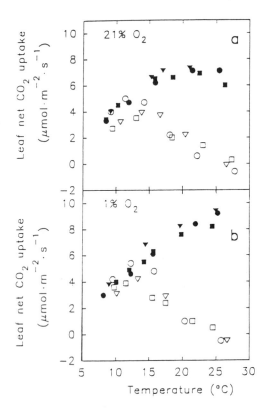

FIGURE 11.4 *Relation between leaf net carbon dioxide uptake and leaf temperature measured on dehydrated 37% leaf water deficit (○, □, ▽) or non-dehydrated (●, ■, ▼) leaves of* Phaseolus *in (a) 21% oxygen, (b) 1% oxygen. In each condition the different symbols represent measurements performed independently on different leaves. (From Cornic and Ghashghaie, 1991.)*

with bigger responses recorded in the afternoons. One explanation for these responses would be a differential effect of temperature between early morning and afternoon hours. Control of stomata via a dynamic change in sensitivity is an attractive possibility because it provides a responsive link between climatic variation and the root signal which does not depend on the *de novo* synthesis of more ABA for a rapid response. As we are suggesting that much ABA synthesis is spatially separated from the responses taking place in the leaves, responsive variation in the rate of synthesis would be difficult to achieve. In fact, we have found that the ABA concentration in the xylem is tightly regulated by the many feedbacks in the system. Although this relatively stable concentration provides a suitable regulator of development, without modulation of the sensitivity of the response, this kind of signal could not provide dynamic control of stomatal behaviour as a function of variation in the climate.

Our recent experiments have failed to establish a temperature modulation of stomatal sensitivity to ABA in maize plants growing in the field. Instead, high stomatal sensitivity to ABA in the afternoon hours has been attributable to the lower leaf water potential recorded at this time (Tardieu and Davies, 1992). Experiments with isolated epidermis incubated at different water potentials show that the effects of ABA are amplified as water potential falls (Fig. 11.5). We have also confirmed this conclusion by feeding ABA to detached leaves of *Phaseolus* held at different water potentials.

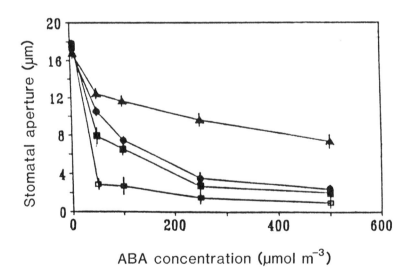

FIGURE 11.5 *Apertures of C.* communis *stomata in detached epidermis as a function of [ABA] in the incubating medium, at four water potentials:* ▲, *−0.3 MPa;* ●, *−0.5 MPa;* ■, *−0.8 MPa;* □, *−1.5 MPa. Each point is the mean of 200–300 measurements from five or six epidermal pieces. Bars, confidence intervals (P = 0.05). (From Tardieu and Davies, 1992.)*

Here, decreases in water potential of only 0.3 MPa can enhance the effectiveness of ABA by around 50%.

11.5. Is there a role for abscisic acid as a general stress hormone?

We have described above how the ABA balance of the plant might be used to regulate the behaviour of stomata as a function of fluctuations in both the soil and the atmospheric environment. The information summarized in Table 11.1 suggests that there is a more general role for ABA as a stress hormone in plants. We now examine another well documented effect of ABA on plants and compare the mode of action of the hormone in this system with that described above.

11.5.1. Abscisic acid and the growth of roots and shoots of plants at low water potential

When soil starts to dry, one of the first responses shown by plants is a reduction in shoot growth. In most plants, root growth is much less sensitive

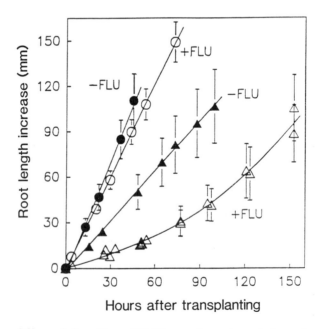

FIGURE 11.6 *Effect of fluoridone (FLU) on elongation of the primary root of maize seedlings growing in high (−0.03 MPa, ○ ●) or low (−1.60 MPa, △ ▲) water potential vermiculite. Data points represent the mean ± s.d. of 20–40 roots. (From Saab et al., 1990.)*

to drying soil. When the water potentials of growing tissues are reduced to a comparable degree, the effects on shoot growth are much greater than the effects on roots (Westgate and Boyer, 1985). In maize, for example, roots will continue to grow at tissue water potentials that cause complete inhibition of shoot growth (Sharp *et al.*, 1988). We have seen above that the accumulation of ABA apparently leads to an inhibition of shoot growth. The effects of ABA on root growth are much more difficult to predict (see Saab *et al.*, 1990, for a literature review), but there are some reports of ABA-stimulated increases in root growth (e.g. Mulkey *et al.*, 1983).

In a recent paper, Saab *et al.* (1990) have manipulated endogenous ABA concentrations using the carotenoid biosynthesis inhibitor fluoridone and a mutant deficient in carotenoid biosynthesis. Both of these approaches show that reduced ABA concentration was associated with inhibition of root elongation and promotion of shoot elongation at *low water potential* (Fig. 11.6). Manipulation of ABA content at high water potential had little effect on elongation rate. There are direct parallels here with the effects of ABA on stomata in that, in both cases, low water potential increases the sensitivity of the target cells to the hormone. A role for ABA in the maintenance of root elongation and in the inhibition of shoot elongation at low water potential accords with a view of the compound as a true stress hormone in plants. Maintenance of root elongation in drying soil can presumably act to sustain water uptake, whereas restrictions in shoot growth will slow the development of a transpiring area.

11.6. A basis for an apparent change in tissue sensitivity to abscisic acid

Much work on the variation in stomatal sensitivity to a range of influences has been performed with epidermal strips detached from leaves (e.g. Snaith and Mansfield, 1981; Weyers *et al.*, 1987). Among the variables that influence stomatal sensitivity to ABA are the concentrations of other hormones present in the incubation medium, the ionic status and pH of the bathing medium, the temperature of the medium, the water potential of the bathing medium and the nutritional status and age of the plant from which the epidermis has been removed.

Firn (1986) and Weyers *et al.* (1987) have emphasized that sensitivity is not a simple unambiguous concept. The latter authors have attempted a quantitative analysis of sensitivity (see also Paterson *et al.*, 1988), which has enabled them to ascribe tentative physicochemical explanations to the changes that they see. Possibilities that they consider are the number or availability of receptors, the affinity of the receptors for the hormone and the binding characteristics of the molecule.

Until the binding sites for ABA are characterized, it is difficult to go much further with the kind of analysis described above. It is important to be aware, however, that with much of the work with ABA and whole plants we have described relationships between stomatal behaviour and ABA concentrations in the xylem. An apparent variation in the sensitivity of this response may reflect a fundamental alteration in membrane properties, but equally it could reflect the influence of another environmental variable on the partitioning of xylem ABA between different compartments within the leaf. Hartung and Davies (1992) have described the influence of pH on the distribution of ABA between compartments and it is clear, for example, that any treatment that increased the flux of ABA from the free space into the cell would be likely to decrease the apparent sensitivity of the stomata to an ABA signal in the xylem.

Recent experiments reported by Trejo et al. (1993) suggest that metabolism of ABA in the mesophyll can have a strong controlling influence on the concentration of ABA in the apoplast of the epidermis. We know that the rate of ABA metabolism is highly sensitive to a range of different variables, not least the water status of the leaf (e.g. Zeevaart and Creelmann, 1988). It seems likely that any treatment that moderates the rate at which ABA is broken down in the leaf will have a substantial effect on the apparent sensitivity of leaf processes to ABA.

11.7. Conclusions

It is now clear that chemical regulation is an important component of the plant's responses to soil drying. We have a good understanding of how the delivery of the hormone to the shoots is modified by changing atmospheric and edaphic conditions, and also some understanding of how the plant responds to the hormone. All of this has come about as a result of a combination of laboratory and field work. More detailed laboratory work on mechanisms of action is clearly needed, but it has become apparent that, if we are to understand how plants work in the field, experiments must also be conducted in the field. Our framework of analysis can now be applied to different crops. It will be important to investigate the impact of genetic variation in some of the characters that we have identified as important. There is a particular need for a greater understanding of the control of growth processes in droughted plants in the field. Our laboratory analysis has led us to believe that ABA must play an important regulatory role.

REFERENCES

Bray, E.A. (1991). Regulation of gene expression by endogenous ABA during drought stress. In: *Abscisic Acid: Physiology and Biochemistry*, pp. 81–98, eds. W.J. Davies and H.G. Jones. Oxford: Bios.

Burschka, C., Tenhunen, J.D. and Hartung, W. (1983). Diurnal variations in abscisic acid content and stomatal response to applied abscisic acid in leaves of irrigated and non irrigated *Arbutus unedo* plants under naturally fluctuating environmental conditions. *Oecologia* **58**, 1128–1134.

Cornic, G. and Ghashghaie, J. (1991). Effect of temperature on net CO_2 assimilation and photosystem II quantum yield of electron transfer of French bean (*Phaseolus vulgaris* L.) leaves. *Plant Physiol.* **185**, 255–260.

Firn, R.D. (1986). Growth substance sensitivity: the need for clearer ideas, precise terms and purposeful experiments. *Physiol. Plant.* **67**, 267–272.

Gollan, T., Schurr, U. and Schulze, E.-D. (1992). Stomatal response to drying soil in relation to changes in the xylem sap composition of *Helianthus annuus*. 1. The concentration of cations, anions, amino acids in, and pH of, the xylem sap. *Plant Cell Environ.* **15**, 551–560.

Gowing, D.G.C., Jones, H.G. and Davies, W.J. (1990). A positive root-sourced signal as an indicator of soil drying in apple, *Malus x domestica* Borkh. *J. Exp. Bot.* **41**, 1535–1540.

Hartung, W. and Davies, W.J. (1992). Drought-induced changes in physiology and ABA. In: *Abscisic Acid: Physiology and Biochemistry*, pp. 63–79, eds. W.J. Davies and H.G. Jones. Oxford: Bios.

Mulkey, T.J., Evans, M.L. and Kuzmanoff, K.M. (1983). The kinetics of abscisic acid action on root growth and gravitropism. *Planta* **157**, 150–157.

Munns, R. and King, R.W. (1988). Abscisic acid is not the only stomatal inhibitor in the transpiration stream. *Plant Physiol.* **88**, 703–708.

Passioura, J.B. (1988). Root signals control leaf expansion in wheat seedlings growing in drying soil. *Aust. J. Plant Physiol.* **15**, 687–693.

Paterson, N.W., Weyers, J.D.B. and Brook, R.A. (1988). The effect of pH on stomatal sensitivity to ABA. *Plant Cell Environ.* **11**, 83–90.

Quarrie, S.A. (1991). Implications of genetic differences in ABA accumulation for crop production. In: *Abscisic Acid: Physiology and Biochemistry*, pp. 227–243, eds. W.J. Davies and H.G. Jones. Oxford: Bios.

Saab, I., Sharp, R.E., Pritchard, J. and Voetberg, G.S. (1990). Increased endogenous ABA maintains primary root growth and inhibits shoot growth of maize seedlings at low water potential. *Plant Physiol.* **93**, 1329–1336.

Schurr, U., Gollan, T. and Schulze, E.-D. (1992). Stomatal response to soil drying in relation to changes in xylem sap composition of *Helianthus annuus*. 11. Stomatal sensitivity to abscisic acid imported from the xylem sap. *Plant Cell Environ.* **15**, 561–568.

Sharp, R.E., Silk, W.K. and Hsiao, T.C. (1988). Growth of the maize primary root at low water potential. 1. Spatial distribution of expansive growth. *Plant Physiol.* **87**, 50–57.

Sinnott, E.W. (1960). *Plant Morphogenesis*. New York: McGraw-Hill.

Snaith, P.J. and Mansfield, T.A. (1981). Stomatal sensitivity to abscisic acid: can it be defined? *Plant Cell Environ.* **5**, 309–311.

Tardieu, F. and Davies, W.J. (1992). Stomatal response to ABA is a function of current plant water status. *Plant Physiol.* **98**, 540–545.

Tardieu, F., Katerji, N., Bethenod, O., Zhang, J. and Davies, W.J. (1991). Maize stomatal conductance in the field, its relationship with soil and plant water potentials, mechanical constraints and ABA concentration in the xylem sap. *Plant Cell Environ.* **14**, 121–124.

Tardieu, F., Zhang, J. and Davies, W.J. (1992a). What information is conveyed by an ABA signal from maize roots in drying field soil? *Plant Cell Environ.* **15**, 185–192.

Tardieu, F., Zhang, J., Katerji, N., Bethenod, O., Palmer, S. and Davies, W.J. (1992b). Xylem ABA controls the stomatal conductance of field-grown maize subjected to soil compaction and soil drying. *Plant Cell Environ.* **15**, 193–198.

Tardieu, F., Zhang, J. and Gowing, D.J.C. (1993). A model of stomatal control by both ABA concentration in the xylem sap and leaf water status: test of the model and of alternative mechanisms for droughting and ABA-fed field grown maize. *Plant Cell Environ.* in press.

Trejo, C.L. and Davies, W.J. (1991). Drought-induced closure of *Phaseolus vulgaris* stomata precedes leaf water deficit and any increase in xylem ABA concentration. *J. Exp. Bot.* **42**, 1507–1516.

Trejo, C.L., Ruiz, L.M.P. and Davies, W.J. (1993). Sensitivity of stomata to ABA: an effect of the mesophyll. *Plant Physiol.* in press.

Trewavas, A.J. (1981). How do plant growth substances work? *Plant Cell Environ.* **4**, 203–228.

Trewavas, A.J. and Jones, H.G. (1991). An assessment of the role of ABA in plant development. In: *Abscisic Acid: Physiology and Biochemistry*, pp. 169–188, eds. W.J. Davies and H.G. Jones. Oxford: Bios.

Westgate, M.E. and Boyer, J.S. (1985). Osmotic adjustment and the inhibition of leaf, root, stem and silk growth at low water potentials in maize. *Planta* **164**, 540–549.

Weyers, J.B., Paterson, N.W. and Brook, R.A. (1987). Towards a quantitative definition of plant hormone sensitivity. *Plant Cell Environ.* **10**, 1–10.

Zeevaart, J.A.D. and Creelmann, R.A. (1988). Metabolism and physiology of abscisic acid. *Annu. Rev. Plant Physiol. Mol. Biol.* **39**, 439–473.

Zhang, J. and Davies, W.J. (1989). Abscisic acid produced in dehydrating roots may enable the plant to measure the water status of the soil. *Plant Cell Environ.* **12**, 73–81.

Zhang, J. and Davies, W.J. (1990a). Does ABA in the xylem sap control the rate of leaf growth in soil dried maize and sunflower plants? *J. Exp. Bot.* **41**, 1125–1132.

Zhang, J. and Davies, W.J. (1990b). Changes in the concentration of ABA in xylem sap as a function of changing soil water status can account for changes in leaf conductance and growth. *Plant Cell Environ.* **13**, 271–285.

Zhang, J. and Davies, W.J. (1991). Antitranspirant activity in the xylem sap of maize plants. *J. Exp. Bot.* **42**, 317–321.

CHAPTER 12

Perception and transduction of stress by plant cells

R. A. LEIGH

12.1. Introduction

To adapt to stress, plants must be able to perceive that environmental conditions have changed and then transduce that perception into an appropriate response that makes them more able to grow and reproduce in the presence of the change. This chapter deals with some of the ways that stresses can be perceived by plant cells and the mechanisms by which the signals are transduced into an appropriate response. The discussion is broken down into three main areas:

1. Situations where the same protein perceives and responds to the stress
2. Situations where perception and response involve different proteins in the same cell, and so intracellular signalling pathways must be involved
3. The nature of the signals used to transfer information between different parts of intact plants, particularly in response to physical wounding

The account is not exhaustive but aims to describe recent results that have shed light on the transduction processes that operate at these different levels.

12.2. Perception and response by the same protein

This is clearly the simplest situation, and two particular examples will be discussed: the response of the plasma membrane H^+ pump to changes in cytoplasmic pH and the mediation of turgor changes by stretch-activated ion channels.

12.2.1. *Response of the plasma membrane H⁺ pump to cytoplasmic pH*

The pH of the cytoplasm of plant cells is normally just above neutrality and is maintained at about this level to allow the optimal operation of metabolism (Kurkdjian and Guern, 1989). A number of stresses can induce acidification of the cytoplasm, including anaerobiosis (Roberts *et al.*, 1984) and increases in the concentration of pollutant gases such as sulphur dioxide (Pfanz *et al.*, 1987). In addition, normal metabolic activities involve the production of H^+ or OH^- (e.g. assimilation of nitrogen from different sources; see Raven and Smith, 1976) and these ions must be removed from the cytoplasm if metabolism is to be maintained.

Several mechanisms can contribute to pH regulation, including production and consumption of malate *via* the biochemical 'pH stat', transport of H^+ to and from the vacuole, and transport of H^+ at the plasma membrane. The first two mechanisms have only a limited capacity and therefore it is generally assumed that long-term regulation of cytoplasmic pH must involve net efflux of H^+ from the cell, which is an active transport process and is catalysed by the plasma membrane H^+ pump (Kurkdjian and Guern, 1989).

Changes in H^+-pump activity in response to pH changes in the cytoplasm have been demonstrated using electrophysiological techniques. For instance, measurement of the membrane potential shows that acidification of the cytoplasm causes an initial depolarization of the membrane potential (explained by influx of positively charged H^+) followed by repolarization as the H^+ pump is activated (Fig. 12.1; Bates and Goldsmith, 1983; Frachisse *et al.*, 1988). More detailed descriptions of the underlying changes have been

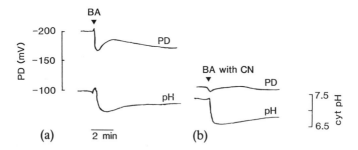

FIGURE 12.1 *The effect of butyric acid on the cytoplasmic pH and membrane potential of rhizoid cells of* Riccia fluitans. *(a) Butyric acid (BA, 0.7 mM) was added at the time indicated. Undissociated butyric acid crosses the plasma membrane and dissociates in the cytoplasm releasing H^+ and acidifying the cytosol (cyt) (lower trace). This causes an initial depolarization of the membrane potential difference (PD, upper trace) to less negative values followed by a hyperpolarization back to more negative values as the plasma membrane H^+ pump is activated. (b) Same concentration of butyric acid added in the presence of 1 mM sodium cyanide (CN). The cyanide prevents ATP formation and so inhibits the pump and prevents its activation although the cytoplasm still acidifies. (After Frachisse et al., 1988.)*

explored by analysis of current–voltage relationships and the partitioning of changes in current movement across the plasma membrane between pump activity and other transporters (e.g. Sanders *et al.*, 1981). The results show that cytoplasmic acidification strongly activates the pump (Sanders *et al.*, 1981) and also induces a large 'leak' conductance. The latter is now thought to be activation of K^+–H^+ exchange (Blatt and Slayman, 1987), which recirculates current across the membrane, preventing the membrane potential from becoming too negative and thermodynamically 'stalling' the pump. Alkalinization of the cytoplasm (caused by weak bases such as procaine) inhibits H^+-pump activity (e.g. Felle and Bertl, 1986).

The effects of cytoplasmic pH on H^+-pump activity are explained by the pump's pH optimum of about 6.5, which means that it is strongly activated by decreases in pH from neutrality and by the increase in the concentration of the transport substrate, cytoplasmic H^+. Although the pump has the capacity to remove large amounts of H^+ from the cell, its role in tightly regulating cytoplasmic pH remains unclear because inhibition of the pump does not always prevent regulation of pH. Therefore, the pump may be only part of an ensemble of pH-regulating mechanisms, which probably vary in importance with both cell type and the size of the stress imposed (Kurkdjian and Guern, 1989).

12.2.2. *Role of stretch-activated ion channels in turgor regulation*

The application of the patch clamping technique (Hedrich *et al.*, 1987; Hedrich and Schroeder, 1989; Cahalan and Neher, 1992) has led to the recognition that passive ion movements across membranes are mediated by ion channels. These channels can show considerable ion specificity and act as regulated pores which, when open, allow ions to move passively into or out of the cell down electrochemical potential gradients. Patch clamp studies of plant protoplasts and vacuoles have revealed that both the plasma membrane and the tonoplast contain a spectrum of ion channels with a range of ion specificities and regulatory characteristics (Hedrich and Schroeder, 1989). Like ion channels in animal cells, plant ion channels are regulated (gated) by a variety of factors, including voltage, nucleotides and ions (e.g. Ca^{2+}). These regulators ensure that each channel opens only under a particular set of permissive conditions (e.g. when the cytoplasmic Ca^{2+} level is elevated or the membrane potential is in a particular voltage range) thus preventing uncontrolled leakage of ions through the channel.

In animal cells, some channels are regulated by the physical state of the membrane and open or close when the membrane is stretched by applying suction to the membrane via the patch clamp apparatus (Sachs, 1991). These channels are thought to be involved in the regulation of cell volume by modulating intracellular ion concentrations and osmotic pressure and so controlling water movement into or out of the cell. They allow changes in the physical state of the cell resulting from a variety of sources (deformation, gravity, friction, osmotic pressure, etc.) to be transduced into changes in cellular ionic composition. Such channels are clearly potential candidates for sensing and transducing changes of turgor in plant cells.

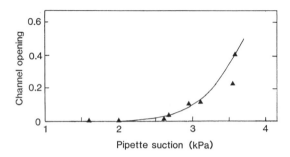

FIGURE 12.2 *The relationship between opening of an ion channel in the plasma membrane of* Vicia faba *guard cell protoplasts and the suction applied to the protoplast through the patch clamp pipette. The results indicate that the channel is stretch activated. Channel opening is expressed as the product of channel number and open probability. (After Cosgrove and Hedrich, 1991.)*

There are several reports of stretch-activated ion channels in plant cells. The first such channel was described in protoplasts from tobacco suspension cells and was selective for anions over cations (Falke *et al.*, 1988). More recently, Cosgrove and Hedrich (1991) have described three stretch-activated channels in guard cell protoplasts from *Vicia faba*. These were permeable to Cl^-, K^+ and Ca^{2+}, respectively. In both *Nicotiana* and *Vicia* protoplasts, channel opening increased with the suction applied to the patch pipette (Fig. 12.2). In addition, a stretch-activated channel with greater specificity to anions than cations has been described in vacuoles from red beet (Alexandre and Lassalles, 1991). This channel was also activated by the osmotic gradient across the membrane and so appears to be responsive to both hydrostatic and osmotic gradients.

The role of these channels in physiological responses to changes in cell water relations has not yet been proven, so their role as both sensors and modulators of changes in turgor remains speculative. Similarly, the mechanisms by which changes in the physical state of the membrane alter ion fluxes through the channels are not known. Zimmermann (1977) put forward an 'electromechanical' model of turgor sensing, which proposed that changes in membrane compression affected the activity of transport proteins embedded in the membrane. It is tempting, therefore, to speculate that stretch-activated ion channels are the physical reality of Zimmermann's theoretical model. However, in animal cells, stretch-activated channels may also interact with the cytoskeleton, and their activity may be controlled by changes in tension mediated via cytoskeletal elements. Intriguingly, a K^+ channel recently cloned from *Arabidopsis* has a sequence in its C-terminus that has homology with a highly repeated motif of the erythrocyte protein ankyrin. Ankyrin has been shown to be involved in the tethering of proteins, including ion channels, to the cytoskeleton (Sachs, 1991). It is possible that this is the function of the motif identified in the *Arabidopsis* K^+ channel and that such anchoring plays some role in modulation of the channel's activity. At present, however, the gating properties of the *Arabidopsis* K^+ channel have not been determined.

12.3. Transduction pathways involving intracellular signalling

12.3.1. The role of calcium ions and inositol phosphates

In most situations, the proteins perceiving and responding to changes in a cell's environment will be different. Thus information must pass between these proteins, and there is currently much interest in the pathways of such stimulus–response coupling in plant cells. This research has largely been driven by the insight that has been gained into the role of Ca^{2+} and inositol phosphates in intracellular signalling in animal cells (Berridge and Irvine, 1989). The descriptions in plant cells are not as well developed as those in animals but there is good evidence that similar pathways are used by plants for responding to environmental stimuli.

An example of the involvement of Ca^{2+} in stress transduction came from work on *Poterioochromonas malhamensis*. In this alga, volume regulation following osmotically induced shrinkage involves the accumulation of a glycerol metabolite, isofluoridoside, to promote osmotically driven water uptake. Synthesis of isofluoridoside involves the proteolytic activation of the key enzyme, isofluoridoside phosphate synthase, by a Ca^{2+}-dependent protease (Kauss, 1983). The minimum chain of events leading to isofluoridoside accumulation is:

1. Sensing of the change in volume probably via a mechanism involving changes in the physical state of the membrane
2. Increase in the cytoplasmic Ca^{2+} concentration, probably by transport from either the external medium or internal stores
3. Activation of the protease by Ca^{2+}
4. Activation of isofluoridoside phosphate synthase by the protease (Kauss, 1983)

Although the full chain of events has not been elucidated, this example shows how even an apparently simple response can involve a multi-step transduction pathway. However, more complete insight into the role of Ca^{2+} and inositol phosphates in intracellular signalling in plants has come from work on stomatal guard cells, particularly attempts to understand the basis of stomatal closure induced by abscisic acid (ABA).

Stomatal closure in response to decreases in soil water supply is mediated by ABA, which is thought to be produced in the roots and then transported to the leaves via the xylem (Davies and Zhang, 1991). Hartung (1983) demonstrated that ABA uptake by guard cells is not required for closure, and so ABA probably binds to a receptor in the guard cell plasma membrane. This receptor has not been positively identified, although ABA binding to the guard cell plasma membrane has been demonstrated and three possible binding proteins separated (Hornberg and Weiler, 1984). Closure of stomata involves the net loss of salt from guard cells, and ABA induces a transient increase in the efflux of ions from guard cells (MacRobbie, 1981). Electrophysiological studies (Hedrich and Schroeder, 1989; Blatt, 1991; MacRobbie, 1992) have demonstrated that guard cells contain a number of ion channels that could mediate these ion movements including:

1. An inwardly rectifying K^+ channel that is activated at membrane potentials more negative than -120 mV and which is inactivated by increases in cytoplasmic Ca^{2+}
2. An outwardly directed anion channel that requires cytoplasmic Ca^{2+} and ATP for opening, and which opens when the membrane potential is between 0 and -80 mV
3. An outwardly directed K^+ channel that opens when the membrane potential is more positive than -20 mV
4. Several stretch-activated channels (see above)

ABA inhibition of stomatal opening requires Ca^{2+} (Mansfield et al., 1990), and the sensitivity of some of the channels to this ion have led to intensive investigation of the possibility that stomatal closure involves a Ca^{2+}-based intracellular signalling pathway. The involvement of such a pathway is supported by the observation that Ca^{2+}-channel blockers will partially inhibit ABA-induced closure (De Silva et al., 1985; McAinsh et al., 1991) and that ABA increases the turnover of inositol phosphates in guard cells (MacRobbie, 1992).

Direct investigations of the changes in cytoplasmic Ca^{2+} levels of guard cells during stomatal closure have been undertaken using Ca^{2+}-sensitive fluorescent dyes injected into the cells. Using this approach, it was shown that elevation of cytoplasmic Ca^{2+} in guard cells by photoinduced release of Ca^{2+} from injected 'caged' Ca^{2+} led to stomatal closure (Gilroy et al., 1990). Release of inositol 1,4,5-trisphosphate ($InsP_3$) from caged $InsP_3$ also caused closure, which was accompanied by elevation of cytosolic Ca^{2+}, inhibition of the inward-rectifying K^+ channel, and activation of an unidentified leak current (Gilroy et al., 1990; Blatt et al., 1990). These results indicate that increases in cytoplasmic Ca^{2+} concentration mediated by $InsP_3$ do lead to closure. However, attempts to demonstrate that ABA induces a rise in cytoplasmic Ca^{2+} that precedes closure have been less successful and no clear picture has emerged. McAinsh et al. (1990) found that ABA induced a rise of Ca^{2+} concentrations from about 100 nM to about 700 nM in eight of ten Commelina guard cells, but Gilroy et al. (1991), also working with Commelina, found that only four of 54 showed a definite increase in cytoplasmic Ca^{2+}, although ABA induced stomatal closure in all cases. However, other closing treatments investigated by Gilroy et al. (1991) all caused a rise in cytoplasmic Ca^{2+} prior to closure. A variable response of guard cell cytoplasmic Ca^{2+} to ABA was also found by Irving et al. (1992) using Phaphiopedilum tonsum. Therefore, although it is clear that stomatal closure can be induced by elevation of cytoplasmic Ca^{2+} levels, the role of this phenomenon in ABA-induced closure is not proven.

The reason why a rise in cytoplasmic Ca^{2+} has not been universally found in ABA-induced stomatal closure remains unclear. It could be caused by the physiological state of the plants or the guard cells used in the different studies, or perhaps by growth conditions or some other factor. For instance, the stomata studied by McAinsh *et al.* (1990) had starting apertures of $18\,\mu m$, whereas those used by Gilroy *et al.* (1991) had apertures only $12\,\mu m$ wide. It is possible therefore that lower levels of cytoplasmic Ca^{2+} may be needed to induce closure from the more closed cells and that these are not detectable by the Ca^{2+}-sensitive dyes. However, this does not explain why other closing treatments used by Gilroy *et al.* (1991) did lead to a detectable rise in cytoplasmic Ca^{2+} concentration. Alternatively, in some cells the rise in Ca^{2+} level may be very localized and not amenable to study by the methods used. Gilroy *et al.* (1991) reported that fluorescence from very thin areas of cytoplasm could not be measured or imaged. Therefore if Ca^{2+} was released very close to the plasma membrane it could still induce closure but remain undetectable. Finally, it is possible that the lack of a rise in cytoplasmic Ca^{2+} concentration may be physiologically meaningful and indicate that there are Ca^{2+}-dependent and Ca^{2+}-independent routes by which ABA causes closure (Gilroy *et al.*, 1991). One possible explanation is that cytoplasmic pH may also be a signal acting either independently or in concert with the Ca^{2+} pathway. Irving *et al.* (1992) have shown that whereas both opening and closing agents cause a rise in cytoplasmic Ca^{2+} level, cytoplasmic pH was decreased by agents which open stomata (e.g. IAA, kinetin, fusicoccin), but increased by closing agents (ABA, procaine). Therefore it is possible that cytoplasmic pH is part of a transduction pathway that affects stomatal aperture and can cause closure independently of an increase in cytoplasmic Ca^{2+} concentration.

The source of the Ca^{2+} that causes the rise in the cytoplasmic concentration of this ion is unclear. The inhibition of closure by externally applied Ca^{2+}-channel blockers suggests that Ca^{2+} influx at the plasma membrane may be important (De Silva *et al.*, 1985). However, these agents only partially inhibit closure and therefore it seems likely that intracellular Ca^{2+} stores may also be involved (McAinsh *et al.*, 1991). This is also supported by studies with Ca^{2+}-sensitive fluorescent dyes, which show that rises in cytoplasmic Ca^{2+} concentration can still occur in the presence of Ca^{2+}-channel blockers and that the highest rises are located around the vacuole (Gilroy *et al.*, 1991). It therefore seems most likely that the vacuole contributes to the rise in cytoplasmic Ca^{2+} concentration and in some conditions may be the sole source of Ca^{2+}. The event that initiates mobilization of vacuolar Ca^{2+} has not been established. Two Ca^{2+} channels, mediating Ca^{2+} efflux from the vacuole, have been demonstrated in beet tonoplast, one gated by voltage, the other by $InsP_3$ (Johannes *et al.*, 1992). Therefore, either or both of these could be involved in release of vacuolar Ca^{2+} in closing guard cells.

It is also likely that other tonoplast transport systems are activated during closure, as the loss of guard cell turgor is ultimately due to loss of vacuolar salts (MacRobbie, 1981). Thus transport from the vacuole to the cytoplasm will control the rate of ion loss across the plasma membrane and so will determine the rate of closure. At present, we have very little knowledge of the tonoplast transport processes involved in closure or the way they are controlled. Hedrich and Neher (1987) demonstrated Ca^{2+}-regulated ion channels in the tonoplast of beet, and it may be that the rise in cytoplasmic Ca^{2+} concentration during stomatal closure mediates the activity of such channels in the guard cell tonoplast. This is clearly an area in which there are opportunities for further research.

In summary, the sequence of events that leads to closure of stomata seems to involve Ca^{2+} and $InsP_3$, but other signalling pathways may also play a role. Increases in cytoplasmic Ca^{2+} concentration definitely cause closure, and the regulatory properties of the ion channels in the plasma membrane and tonoplast control loss of ions. Many of the events still remain to be elucidated, but based on current knowledge the effects on plasma membrane channels can be described. Thus the increase in cytoplasmic Ca^{2+} concentration will prevent K^+ influx via the Ca^{2+}-inhibited inwardly rectifying K^+ channel. In cells with a large negative membrane potential, some event must depolarize the membrane potential to between 0 and $-80\,mV$ so that the Ca^{2+}-activated anion channel can open. This event has not been positively identified, although it could be Ca^{2+} influx across the plasma membrane (MacRobbie, 1992). When the voltage is within the necessary range, the rise in Ca^{2+} concentration will activate the Ca^{2+}-dependent anion channel, thus permitting loss of anions from the cell and further depolarizing the membrane potential towards more positive values. This will then activate K^+ loss via the outward-rectifying K^+ channel. The net result will be the loss of K^+ and anions, which decreases turgor and closes the stomatal pore.

12.3.2. Jasmonate-based signalling in pest and pathogen attack

Plants have elaborate defence mechanisms that help protect them from attack by pests, pathogens and herbivores. These include the induction of new proteins such as proteinase inhibitors (Green and Ryan, 1972), and the accumulation of secondary metabolites (Franceschi and Grimes, 1991; Gundlach et al., 1992). Attack of one part of the plant can lead to systemic induction of these protective mechanisms, and therefore signals must pass from the site of attack to other parts of the plant. Recently it has been shown that jasmonic acid and methyl jasmonate can cause increases in the synthesis of protective compounds (Farmer and Ryan, 1990; Franceschi and Grimes, 1991; Gundlach et al., 1992; R.M. Wallsgrove, personal communication). Therefore, it has been proposed that jasmonic acid and methyl jasmonate act as part of a signalling pathway that is an important component of the plant's defences, although full details of the pathway have yet to be elucidated (Farmer and Ryan, 1992).

Jasmonic acid is synthesized from linolenic acid by a pathway that involves a lipoxygenase, a dehydrase and β-oxidation (Vick and Zimmerman, 1984) and intermediates in the pathway have been shown to be effective inducers of proteinase inhibitors (Farmer and Ryan, 1992). Therefore the latter authors proposed that attack leads to the production of various signal molecules by damaged cells, and that these bind to receptors in the plasma membrane of non-wounded cells and activate a lipase activity. This lipase then releases linolenic acid from membrane lipids, which is converted to jasmonic acid (Farmer and Ryan, 1992). Jasmonic acid or methyl jasmonate then induces changes in gene expression that ultimately mediate the changes in the concentrations of protective compounds. It remains unclear at present whether other signalling pathways, such as elevation of cytoplasmic Ca^{2+} concentration, are also involved in the responses induced by jasmonates. However, one interesting feature of the mode of action of jasmonates is that methyl jasmonate is effective when supplied as vapour, and so potentially can act as a signal not only for the plant under attack but also for neighbouring plants (presumably whether of the same species or not). Thus it may have a role as both a long-distance and an intracellular signalling molecule.

12.3.3. Other intracellular signalling pathways

Presumably there are many other cell signalling systems that remain to be discovered in plants. In the past there was much interest in the role of cyclic AMP (cAMP) as a signal molecule, but failure to prove unequivocally the presence of the compound in plants has led to the conclusion that it probably does not have an important role (Spiteri et al., 1989). However, interest in this topic may be aroused again following the cloning of the AKT1 K^+ channel gene from *Arabidopsis* (Sentenac et al., 1992). Sequence analysis of this gene showed that the C-terminus contains a motif with high homology to mammalian channels and transporters that are regulated by cAMP or cGMP. This raises the possibility that cyclic nucleotides do play a role in regulation of activities in plants. However, it remains to be shown that the sequence homology found by Sentenac et al. (1992) also indicates functional similarity.

12.4. Long-distance signalling in plants

The idea that plants signal between tissues to integrate their activities is, of course, not new; the study of plant hormones is based precisely on this idea. However, recent research has highlighted the variety of new and different signals that plants appear to use for long-range integration and for the transfer of information between organs. These are peptide, electrical and hydraulic signals.

12.4.1. Systemin: a long-distance signalling peptide

Studies of the induction of plant defence mechanisms, particularly synthesis of proteinase inhibitors, have led to the identification of a peptide hormone, systemin, in plants (Farmer and Ryan, 1992). Proteinase inhibitors are part of a plant's defence mechanism and are produced systemically in response to wounding (Green and Ryan, 1972), and it was thought that the systemic signal was an oligogalacturonide. However, it was found that oligogalacturonides of the required length were not mobile and so can act only as local signals (Pearce et al. (1991) and references therein). Therefore Pearce et al. (1991) tried purifying peptides that would induce proteinase inhibitors and identified

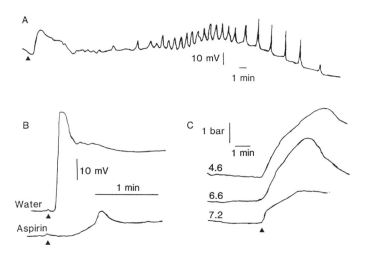

FIGURE 12.3   *The effects of wounding on electrical and pressure signals in plants. In all cases the plant was wounded by applying localized heat to the lamina of a leaf at the times indicated by the arrow heads. (A) A train of electrical pulses detected in the petiole of a wounded tomato leaf. (After Wildon et al., 1989.) (B) Inhibition of the electrical signal in tomato plants by aspirin. (After Wildon et al., 1989.) (C) Changes in turgor pressure measured in epidermal cells of a wheat leaf following wounding of another leaf on the plant. The numbers on each trace indicate the turgor before the wounding treatment was given. (After Malone and Stanković, 1991.)*

an 18-amino-acid peptide that they named systemin. Systemin has the following sequence (N-terminus on the left):

AlaValGlnSerLysProProSerLysArgAspProProLysMetGlnThrAsp

Pearce *et al.* (1991) chemically synthesized systemin and showed that synthetic systemin is able to induce the synthesis of proteinase inhibitors and, when applied at sites of wounding, moves quickly throughout tissues and into the phloem. Thus this peptide has all the characteristics required of a compound that is produced on wounding and then moves to other parts of the plant to induce defence compounds. Systemin is derived from a 200 amino acid polypeptide, prosystemin, which is synthesized in response to wounding (McGurl *et al.*, 1992). It is assumed that induction of proteinase inhibitors involves binding of systemin to a receptor leading to activation of jasmonate synthesis (Farmer and Ryan, 1992).

12.4.2.   *Electrical long-distance signalling*

There have been many reports that wounding of plants can induce changes in electrical activity in plants (see Malone and Stanković, 1991), but little evidence that they can relay information that induces changes in remote tissues. However, Wildon *et al.* (1989) examined the relationship between wounding-induced electrical activity and the induction of proteinase inhibitors. They used heat treatment of leaflets on a tomato plant to induce proteinase inhibitors in other leaves and showed that the heat also caused propagation of an electrical signal that could be measured with external electrodes attached to the petiole and stem below the treated leaf (Fig. 12.3A). They also demonstrated that aspirin, which prevented induction of proteinase inhibitors, also inhibited the electrical signal (Fig. 12.3B). In more recent work, the same group has shown that preventing phloem transport by cold-girdling of the petiole of the treated leaf has no effect on inhibitor production in the remote leaf or on the propagation of the electrical signal, suggesting that phloem translocation is not needed for the inducing signal to move (Wildon *et al.*, 1992). In contrast, steam-girdling, which kills living cells, did prevent both inhibitor production and the electrical activity, suggesting that living cells were required for transmission of the signal. The latter result also suggests that transmission via the non-living xylem vessels is not involved. The cells that form the pathway for electrical signalling have not been identified, nor has it yet been possible to show that giving an electrical impulse of an appropriate type to one leaf will induce proteinase inhibitors in another (J. F. Thain, personal communication). The latter experiment is obviously a crucial test of the electrical signalling hypothesis. Nevertheless, the work strongly suggests that long-distance electrical signalling may be an important phenomenon in plants and it will be interesting to see whether it is involved in other integrative activities between organs.

12.4.3. *Pressure as a long-distance signal*

Pressure waves transmitted via the xylem may also be long-distance signals in plants. Palta *et al.* (1987) showed that in sugar beet plants changes in stomatal conductance led to changes in storage root cell turgor pressure and root diameter. They suggested that hydraulic changes induced by alterations in the rate of transpiration could be a system of communication between root and shoot and thus play a role in integrating activities of these organs. Malone and Stanković (1991) investigated the effect of wounding on cell turgor pressure in a leaf remote from the wound site and showed that it increased 1–2 min after the wound was inflicted (Fig. 12.3C). They tested a number of possible explanations for this effect and concluded that the most likely is a pressure pulse that results from release of water into the xylem from the damaged cells. Such a pulse would move through the xylem at the speed of sound and so would provide a very rapid means of transmitting information between different tissues. More recent work, however, has shown that the pressure pulse does not induce proteinase inhibitors, and the significant event may be the slower mass flow of water and solutes through the xylem that wounding induces. This flow could carry systemin and other signals from the wound site to other parts of the plant (Malone and Stanković, personal communication). Therefore, a positive role for pressure signals has still to be demonstrated.

12.5. Concluding remarks

Knowledge of the pathways of signal transduction in plants is still in its infancy and in no case do we have a full understanding of all of the steps from initial perception to final response. If experience in animals is a clue, it is likely that the full pathways are much more complicated than the relatively simple schemes that have thus far been demonstrated in plants (Berridge and Irvine, 1989). Nevertheless, there are strong indications that plants use a variety of pathways to communicate information both within and between cells. The diversity of potential pathways being suggested indicates that there is a large number of possible ways in which plants can perceive and transduce environmental stresses and changes, be they biological, chemical or physical in nature.

ACKNOWLEDGEMENTS

I wish to thank the following for helpful discussions during the preparation of this chapter: Professor E.A.C. MacRobbie (University of Cambridge), Dr J. F. Thain (University of East Anglia), Dr M. Malone (Horticulture Research International) and Dr A. D. Tomos (University of Wales, Bangor).

REFERENCES

Alexandre, J. and Lassalles, J.-P. (1991). Hydrostatic and osmotic pressure activated channel in plant vacuole. *Biophys. J.* **60**, 1326–1336.

Bates, G.W. and Goldsmith, M.H.M. (1983). Rapid response of the plasma membrane potential in oat coleoptiles to auxin and other weak acids. *Planta* **159**, 231–237.

Berridge, M.J. and Irvine, R.F. (1989). Inositol phosphates and cell signalling. *Nature* **341**, 197–205.

Blatt, M.R. (1991). Ion channel gating in plants: physiological implications and integration for stomatal function. *J. Membrane Biol.* **124**, 95–112.

Blatt, M.R. and Slayman, C.L. (1987). Role of 'active' potassium transport in the regulation of cytoplasmic pH by non-animal cells. *Proc. Natl. Acad. Sci. USA* **84**, 2737–2741.

Blatt, M.R., Thiel, G. and Trentham, D.R. (1990). Reversible inactivation of K^+ channels of *Vicia* stomatal guard cells following the photolysis of caged inositol 1,4,5-trisphosphate. *Nature* **346**, 766–769.

Cahalan, M. and Neher, E. (1992). Patch clamp techniques: an overview. *Methods Enzymol.* **207**, 3–14.

Cosgrove, D.J. and Hedrich, R. (1991). Stretch-activated chloride, potassium, and calcium channels coexisting in plasma membranes of guard cells of *Vicia faba* L. *Planta* **186**, 143–151.

Davies, W.J. and Zhang, J. (1991). Root signals and the growth and development of plants in drying soil. *Annu. Rev. Plant Physiol. Plant Mol. Biol.* **42**, 55–76.

De Silva, D.L.R., Cox, R.C., Hetherington, A.M. and Mansfield, T.A. (1985). Suggested involvement of calcium and calmodulin in the responses of stomata to abscisic acid. *New Phytol.* **101**, 555–563.

Falke, L.C., Edwards, K.L., Pickard, B.G. and Misler, S. (1988). A stretch-activated anion channel in tobacco protoplasts. *FEBS Lett.* **237**, 141–144.

Farmer, E.E. and Ryan, C.A. (1990). Interplant communication: airborne methyl jasmonate induces synthesis of proteinase inhibitors in plant leaves. *Proc. Natl. Acad. Sci. USA* **87**, 7713–7716.

Farmer, E.E. and Ryan, C.A. (1992). Octadecanoid precursors of jasmonic acid activate the synthesis of wound-inducible proteinase inhibitors. *Plant Cell* **4**, 129–134.

Felle, H. and Bertl, A. (1986). Light-induced cytoplasmic pH changes and their interrelation to the activity of the electrogenic proton pump in *Riccia fluitans*. *Biochim. Biophys. Acta* **848**, 176–182.

Frachisse, J.-M., Johannes, E. and Felle, H. (1988). The use of weak acids as physiological tools: a study of the effects of fatty acids on intracellular pH and electrical plasmalemma properties of *Riccia fluitans* rhizoid cells. *Biochim. Biophys. Acta* **938**, 199–210.

Franceschi, V.R. and Grimes, H.D. (1991). Induction of soybean vegetative storage proteins and anthocyanins by low-level atmospheric methyl jasmonate. *Proc. Natl. Acad. Sci. USA* **88**, 6745–6749.

Gilroy, S., Read, N.D. and Trewavas, A.J. (1990). Elevation of cytoplasmic calcium by caged calcium or caged inositol trisphosphate initiates stomatal closure. *Nature* **346**, 769–771.

Gilroy, S., Fricker, M.D., Read, N.D. and Trewavas, A.J. (1991). The role of calcium in signal transduction of *Commelina* guard cells. *Plant Cell* **3**, 333–344.

Green, T.R. and Ryan, C.A. (1972). Wound-induced proteinase inhibitor in plant leaves: a possible defence against insects. *Science* **175**, 776–777.

Gundlach, H., Mürrer, M.J., Kutchan, T.M. and Zenk, M.H. (1992). Jasmonic acid is a signal transducer in elicitor-induced plant cell culture. *Proc. Natl. Acad. Sci. USA* **89**, 2389–2393.

Hartung, W. (1983). The site of action of abscisic acid at the guard cell plasmalemma of *Valerianella locusta*. *Plant Cell Environ.* **6**, 427–428.

Hedrich, R. and Neher, E. (1987). Cytoplasmic calcium regulates voltage-dependent ion channels in plant vacuoles. *Nature* **329**, 833–836.

Hedrich, R. and Schroeder, J.I. (1989). The physiology of ion channels and electrogenic pumps in higher plants. *Annu. Rev. Plant Physiol. Plant Mol. Biol.* **40**, 539–569.

Hedrich, R., Schroeder, J.I. and Fernandez, J.M. (1987). Patch-clamp studies on higher plant cells: a perspective. *Trends Biochem. Sci.* **12**, 49–52.

Hornberg, C. and Weiler, E.W. (1984). High-affinity binding sites for abscisic acid on the plasmalemma of *Vicia faba* guard cells. *Nature* **310**, 321–325.

Irving, H.R., Gehring, C.A. and Parish, R.W. (1992). Changes in cytosolic pH and calcium of guard cells precede stomatal movements. *Proc. Natl. Acad. Sci. USA* **89**, 1790–1794.

Johannes, E., Brosnan, J.M. and Sanders, D. (1992). Parallel pathways for intracellular Ca^{2+} release from the vacuole of higher plants. *Plant J.* **2**, 97–102.

Kauss, H. (1983). Volume regulation in *Poterioochromonas*. Involvement of calmodulin in the Ca^{2+}-stimulated activation of isofluoridoside phosphate synthase. *Plant Physiol.* **71**, 169–172.

Kurkdjian, A. and Guern, J. (1989). Intracellular pH: measurement and importance in cell activity. *Annu. Rev. Plant Physiol. Plant Mol. Biol.* **40**, 271–303.

MacRobbie, E.A.C. (1981). Effects of ABA in 'isolated' guard cells of *Commelina communis* L. *J. Exp. Bot.* **32**, 563–572.

MacRobbie, E.A.C. (1992). Calcium and ABA-induced stomatal closure. *Phil. Trans. R. Soc. Lond.* B **338**, 5–18.

Malone, M. and Stanković, B. (1991). Surface potentials and hydraulic signals in wheat leaves following localized wounding by heat. *Plant Cell Environ.* **14**, 431–436.

Mansfield, T.A., Hetherington, A.M. and Atkinson, C.J. (1990). Some current aspects of stomatal physiology. *Annu. Rev. Plant Physiol. Plant Mol. Biol.* **41**, 55–75.

McAinsh, M.R., Brownlee, C. and Hetherington, A.M. (1990). Abscisic acid-induced elevation of guard cell cytosolic Ca^{2+} precedes stomatal closure. *Nature* **343**, 186–188.

McAinsh, M.R., Brownlee, C. and Hetherington, A.M. (1991). Partial inhibition of ABA-induced stomatal closure by calcium-channel blockers. *Proc. R. Soc. Lond.* B **243**, 195–201.

McGurl, B., Pearce, G., Orozco-Cardenas, M. and Ryan, C.A. (1992). Structure, expression and antisense inhibition of the systemin precursor gene. *Science* **255**, 1570–1573.

Palta, J.A., Wyn Jones, R.G. and Tomos, A.D. (1987). Leaf diffusive conductance and tap root cell turgor pressure of sugarbeet. *Plant Cell Environ.* **10**, 735–740.

Pearce, G., Strydom, D., Johnson, S. and Ryan, C.A. (1991). A polypeptide from tomato leaves induces wound-inducible proteinase inhibitor proteins. *Science* **253**, 895–898.

Pfanz, H., Martinoia, E., Lange, O.-L. and Heber, U. (1987). Flux of SO_2 into leaf cells and cellular acidification by SO_2. *Plant Physiol.* **85**, 928–933.

Raven, J.A. and Smith, F.A. (1976). Nitrogen assimilation and transport in vascular land plants in relation to intracellular pH regulation. *New Phytol.* **76**, 415–431.

Roberts, J.K.M., Callis, J., Wemmer, D., Walbot, V. and Jardetzky, O. (1984). Mechanism of cytoplasmic pH regulation in hypoxic maize root tips and its role in survival under hypoxia. *Proc. Natl. Acad. Sci. USA* **81**, 3379–3383.

Sachs, F. (1991). Mechanical transduction by membrane ion channels: a mini review. *Mol. Cell. Biochem.* **104**, 57–60.

Sanders, D., Hansen, U.-P. and Slayman, C.L. (1981). Role of the plasma membrane proton pump in pH regulation in non-animal cells. *Proc. Natl. Acad. Sci. USA* **78**, 5903–5907.

Sentenac, H., Bonneaud, N., Minet, M., Lacroute, F., Salmon, J.-M., Gaymard, F. and Grignon, C. (1992). Cloning and expression in yeast of a plant potassium ion transport system. *Science* **256**, 663–665.

Spiteri, A., Viratelle, O.M., Raymond, P., Rancillac, M., Labouesse, J. and Pradet, A. (1989). Artefactual origins of cyclic AMP in higher plant tissues. *Plant Physiol.* **91**, 624–628.

Vick, B.A. and Zimmerman, D.C. (1984). Biosynthesis of jasmonic acid by several plant species. *Plant Physiol.* **75**, 458–461.

Wildon, D.C., Doherty, H.M., Eagles, G., Bowles, D.J. and Thain, J.F. (1989). Systemic responses arising from localized heat stimuli in tomato plants. *Ann. Bot.* **64**, 691–695.

Wildon, D.C., Thain, J.F., Minchin, P.E.H., Gubb, I.R., Reilly, A.J., Skipper, Y.D., Doherty, H.M., O'Donnell, P. and Bowles, D.J. (1992). Electrical signalling and systemic proteinase inhibitor induction in the wounded plant. *Nature* **360**, 62–65.

Zimmermann, U. (1977). Cell turgor pressure regulation and turgor pressure-mediated transport processes. *Symp. Soc. Exp. Biol.* **31**, 117–154.

CHAPTER 13

Molecular and genetic analysis of the heat-shock response in transgenic plants

F. SCHÖFFL, E. KLOSKE, A. HÜBEL, K. SEVERIN
and A. WAGNER

13.1. Heat-shock response

One of the best-characterized environmental stress responses is that to high temperature or heat stress. At present there is only circumstantial evidence that the molecular or cellular response, e.g. the synthesis of heat-shock proteins (HSPs), is the basis for the organismal response; that is, the increase in thermotolerance. Both phenomena are observed in almost every organism, but in plants the heat-shock (*hs*) response may be particularly important for survival in nature.

A comparison of the *hs* response in different organisms including bacteria, yeast, animals and humans reveals conserved features: *de novo* synthesis of HSPs and conservation of their primary sequences, the molecular mechanisms of transcriptional regulation of the *hs* genes (encoding HSPs), induction of the response by certain physical and chemical stressors (Nover, 1991) and developmental regulation of HSP expression (Hightower and Nover, 1991).

13.1.1. Roles of heat-shock proteins

HSPs are defined as proteins that are abundantly synthesized following heat shock. They are usually designated according to their approximate molecular masses in kilodaltons, although for some newly identified HSPs their original designations have been kept (see Table 13.1). For many of the HSPs, homologous proteins that are constitutively expressed exist in the cell. Much of what is known about the biochemical and physiological properties and the functional role of HSPs is extrapolated from studies in *Escherichia coli*, yeast and *Drosophila*, and from their constitutive cellular counterparts, the so-called constitutive or cognate heat-shock proteins/genes (HSC, *hsc*). Certain HSPs, e.g. HSP70, HSP60 and HSP20, are not confined to a single cellular compartment. Because of their induction by a number of other stressors, HSPs are more properly referred to 'stress proteins'. Not all of the numerous non-*hs* stressors induce a high level of the *hs* response and, in plants, only arsenite has been shown to elicit a significant expression of all HSPs. This raises questions about the specificity of HSPs, the pathway of signal transduction that leads to the elicitation of the *hs* response, and specific aspects of the *hs* response in plants.

The HSPs listed in Table 13.1 are not exclusively synthesized in plants, but certain aspects of the expression and localization of some groups of HSPs are phytospecific. For more details see recent reviews by Vierling (1991) and Nover (1991).

HSP100, sometimes also referred to HSP110, is expressed in plants and many other organisms, with the exception of *Drosophila*. HSP104 in yeast, the best studied HSP of this group, is not essential for growth but is evidently required for the development of thermotolerance by this organism.

HSP90 has been much studied in yeast, *Drosophila* and vertebrate cells, and the current model of its function suggests a reversible interaction with hormone receptors and other gene regulatory proteins in the cell. HSP90 seems to maintain its target proteins in an inactive but receptive (to activation) conformation.

TABLE 13.1. *Biochemical and functional properties of heat-shock proteins (HSPs).*

HSP family*	Properties
HSP100	HSP 104 important for thermotolerance in yeast
HSP90[†]	Associated in the cytoplasm with regulatory proteins (inactive forms) in animal cells (hormone receptors, protein kinases, etc.)
HSP70[†]	Most conserved HSP, ATPase, reversible interaction with the nucleolus, dissociation from other (denatured) proteins requires ATP, possibly a negative regulator of the heat-shock response, may interact with HSF
HSP60[†]	Molecular 'chaperone' for proper assembly of multimeric protein complexes in mitochondria, chloroplasts, cytoplasm?
HSP20[‡]	Heat-shock-dependent aggregation of these HSPs in the cytoplasm, formation of granules in plants, chaperone function, mRNA preservation?
Ubiquitin	8 kDa protein involved in recognition of proteins for proteolytic degradation
HSF[†]	*hs* factor, positive regulator of the transcription of *hs* genes

*Classified by molar mass in kilodaltons; different sizes in many organisms; structural and functional conservation within families.

[†]Multiple isoforms; constitutive and heat-inducible proteins.

[‡]Multiple HSPs, HSP families in plants (members are also transported into the chloroplast and to endoplasmic reticulum); only one HSP in humans; structural relationship with the lens cristallins of the vertebrate eye.

Protein–protein interaction is a key feature of HSP70 and HSP60 proteins, and both are the subject of intensive studies in plants, which also extend to the constitutive HSP70 homologues, BiP (binding protein) and the endoplasmic reticulum (ER)-targeted GRP (glucose-regulated protein). The common feature of HSP70 proteins is their capacity to bind to and to hydrolyse ATP. This property is obviously required to dissociate HSP70 from complexes with other proteins in the cytoplasm or in the nucleus. One important role of cytoplasmic HSP70 homologues may be the binding of nascent proteins to keep them in a suitable conformation for a subsequent transport to, or through, biological membranes (chloroplasts, mitochondria, endoplasmic reticulum). Constitutive HSP70-like proteins are also found in chloroplasts and mitochondria of plants, where they may be involved in the reception and internal shuttling of incoming proteins. Interestingly, these HSP70 variants are obviously also encoded by the nucleus, but their sequences resemble more closely the prokaryotic type of HSP70, the *E. coli* DnaK protein. ER-localized BiP or GRP-like proteins also seem to be involved in proper folding of newly synthesized proteins. There is some evidence that HSP70 and its constitutive homologues are able to associate with the heat-shock transcription factor (HSF). A mutational analysis of *E. coli* and yeast HSP70 genes revealed a negative (suppression) effect of HSP70 on HSF activity.

HSP60 proteins are nuclear encoded, but are abundant in mitochondria and chloroplasts. There is some evidence that HSP60 may also be present in the cytoplasm of plant cells. HSP60 is thought to be involved in the assembly of multimeric protein complexes, as deduced from the activity of the HSP60 homologous ribulose bisphosphate carboxylase-binding protein and the prokaryotic *GroEL* gene product. The term 'molecular chaperone' has been coined for this function of HSP60, and the importance of this property is demonstrated by conditional lethal mutants of HSP60 in yeast.

HSP20 represents a group of proteins with molecular masses usually between 17 and 30 kDa, encoded by multigene families in plants but by far fewer genes in other organisms. Moreover, these plant HSPs appear to be more abundantly expressed following heat shock than other HSPs. Some members of this group of proteins are transported into the chloroplasts and others into other endomembrane systems in plants. There is no evidence for either organelle or endomembrane localized HSP20 outside the plant kingdom. HSP20 proteins are strictly heat-inducible and there is limited evidence for the existence of homologous HSCs, with the exception of some other stress proteins and the eye lens α-crystallins, which share domains of limited structural conservation with HSP20 and new evidence suggests a chaperone function (for review, see Jaenick and Creighton, 1993). HSP20 function is probably only required for higher eukaryotes and most critically in plants. Most prokaryotes seem to lack HSP20 and, in yeast, the function of the single HSP20 gene is dispensable. HSP20 proteins aggregate in the cytoplasm to form granules in response to heat shock in plants. These granules may contain additional HSPs, and granules purified from tomato cells were associated with normal cellular mRNAs but not with *hs*-mRNAs. This finding suggests a structural role of HSP20 for mRNA preservation, but it cannot be excluded entirely that HSP20s have an as yet unidentified enzymatic role.

It is not surprising that ubiquitin is also expressed at an enhanced level following heat shock. The function of ubiquitin is to attach to denatured proteins destined to be degraded by the proteolytic pathway. Increased levels of denatured cellular proteins require an appropriate adjustment of the ubiquitin level. However, ubiquitin may also serve other regulatory roles in the cell.

Recently it was shown that the transcript level of one HSF-like gene from tomato is significantly higher after heat shock than at normal temperatures (Scharf et al., 1990). The heat inducibility of this protein has implications for the possible functional role of HSF, and this aspect is discussed later.

13.1.2. Developmental regulation of heat-shock proteins

It is no surprise that most investigations of HSP expression have been related to stress induction because the hs response seems to be beneficial to the cell and organism under unfavourable environmental conditions. However, evidence is accumulating for developmental induction of the synthesis of HSPs in the absence of heat stress (for a recent review see Vierling, 1991; Hightower and Nover, 1991).

It has been well established that HSP synthesis can be induced in almost all vegetative tissues of plants, but with different efficiencies. Heat-induced expression seems to be significantly impaired during early embryo development and pollen germination. The lack of heat-inducibility in embryos may be countered by constitutive HSPs. During late embryogenesis and seed development, accumulation of the HSP20 was observed at normal temperatures without heat shock.

In our laboratory we are using hs promoter-glucuronidase (GUS) reporter gene constructs to trace the tissue specificity and developmental regulation of the heat-shock response. The soybean *Gmhsp17.3-B* heat-shock promoter drives heat-inducible GUS activity in leaves of transgenic tobacco (Schöffl et al., 1991, 1992) and *Arabidopsis* plants (Schöffl et al., unpublished data). GUS activity was not evenly distributed in tissues on histochemical staining using the method of De Block and Debrouwer (1992). Most intensely stained were the meristematic and differentiating cells of the shoot, root and leaves (Kloske and Schöffl, unpublished data). The general staining pattern is in accordance with the uneven distribution of hs mRNA levels in different organs of soybean (Schöffl, 1984). These differences are suggestive of a higher demand of HSPs in physiologically active and differentiating cells either to protect heat-sensitive tissues or to serve other functions. The high level of constitutive GUS activity resulting from the *Gmhsp17.3-B* promoter-GUS gene construct in seeds of transgenic tobacco (Kloske and Schöffl, unpublished data) indicate that hs genes are also developmentally regulated during embryogenesis and germination. Heat shock has no effect on the level of GUS activity for several days during imbibition and germination, and the tissue may already be well protected by the 'developmental heat-shock response'. However, the role of HSPs in development is still a matter of speculation and it is not known whether the developmental signal is transmitted via HSF or other *cis* elements and *trans*-acting factors. Embryogenesis and differentiation may either cause cellular stresses countered by HSP synthesis or, alternatively, HSPs may be required for embryogenesis and determination or differentiation of cells and tissues, and hence their synthesis would be under developmental control.

13.2. Molecular model of transcriptional regulation

Since it was discovered in 1962 in *Drosophila*, the expression of HSPs has been one of the most studied models for transcriptional regulation and gene expression (for a review see Schlesinger *et al.*, 1982). In plants, during 10 years of investigation of the heat-shock response since the first plant *hs* genes were cloned from soybean (Schöffl and Key, 1982, 1983), significant contributions have been made to the understanding of the molecular mechanisms of transcriptional regulation of *hs* genes. A simplified model describing the molecular mechanism of the transcriptional activation of *hs* genes is depicted in Fig. 13.1, and recent reviews have been published by Sorger (1991), Vierling (1991) and Schöffl *et al.* (1992).

13.2.1. *Heat-shock promoter and upstream elements*

The structural features of the heat-shock promoter are binding sites for the heat-shock factor (HSF). The *cis*-active heat-shock elements (HSE) have the consensus sequence of –GAA–TTC–, but at least one and a half units of

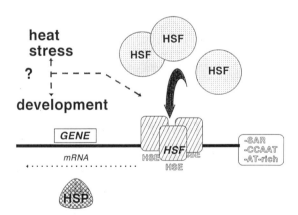

FIGURE 13.1 *Model of transcriptional activation of* hs *genes. HSF, heat-shock transcription factor; HSE, heat-shock element (specifies heat-inducible transcription as a binding site for HSF); HSP, heat-shock protein; SAR, scaffold attachment region; CCAAT, CAAT box element; AT-rich, simple sequences. SAR, CCAAT and AT-rich are upstream promoter elements for enhanced transcription.*

HSE are minimally required for heat inducibility of transcription. The multiplicity of HSE elements in *hs* promoter regions was recognized for the first time in a plant *hs* gene (Schöffl *et al.*, 1984) and emphasis was given to the overlap of HSE sequences upstream from the TATA box, which represent the minimally active consensus sequence –GAA–TTC–GAA–. Sequence analysis of many other plant and animal *hs* genes has confirmed these structures. Functional analyses of soybean *Gmhsp17.3-B* promoter deletions in transgenic tobacco (Baumann *et al.*, 1987; Schöffl *et al.*, 1989) and the use of synthetic HSE consensus sequences (Schöffl *et al.*, 1989) revealed the functional role of HSEs as *hs*-specific sites recognized by the plant HSF (Scharf *et al.*, 1990). The amplitude of *hs* gene transcription seems to be affected by several additional *cis*-active promoter upstream sequences. Additional HSE, but also CCAAT box, elements may interact synergistically for both heat-inducibility and enhancement of transcription (Rieping and Schöffl, 1992). Upstream A- and T-rich simple sequences have been implicated as other enhancer-like sequences of *hs* genes (Rieping and Schöffl, 1992), which also bind nuclear proteins (Schöffl *et al.*, 1990). Another type of sequence, involved in an enhanced and gene dosage correlated expression of soybean *hs* promoter-GUS reporter genes, is the scaffold attachment region (SAR) (Schöffl *et al.*, 1993). SAR sequences have been identified upstream and/or downstream from a number of soybean *hs* genes (Schöffl *et al.*, 1992) and a five- to nine-fold increase of heat-inducible GUS activity was observed in transgenic tobacco plants containing constructs with a SAR_L fragment (derived from the *Gmhsp17.6-L* gene) either at both sides of (or at least one copy located 5′ upstream from) the test gene (Schöffl *et al.*, 1993). In contrast to *hs* genes without SAR enhanced expression, the gene copy number of SAR_L enhanced constructs is positively correlated with the level of induced reporter enzyme activity; however, positional effects are not entirely eliminated. The molecular mechanism by which SAR sequences influence gene expression is unknown and it has not yet been tested whether *hs*-gene-derived SAR sequences influence the co-expression of other genes. If SARs regulate opening and closing of chromatin, according to the chromatin switch model (Laemmli *et al.*, 1992), SAR could either repress or stimulate transcription of the nearby located gene, depending on the ratio of certain chromatin proteins.

13.2.2. *Heat-shock factor*

HSF and its genes were originally isolated from *Drosophila* and yeast (for review see Nover, 1991; Sorger, 1991). HSF is constitutively synthesized in these organisms, but differs fundamentally in the capacity to bind DNA. Both types of HSF bind to HSE sequences of *hs* promoters, but HSF from yeast can bind at non-stress temperatures. However, the transcription of RNA polymerase II-initiated TATA-box complexes of *hs* genes is only supported following heat shock. In contrast, *Drosophila* and mammalian HSFs are unable to bind HSEs at normal temperatures, but binding is induced on heat shock, followed by transcriptional activation of the basal transcription factor/RNA polymerase II complex. These differences suggest that HSF activation, HSE recognition or binding, and stimulation of transcription probably occur in two separable steps. The HSF–HSE association is largely stimulated by trimerization of HSF, provided that at least three alternating 5 bp –GAA– boxes are present at the target site. Hence, trimerization of HSF is crucial to DNA binding but not sufficient for transcriptional activation. The trimerization of HSF is attributed to several leucine zipper motifs of the oligomerization domain (OD). The OD of one HSF may interact *via* a coiled coil formation with (probably) two other HSF monomers to form a trimer. The binding of trimeric HSF to DNA occurs *via* the DNA binding domains (DBD), which seem to be properly spaced and activated in trimeric complexes. Despite the implications for a strong conservation of the molecular mechanisms of transcriptional activation, only the DBD and the OD are significantly conserved in HSFs from different organisms. Other parts of HSFs, including the C-terminus that seems to be responsible for transcriptional activation in yeast and mammalian cells, are not conserved.

The first plant HSF genes have been isolated from tomato cells *via* DNA-ligand south-western screening of a cDNA expression library using a consensus HSE oligonucleotide with several units of alternating –GAA– and –TTC– (Scharf *et al.*, 1990). In contrast to the single HSF genes in yeast and *Drosophila*, tomato contains at least three different genes represented by cDNAs. Interestingly, only the DBD and OD regions are conserved, but the sequences diverge in other parts of the gene. The role of the different HSF genes/proteins in tomato is not known, but their differences in expression (one of them shows a clearly heat-inducible mRNA level) suggest that different HSFs may be required at different stages of the heat-shock reponse, and that they may serve different functions.

The HSF genes from tomato cross-hybridize with genomic DNA from different plants (Schöffl, unpublished data). Three different clones were isolated from *Arabidopsis thaliana* genomic libraries using tobacco HSF cDNA probes. To date, two different types of HSF-like sequences have been verified by DNA sequencing, but the conservation is, like in tomato, restricted to the DNA-binding and oligomerization domains (Schöffl *et al.*, unpublished data). The genomic gene structure and the characterization of the corresponding cDNAs is under investigation.

13.3. Genetic approach

The selection of regulatory mutants affecting the entire heat-shock response is required to identify components of the signal transduction pathway, from heat-shock to the activation of HSF, and to study the role of HSPs in thermotolerance and development.

Regulatory mutants which either constitutively synthesize HSPs at all temperatures or, alternatively, are unable to induce their synthesis even after heat shock would be the expected changes at the cellular level if the central regulator HSF or other components regulating its activity were mutated. In animal systems, increased intracellular calcium concentration is a prerequisite for HSF–DNA binding (Price and Calderwood, 1991). Active HSF is phosphorylated in yeast (Sorger and Pelham, 1988) and in human cells (Larson et al., 1988), and inactive HSF seems to be associated with HSP70 (possibly a negative regulator of HSF) in bacteria, yeast (for review see Craig and Gross, 1991) and human cells (Abravaya et al., 1992). This suggests that several steps of a signal transmission pathway may exist that could become suitable targets for trans-active second-site mutations. Such mutations would affect the heat-shock response either positively (constitutive HSP synthesis) or negatively (lack or inactivity of HSF). However, it would not be possible to select such mutants in plants because of their unknown phenotypic properties. Therefore it has been necessary to generate an appropriate genetic background for mutant selection. Arabidopsis thaliana was chosen as a model plant for this purpose because of its superior capacity to serve both genetic and molecular analyses.

13.3.1. Use of heat-inducible reporter genes

Two different hs promoter–reporter gene constructions have been used to create a genetic background in transgenic Arabidopsis for a positive selection of mutants of the heat-shock response. These reporter genes are used respectively as selectable markers for the induction or inhibition of HSP synthesis.

For the selection of regulatory mutations causing a constitutive synthesis of HSPs, a heat-inducible hygromycin-resistance gene (hygromycin phosphotransferase, hs-hpt) (Severin and Schöffl, 1990) has been introduced into Arabidopsis plants. Tansgenic seeds are unable to grow on hygromycin-containing media at normal temperature (25 °C) unless they are heat-treated for 1 h at least every second day (Schöffl et al., 1992). Mutations, such as those causing a constitutively active HSF and thus eliciting a constitutive heat-shock response, would also grow on hygromycin at room temperature.

Reverse mutations, that is, those exhibiting an HSF null phenotype, require a different genetic background for selection. A soybean hs promoter-Adh gene (Severin and Schöffl, unpublished data) has been constructed and introduced via Agrobacterium tumefaciens-mediated transfection into an Arabidopsis thaliana line carrying an Adh^- mutation isolated by Jacobs et al. (1988). In these lines the hs-adh gene is only expressed after a heat stress, as indicated by the enzymatic assay for alcohol dehydrogenase (ADH) according to Dolferus and Jacobs (1984). The heat-inducibility of ADH activity is dependent on the developmental stage of the germinating seed, and maximum levels are expressed after several days of imbibition (Wagner and Schöffl, unpublished data). The window of Adh activity coincides with the maximum susceptibility of heat-treated seedlings to allyl alcohol. Either of the two treatments alone, heat shock or allyl alcohol, is ineffective. The killing of plants by the allyl alcohol treatment is attributed to the enzymatic conversion of this substrate to acrolein via ADH:

$$CH_2=CH-CH_2OH + NAD \xrightarrow{ADH} CH_2=CH-CHO + NADH + H^+$$
allyl alcohol   acrolein

Because of the cytotoxic effect of acrolein, allyl alcohol can be used to counterselect cells exhibiting a regular heat-shock response.

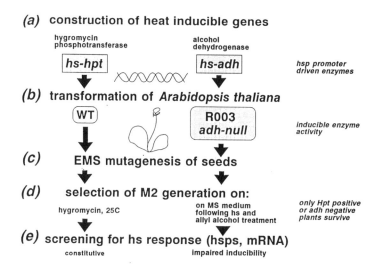

(a) **construction of heat inducible genes**

hygromycin
phosphotransferase

alcohol
dehydrogenase

| hs-hpt | | hs-adh |

*hsp promoter
driven enzymes*

(b) **transformation of *Arabidopsis thaliana***

WT

R003
adh-null

*inducible enzyme
activity*

(c) EMS mutagenesis of seeds

(d) selection of M2 generation on:

hygromycin, 25C

on MS medium
following hs and
allyl alcohol treatment

*only Hpt positive
or adh negative
plants survive*

(e) screening for hs response (hsps, mRNA)

constitutive

impaired inducibility

FIGURE 13.2 *Schemes for positive selection of regulatory mutants of the heat-shock response in* Arabidopsis thaliana.

13.3.2. *Schemes for a positive selection of regulatory mutants*

Transgenic lines or *Arabidopsis thaliana* containing either *hs-hpt* or *hs-adh* genes are currently used for mutant selection according to the schemes depicted in Fig. 13.2. The construction of heat-inducible genes (a) and *Arabidopsis* lines (b) has been described in Section 13.3.1. Seeds from several generations of transgenic plants exhibiting high levels of reporter gene activity were collected and subsequently subjected to EMS mutagenesis (c) according to Jacobs *et al.* (1988). M2 seeds of mutagenized plants were subjected to selection (d) by either spreading *hs-hpt* transformed lines on hygromycin at 25 °C or seeding *hs-adh* transformed lines on MS medium after seeds had been treated by a mild heat shock followed by a short allyl alcohol treatment. Plants surviving on hygromycin are currently screened for a constitutive heat-shock response; allyl alcohol-resistant lines are tested for their inability to elicit a response to heat shock.

Not unexpectedly, a number of plants selected by the above protocols are obviously not mutated in genes affecting the heat-shock response. The exact nature of resistance is not known, but it is conceivable that, for example, certain *cis*-active mutations in the reporter genes would mimic the expected phenotypes. Screening for HSPs and *hs* mRNAS is one possible means to discriminate false-positive scoring mutants.

13.4. Summary

The molecular model of transcriptional regulation of *hs* genes in plants has implications for molecular and genetic analyses of both the signal transduction pathway and the manipulation of the heat-shock response. Using *Arabidopsis thaliana* as a model for higher plants, transgenic lines containing suitable selection markers have been generated that are currently being tested for overexpression of the central transcriptional activator (HSF) of *hs* genes and for the selection of mutants affecting the heat-shock response in a positive (constitutive HSP synthesis) or negative (lack of HSP synthesis) fashion. It is anticipated that these approaches will provide insights into the pathway of signal transduction for environmental and developmental regulation of the heat-shock response and elucidate the role of HSPs in thermotolerance.

ACKNOWLEDGEMENTS

The research in our laboratory is supported by grants of the Deutsche Forschungsgemeinschaft to F.S.

REFERENCES

Abravaya, K., Meyers, M.P., Murphy, S.P. and Morimoto, R.I. (1992). The human heat shock protein hsp70 interacts with HSF, the transcription factor that regulates heat shock gene expression. *Genes Dev.* **6**, 1153–1164.

Baumann, G., Raschke, E., Bevan, M. and Schöffl, F. (1987). Functional analysis of sequences required for transcriptional activation of a soybean heat shock gene in transgenic tobacco. *EMBO J.* **6**, 1161–1166.

Craig, E.A. and Gross, C.A. (1991). Is hsp70 the cellular thermometer? *Trends Biochem. Sci.* **16**, 135–140.

De Block, M. and Debrouwer, D. (1992). *In-situ* enzyme histochemistry on plastic-embedded plant material: the development of an artefact free β-glucuronidase assay. *Plant J.* **2**, 261–266.

Dolferus, R. and Jacobs, M. (1984). Polymorphism of alcohol dehydrogenase in *Arabidopsis thaliana* (L.) Heynh.: genetical and biochemical characterization. *Biochem. Genet.* **22**, 817–838.

Hightower, L. and Nover, L. (1991). *Heat Shock and Development.* Berlin: Springer Verlag.

Jacobs, M., Dolferus, R. and Van Den Bossche, D. (1988). Isolation and biochemical analysis of ethyl methanesulfonate-induced alcohol dehydrogenase of *Arabidopsis thaliana* (L.) Heynh. *Biochem. Genet.* **26**, 105–122.

Jaenicke, R. and Creighton, T.E. (1993). Junior chaperones. *Curr. Biol.* **3**, 234–235.

Laemmli, U.K., Käs, E., Poljak, L. and Adachi, Y. (1992). Scaffold associated regions: *cis*-acting determinants of chromatin structural loops and functional domains. *Curr. Opin. Genet. Dev.* **2**, 275–285.

Larson, J.S., Schuetz, T.J. and Kingston, R.E. (1988). Activation *in vitro* of sequence specific DNA-binding by a human regulatory factor. *Nature* **335**, 372–375.

Nover, L. (1991). *Heat Shock Response.* Boca Raton, Fla.: CRS Press.

Price, B.D. and Calderwood, S.K. (1991). Ca^{2+} is essential for multistep activation of heat shock factor in permeabilized cells. *Mol. Cell. Biol.* **11**, 3365–3368.

Rieping, M. and Schöffl, F. (1992). Synergistic effect of upstream sequences, CCAAT box elements, and HSE sequences for enhanced expression of chimaeric heat shock genes in transgenic tobacco. *Mol. Gen. Genet.* **231**, 226–232.

Scharf, K.-D., Rose, S., Zott, W., Schöffl, F. and Nover, L. (1990). Three tomato genes code for heat stress transcription factors with a region of remarkable homology to the DNA-binding domain of yeast HSF. *EMBO J.* **9**, 4495–4501.

Schlesinger, M.J., Ashburner, M. and Tissieres, A. (1982). *Heat Shock from Bacteria to Man.* Cold Spring Harbor: Cold Spring Harbor Press.

Schöffl, F. (1984). Aspekte der Temperaturstress-regulierten Genexpression bei höheren Pflanzen, Dechema Monographien, Band 95, pp. 323–338. Weinheim: Verlag Chemie.

Schöffl, F. and Key, J.L. (1982). An analysis of mRNAs for a group of heat shock proteins of soybean using cloned cDNAs. *J. Mol. Appl. Genet.* **1**, 301–314.

Schöffl, F. and Key, J.L. (1983). Identification of a multigene family for small heat shock proteins in soybean and physical characterization of one individual gene coding region. *Plant Mol. Biol.* **2**, 269–278.

Schöffl, F., Raschke, E. and Nagao, R.T. (1984). The DNA sequence analysis of soybean heat shock genes and identification of possible regulatory promoter elements. *EMBO J.* **3**, 2491–2497.

Schöffl, F., Rieping, M., Baumann, G., Bevan, M.W. and Angermüller, S. (1989). The function of plant heat shock promoter elements in the regulated expression of chimaeric genes in transgenic tobacco. *Mol. Gen. Genet.* **217**, 246–253.

Schöffl, F., Rieping, M. and Raschke, E. (1990). Functional analysis of sequences regulating the expression of heat shock genes in transgenic plants. In: *Genetic Engineering of Crop Plants*, pp. 79–94, eds. G.W. Lycett and D. Grierson. London: Butterworths.

Schöffl, F., Rieping, M. and Severin, K. (1991). The induction of the heat shock response: activation and expression of chimaeric heat shock genes in transgenic plants. In: *Plant Molecular Biology*, Vol. 2, pp. 685–694, ed. R. Herrmann. New York: Plenum Press.

Schöffl, F., Diedring, V., Kliem, M., Rieping, M., Schröder, G. and Severin, K. (1992). The heat shock response in transgenic plants: the use of chimaeric heat shock genes. In: *Inducible Plant Proteins*, pp. 247–266, ed. J.L. Wray. Cambridge: Cambridge University Press.

Schöffl, F., Schröder, G., Kliem, M. and Rieping, M. (1993). A SAR sequence containing 395 bp DNA fragment mediates enhanced gene dosage correlated expression of a chimaeric heat shock gene in transgenic tobacco plants. *Transgenic Res.* **2**, 93–100.

Severin, K. and Schöffl, F. (1990). Heat-inducible hygromycin resistance in transgenic tobacco. *Plant Mol. Biol.* **15**, 827–833.

Sorger, P.K. (1991). Heat shock factor and the heat shock response. *Cell* **65**, 363–366.

Sorger, P.K. and Pelham, H.R.P. (1988). Yeast heat shock factor is an essential DNA-binding protein that exhibits temperature-dependent phosphorylation. *Cell* **54**, 855–864.

Vierling, E. (1991). The roles of heat shock proteins in plants. *Annu. Rev. Plant Physiol. Plant Mol. Biol.* **42**, 579–620.

CHAPTER 14

Long-term adaptation and survival

M. A. HUGHES, M. A. DUNN, L. ZHANG, R. S. PEARCE,
N. J. GODDARD and A. J. WHITE

14.1. Introduction

A low-positive-temperature treatment will acclimate many temperate plant species for frost tolerance (ability to withstand a freezing temperature). Using the term 'stress' to represent an environmental factor which causes a change in a biological system that is potentially injurious (Hoffmann and Parsons, 1991), this acclimation can be viewed as an adaptive response of the plant that enables it to survive an environmental stress. The metabolic and cellular changes that occur during acclimation are acknowledged to be complex and can be considered to be quantitative, in that the level of the response may depend on both the temperature scale and the genotype of the plant involved. This property of temperate plants, which allows individuals to acclimate to potentially damaging fluctuations in temperature, is termed phenotypic plasticity. The principal features of this plasticity are as follows:

1. Modification of phenotype may occur in a single genotype.
2. Plasticity is responsive to changes in the environment.
3. The changes have an adaptive value allowing the individual to withstand an environmental stress (Smith, 1990).

Phenotypic plasticity may involve an adjustment to metabolism without changing the development and morphology of the plant.

Temperature will have an effect on all thermosensitive metabolic processes in the plant cell and may lead to a modification of structural components of the cell. In addition, a number of recent studies have demonstrated that acclimating low-temperature treatments can cause changes in gene expression at both the mRNA and the protein level (Guy, 1990; Hughes and Dunn, 1990; Thomashow, 1990). It is not known how plants sense low temperature, nor is the signal transduction pathway understood. The important consequential changes in cell function are also only loosely identified (Hughes and Dunn, 1990). In 1964, Allard and Bradshaw argued that, owing to the complexities of the developmental pathways involved, the interactions between pathways and environment would be so complex that it was unlikely that progress could be made in understanding the basic causes and mechanisms involved in phenotypic plasticity (Allard and Bradshaw, 1964; Bradshaw, 1965). The techniques of molecular genetics, however, represent a powerful tool which can be applied to dissect such complex interactions, and we have begun to use this technology to study acclimation for frost tolerance in the cereal crop barley (*Hordeum vulgare* L.).

14.2. Genetic basis of acclimation for frost tolerance

The degree of expression of phenotypic plasticity is under genetic control; that is, populations or individuals may show differences in the level of plasticity for the same character in response to the same environmental variable (Schlichting, 1986). The wide variation in temperature response that exists between cultivars of barley was the primary reason for choosing this crop as a model species.

The inheritance of the ability to frost harden (acclimate for frost tolerance) has been studied in a wide range of species, with perhaps the most extensive genetic studies being in wheat (Thomashow, 1990; Sutka and Snape, 1989; Roberts, 1990; Whelan, 1990). In barley, analysis of doubled haploid progeny lines produced by crossing the frost-hardy winter cultivar Vogelsanger Gold with each of four frost-sensitive, spring cultivars (Tron, Tystoft Prentice, Clipper and Alf) indicate that good winter hardiness can be inherited with the spring growth habit (Doll *et al.*, 1989). In the context of a molecular approach to the study of acclimation, these genetic studies are important because they demonstrate that it is possible to identify and locate genes that have a major effect on the character.

The validity of this conclusion is also illustrated by the construction of transgenic tobacco plants containing a glycerol-3-phosphate acyltransferase gene from *Arabidopsis*, which have altered chilling sensitivity (Murata *et al.*, 1992). Glycerol-3-phosphate acyltransferase seems to be important for determining the level of phosphatidyl glycerol fatty acid unsaturation. The constitutive expression of this gene isolated from the non-chill-sensitive plant *Arabidopsis* in chill-sensitive tobacco altered both the level of phosphatidyl glycerol unsaturation and the chill sensitivity of the plants.

Our studies are exploiting the genetic variation present in barley both to assign biological and biochemical functions to isolated genes and to use as analytical tools in studies of the mechanisms of low-temperature control of gene expression.

14.3. Cloning low-temperature-responsive genes in barley

A cDNA library was made from mRNA sequences isolated from shoot meristems of the winter barley cultivar Igri, which had been given a low-temperature (6 °C day, 2 °C night, 10 h day) treatment. Plants at the three to four leaf stage were used for mRNA extractions. This temperature treatment has been shown to both acclimate and vernalize young plants of cv. Igri. The shoot meristem was chosen as a source of expressed genes (mRNA) because it is known that, in Gramineae, this part of the plant both perceives low temperature and acclimates for frost tolerance (Peacock, 1975; Thomas and Stoddart, 1984). Furthermore, it is the freezing tolerance of the non-photosynthetic crown tissue of cereals that determines winter survival (Gusta et al., 1982). The 12 low-temperature-induced genes isolated from the cDNA library using a differential screen, which have been subcloned and further characterized, are shown in Table 14.1. The transcript size varies from 500 to 2400 bases, and we have produced some DNA sequences for 11 of the genes to look for homology to previously cloned plant genes. In view of the anticipated complexity of the response of barley to a low-positive-temperature treatment, we have adopted the strategy of producing a foundation of information about a range of genes rather than produce an in-depth study of a single function.

The technique of differential screening identifies genes that are up-regulated at the level of steady-state mRNA in shoot meristems, given a

TABLE 14.1. *Summary of barley low-temperature-responsive cDNA clones that have been subcloned and confirmed in test Northern blots.*

Gene	Transcript size (bases)	Chromosome location	cDNA sequence	Homology[*]
blt 14	500	2[†]	Complete[‡]	*pA086*
blt 4	750	3[†]	Complete[‡]	LTP
blt 101	500	4L	Complete[‡]	None
blt 801	850		Partial	MZGRP
blt 49	850		Partial	None
blt 410	1100§		–	–
blt 1015	1350		Partial	CAB
blt 25	2300§		Partial	None
blt 63	1700	2S,5L[†]	Complete[‡]	EF-1α
blt 411	2200§		Partial	CAT
blt 23	2000		Partial	None
blt 24	2400		Partial	None

[*] Homology with sequences in EMBL or GENEBANK databases; [†] multigene family; [‡] genomic clones isolated; § incomplete CDNA clone.

pA086, low-temperature-induced barley gene with unknown function; LTP, lipid transfer protein; MZGRP, maize glycine-rich protein; CAB, chlorophyll *a/b* binding protein; EF-1α, elongation factor 1α; CAT, catalase.

6 °C/2 °C temperature treatment compared with the 20 °C/15 °C control temperature. There are at least two features of the technique which will affect the conclusions that can be drawn from the range and type of gene isolated. First, the technique will identify genes that are expressed at a high level, but is poor at identifying genes represented by rare transcripts; that is, genes expressed at low levels. The second feature of the technique concerns the identification of different genes within multigene families. It has been proposed that differential regulation of the expression of members of multigene families may represent the molecular basis of phenotypic plasticity (Smith, 1990). The extent to which a differential screen will identify such genes will depend on the extent of homology, at the DNA sequence level, that exists between different members of the multigene family. If considerable homology exists, a labelled sequence of gene A will recognize (hybridize with) the sequence of gene B and it will not be possible using hybridization techniques to distinguish between the expression of gene A versus gene B. The first cDNA library we made contained a large number of partial clones. This meant that our first differential screen identified members of multigene families in which the cloned fragment represented a region of the gene that had diverged from other members of the family.

14.4. Expression studies: multigene families

The list of genes presented in Table 14.1 can provide some evidence to answer the following questions:

1. Does the acclimation response involve the differential expression of protein isoforms that carry out functions maintained at both the control and the cold temperature, as proposed by Smith (1990)?
2. Does the plant respond to a cold treatment by producing novel gene products required for cell function at a low temperature but not necessary at the control temperature?

Evidence for the first type of response comes from *blt*1015, *blt*63 and *blt*411, which have homology with previously cloned chlorophyll *a/b* binding protein, elongation factor 1α and catalase genes, respectively. All these genes are known to be members of multigene families in other plant species, and for *blt*63 (protein synthesis elongation factor 1α) we have shown that a multigene family is distributed between two chromosomes in the barley genome. Catalase catalyses the dismutation of hydrogen peroxide into water and oxygen. In maize, three linked structural genes (*CAT*-1, *CAT*-2 and *CAT*-3) encode three biochemically distinct isozymes. *CAT*-1 and *CAT*-2 are found in glyoxysomes and peroxisomes and *CAT*-3 co-isolates with mitochondria (Roupakias *et al.*, 1980). Catalase is involved in both photorespiration and lipid metabolism. It has recently been shown that the response of these three genes to the fungal toxin cercosporin is very different, with levels of only *CAT*-2 and *CAT*-3 transcript and protein increasing in the presence of the toxin (Williamson and Scandalios, 1992). *blt*1015 has homology to the *CAT*-2 maize gene, and a low-temperature response by a catalase gene is predicted because cold-treatment has been shown to increase the level of hydrogen peroxide in the leaves of winter wheat (Okuda *et al.*, 1991).

Evidence for the second type of response comes from the study of *blt* 101 (Table 14.1). This gene encodes a small transcript, which is undetectable in control (20 °C/15 °C) plants but is a relatively abundant ubiquitous message in cold (6 °C/2 °C) treated material. The gene is of unknown function but exists as a single copy on the long arm of chromosome 4 (Goddard *et al.*, 1992). Because this is a single copy gene which is not detectable in control plants it can be deduced that its expression in cold plants is associated with the production of a protein with a novel low-temperature-specific function. It is perhaps significant that this gene is not up-regulated to any extent in plants subjected to the other stresses we have tested (drought, ABA and nutrient stress).

The gene *blt* 4 (Table 14.1) is also one of a small multigene family, members of which are linked on chromosome 3 (Dunn *et al.*, 1991). We have cloned and studied two members of this multigene family, both of which are up-regulated by a low-temperature treatment. The two genes have considerable homology within the coding region but the transcripts differ significantly in the 3′ non-coding region. These genes have sufficient homology with a group of plant lipid transfer proteins (Hughes *et al.*, 1992) for us to be confident about identifying them as lipid transfer protein genes. Two other barley lipid transfer protein genes have been reported (Jakobsen *et al.*, 1989; Skriver *et al.*, 1992). Both are produced by aleurone cells during germination and are located on chromosome 7. The protein product of one of these aleurone genes (*Ltp*1) has been shown to transfer phosphatidylcholine from liposomes to mitochondria *in vitro* (Breu *et al.*, 1989). There are, therefore, at least two lipid transfer protein gene families in barley, one of which is responsive to low temperature while the other presumably has a function in germination.

There are a number of reports to suggest that an alteration in membrane lipid composition is important in cold acclimation. Steponkus *et al.* (1988), for example, have shown that the cryobehaviour of rye protoplasts can be modified by experimentally manipulating the plasma membrane lipid composition. Thus a lipid transfer function for one of the barley low-temperature-responsive gene families is consistent with the type of modification to cell structure expected during acclimation. In this example we have no evidence for differential low-temperature expression of different members of a single gene family; rather, the plant appears to have different gene families involved in different developmental circumstances. It is interesting that genes from both gene families are induced by abscisic acid (ABA) (Hughes *et al.*, 1992). Thus, although the tissue location and temperature control of gene expression appear to be different, they share an ABA response. Table 14.2 shows that *blt* 4 is also responsive to drought (Dunn *et al.*, 1990). A large number of ABA-responsive plant genes have been reported (e.g. Hong *et al.*, 1992; Marcotte *et al.*, 1992; Sutton *et al.*, 1992) and a proportion of these are also regulated by cold and other stresses (Thomashow, 1990; Hong *et al.*, 1992; Sutton *et al.*, 1992). The relationship between the response of these

TABLE 14.2. *Northern blot analysis of steady-state* blt4 *and* blt14 *mRNA levels in meristems of drought-stressed barley cv. Igri.*

Probe DNA	Integral of autoradiographic peak		
	Moderate temperature	Low temperature	Drought
*blt*4	1057	4368	6281
*blt*14	0	3497	661

10 μg total RNA was applied per well and filters were washed with 0.1 × SSPE (0.18 M NaCl, 0.01 M NaPO$_4$ pH 7.7, 0.001 M EDTA), 0.1% sodium dodecyl sulphate at 50 °C.

genes to the phytohormone and their response to low temperature is not clear, however, because studies of ABA-insensitive mutants in *Arabidopsis* have shown that for two genes the ABA and low-temperature controls are independent (Gilmour and Thomashow, 1991; Nordin *et al.*, 1991).

We have also investigated the low-temperature response of the *blt*14 gene family, which is situated on chromosome 2 (Dunn *et al.*, 1990). *blt*14 encodes a small transcript (500 bases) that has homology with two domains of a low-temperature-induced gene isolated by Cattivelli and Bartels (1990); neither gene has homology with any gene of known function. *blt*14 is not induced by drought, ABA or nutrient stress. In common with the other barley genes we have tested, *blt*14 has homology with sequences present in a number of cereal genomes, including rye (*Secale cereale* L.). In view of the extensive body of work carried out on the North American frost-hardy rye cultivar Puma, we have produced a Puma low-temperature shoot meristem cDNA library and isolated *blt*14 cognate clones from this library (Zhang *et al.*, 1992). Two members of the Puma gene family have been sequenced and their cold-induced expression studied in shoot meristems, mature leaves and roots of both Puma and a less frost-hardy rye cultivar, Rhayader. Total RNA was extracted from different parts of rye plants which had either been grown continuously at 20 °C day/15 °C night or been given a 2 week 6 °C day/2 °C night temperature treatment. RNA was isolated from plants at the three to four leaf stage. Thus control (20 °C/15 °C) plants were 28 days old, whereas low-temperature (6 °C/2 °C) plants were 42 days old. Fig. 14.1 shows the results of probing the same Northern blot of these RNA preparations with the two *blt*14 cognate rye genes, *rlt*1421 and *rlt*1412. Although both genes show higher steady-state levels of mRNA in the cold-treated tissues, Fig. 14.1 shows that *rlt*1412 has highest levels of expression in root tissues whereas *rlt*1421 shows highest levels in mature leaves. This result indicates a tissue- or organ-specific cold expression for different members of the *blt*14 cognate gene family in rye, not a differential response to the same temperature treatment.

It is clear from this discussion that the role of gene families in phenotypic plasticity, as illustrated by the response to an acclimation treatment, is more diverse than that originally proposed. This multiplicity of the molecular response indicates the complexity of the interaction between environmental factor and plant, which may not be evident at the physiological level.

14.5. Expression studies: acclimation response

A number of studies have indicated that the cold-induced shoot meristem barley genes under investigation (Table 14.1) are involved in acclimation for frost hardiness. This evidence comes partly from an association of the level of gene expression and the level of acclimation of the cultivars in

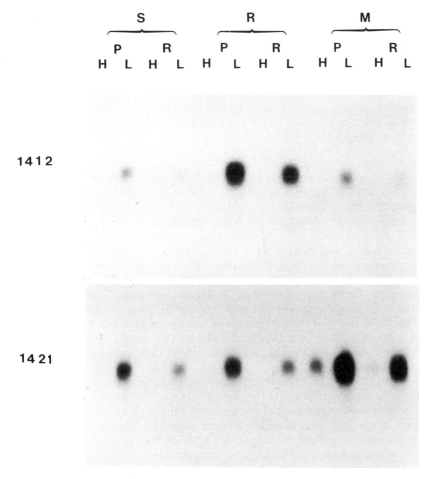

FIGURE 14.1 *Northern blot analysis of the low-temperature-induced expression of* rlt *1412 and* rlt *1421 in rye shoot meristems (S), roots (R') and mature leaves (M). P, cv. Puma; R, cv. Rhayader; H, control temperature (20 °C day/15 °C night); L, low temperature (6 °C day/2 °C night); 10 µg total RNA loaded per well.*

TABLE 14.3. *Comparison of frost hardiness of non-acclimated and acclimated rye plants from the cultivars Puma and Rhayader.*

	Frost hardiness (°C)		Significant difference in hardiness ($P \leq 0.05$)
	Puma	Rhayader	
Control (20 °C/15 °C) non-acclimated plants			
LT_{50} regrowth of crown: 1 week	−2.5	−2.7	n.s.d.
LT_{50} regrowth of crown: 1 month	−2.0	−2.0	n.s.d.
LT_{50} electrolyte leakage: leaf	−7.5	−5.0	s.d.
Low-temperature (6 °C/2 °C) acclimated plants			
LT_{50} regrowth of crown: 1 week	−21.0	−10.2	s.d.
LT_{50} regrowth of crown: 1 month	−10.0	−4.5	s.d.
LT_{50} electrolyte leakage: leaf	−23.0	−11.5	s.d.

n.s.d. no significant difference; s.d., significant difference; LT_{50}, 50% of plants show regrowth or 50% of maximum damage measured by the leakage test. Significant difference between cultivars is based on χ^2 tests of the percentage survival at a range of temperatures. For the leakage test, it is based on the first temperature giving leakage significantly above the control value.

comparative experiments. Fig. 14.1 shows that, for both the rye *blt* 14 cognate genes, the level of expression is higher in all tissues and organs of cv. Puma compared with cv. Rhayader. An analysis of the frost tolerance of plants comparable to those used for RNA extractions is given in Table 14.3. A 2 week treatment at 6 °C/2 °C acclimates both cultivars, as measured by both a regrowth test and electrolyte leakage from mature leaves, such that the LT_{50} values of all tests are lower than those of the non-acclimated plants. Furthermore, all the tests show that, following acclimation, Puma is significantly more hardy than Rhayader. A similar association between the level of cold-induced gene expression and acclimation response has been demonstrated for a member of the *blt* 4 gene family in doubled haploid progeny lines segregating for frost hardiness, which were produced by crossing the winter frost-hardy barley cultivar Vogelsanger Gold with the spring frost-sensitive cultivar Tron.

The other type of experiment that has provided evidence for the role of genes in frost hardiness has used different environmental treatments. The plants from such experiments have been used for both analysis of steady-state mRNA levels and measurements of frost hardiness. Barley plants (cv. Igri) were grown to the third leaf stage at 18 °C day/10 °C night (10 day); some plants continued in the 18 °C/10 °C temperature, while others were transferred to constant 12 °C, constant 9 °C, constant 6 °C or 6 °C day/2 °C night for either 7 or 14 days. Hardiness was measured by a regrowth test and showed that the LT_{50} was inversely related to the treatment temperature. Plants maintained at 18 °C/10 °C showed some hardening over the 14 days of the experiment, whereas plants transferred to constant 12 °C showed some significant dehardening. The steady-state levels of *blt*4 parallel the pattern of acclimation response, with the levels in the 12 °C treatment lower than that at 18 °C/10 °C, and the up-regulation of *blt*4 gradually increasing with the reduction in temperature. This experiment has also been used to analyse the expression of *blt*14 and *blt*101. Both genes are not detectable in the 18 °C/10 °C or constant 12 °C material but show induction at 9 °C, with levels of expression increasing with decreasing temperature. This demonstrates that a difference of just 3 °C can differentiate between the response of *blt*4 and the other two genes.

Differences between *blt*4 and *blt*101 or *blt*14 expression are also revealed by transfer of the 6 °C/2 °C treated plants to a lower temperature; 4 °C day/−4 °C night. This results in a further increase in the level of acclimation of the plants and a further increase in the level of expression of *blt*14 and *blt*101, whereas levels of *blt*4 reach a maximum at 6 °C/2 °C. Interestingly, there is genetic evidence that different gene functions are important for different levels of hardiness. Monosomic substitution lines of chromosome 5A from different winter wheat cultivars into an identical spring wheat background have identified allelic variants of the gene *Fr*1, which is believed to be a major component of frost tolerance in wheat (Sutka and Veisz, 1988). Raising the freezing temperature from −14 °C to −10 °C results in a change such that frost sensitivity is dominant at −14 °C and frost resistance is dominant at −10 °C. It is possible that this effect results from two mechanisms of frost tolerance being important in this material; thus, *Fr*1 is involved in one mechanism conferring frost tolerance at −10 °C but has no role in the other mechanism, resulting in a recessive mode of action at −14 °C.

14.6. Expression studies: mechanism of low-temperature control

The complexity of the molecular response of barley to low temperature is further illustrated by run-on transcription analysis of the genes in Table 14.1. This technique distinguishes between control of gene expression at the level of transcription and controls at a post-transcriptional level. Nuclei were isolated from shoot meristems from either control (20 °C/15 °C) grown plants or from plants given a low temperature (6 °C/2 °C) treatment. Incubation of the nuclei with [^{32}P]UTP produces radiolabelled mRNA in quantities that reflect the rate of transcription of each gene. Each mRNA is identified and its level quantified by hybridizing the RNA from each nuclear preparation to cloned DNA immobilized onto a nitrocellulose membrane. This approach shows that six of the low-temperature-responsive genes (including *blt*4 and *blt*101) are transcriptionally controlled, whereas the others (including *blt*14) have a post-transcriptional control. It should be noted that, although *blt*14 and *blt*101 have very similar patterns of induction (i.e. low-temperature specific; not induced above 9 °C; transcripts at an undetectable level in control (20 °C/15 °C) grown shoot meristems, further induced by transfer from 6 °C/2 °C to 4 °C/−4 °C; time course of induction during acclimation treatment and down-regulation during a deacclimation treatment), one gene is transcriptionally controlled whereas the other appears to be controlled at the level of mRNA stability. Interestingly, although *blt* 14 is transcribed at relatively low levels in both temperature regimens, the transcript accumulates to about 0.2% of the total mRNA in cold shoot meristems. The importance of post-transcriptional mechanisms of regulation of gene expression in acclimation to temperature is reinforced by the finding of Hajela *et al.* (1990) that three of the four cold-induced *Arabidopsis* genes cloned by differential screening are also post-transcriptionally controlled.

14.7. Summary

In summary, the following conclusions can be drawn from this integrated physiological and molecular study of the effects of cold on cereal gene expression. The response of multigene families to low temperature is diverse. The temperature at which gene expression is up-regulated and the temperature inducing the maximum response vary between genes. The response gradually increases with decreasing temperature (it is not an all-or-nothing response). Both transcriptional and post-transcriptional controls operate. Some genes are also responsive to other stresses, whereas others appear to be low-temperature specific.

REFERENCES

Allard, R.W. and Bradshaw, A.D. (1964). The implications of genotype–environment interactions in applied plant breeding. *Crop Sci.* **4**, 503–508.
Bradshaw, A.D. (1965). Evolutionary significance of phenotypic plasticity in plants. *Adv. Genet.* **13**, 115–153.
Breu, V., Guerbette, F., Kader, J.-C., Kannangara, C.G., Svensson, B. and Von Wettstein-Knowles, P. (1989). A 10KD barley basic protein transfers phosphatidylcholine from liposomes to mitochondria. *Carlsberg Res. Commun.* **54**, 81–84.
Cattivelli, L. and Bartels, D. (1990). Molecular cloning and characterisation of cold-regulated genes in barley. *Plant Physiol.* **93**, 1504–1510.
Doll, H., Haahr, V. and Sogaard, B. (1989). Relationship between vernalisation requirement and winter hardiness in doubled haploids of barley. *Euphytica* **42**, 209–213.
Dunn, M.A., Hughes, M.A., Pearce, R.S. and Jack, P.L. (1990). Molecular characterisation of a barley gene induced by cold treatment. *J. Exp. Bot.* **41**, 1405–1413.
Dunn, M.A., Hughes, M.A., Zhang, L., Pearce, R.S., Quigley, A.S. and Jack, P.L. (1991). Nucleotide sequence and molecular analysis of the low temperature induced cereal gene, *blt* 4. *Mol. Gen. Genet.* **229**, 389–394.
Gilmour, S.J. and Thomashow, M.F. (1991). Cold acclimation and cold regulated gene expression in ABA mutants of *Arabidopsis thaliana. Plant Mol. Biol.* **17**, 1233–1240.
Goddard, N.J., Dunn, M.A. and Hughes, M.A. (1992). Molecular analysis of the barley low temperature induced gene *blt* 101. *J. Exp. Bot.* **43**, s57.

Gusta, L.V., Fowler, D.B. and Tyler, N.J. (1982). Factors influencing hardiness and survival in winter wheat. In: *Plant Cold Hardiness and Freezing Stress*, Vol. 2, p. 23, eds. P.H. Li and A. Sakai.

Guy, C.L. (1990). Cold acclimation and freezing stress tolerance: role of protein metabolism. *Annu. Rev. Plant Physiol. Mol. Biol.* **41**, 187–223.

Hajela, R.K., Horvath, D.P., Gilmour, S.J. and Thomashow, M.F. (1990). Molecular cloning and expression of *cor* (cold-regulated) genes in *Arabidopsis thaliana*. *Plant Physiol.* **93**, 1246–1252.

Hoffmann, A.A. and Parsons, P.A. (1991). *Evolutionary Genetics and Environmental Stress*. Oxford: Oxford University Press.

Hong, B., Borg, R. and Ho, D.T.-H. (1992). Developmental and organ specific expression of an ABA- and stress-induced protein in barley. *Plant Mol. Biol.* **18**, 663–674.

Hughes, M.A. and Dunn, M.A. (1990). The effect of temperature on plant growth and development. *Biotechnol. Genet. Eng. Rev.* **8**, 161–188.

Hughes, M.A., Dunn, M.A., Pearce, R.S., White, A.J. and Zhang, L. (1992). An abscisic acid-responsive, low temperature barley gene has homology with a maize phospholipid transfer protein. *Plant Cell Environ.* **15**, 861–865.

Jakobsen, K., Klemsdal, S.S., Aalen, R.B., Bosnes, M., Alexander, D. and Osen, O.-A. (1989). Barley aleurone cell development: molecular cloning of aleurone-specific cDNAs from immature grains. *Plant Mol. Biol.* **12**, 285–293.

Marcotte, N.R., Mark, J.I., Guiltinan, M.J. and Quatrano, R.S. (1992). ABA-regulated gene expression: *cis*-acting sequences and *trans*-acting factors. *Biochem. Soc. Trans.* **20**, 93–97.

Murata, N., Ishizaki-Nishizawa, O., Higashi, S., Hayashi, H., Tasaka, Y. and Nishida, I. (1992). Genetically engineered alteration in the chilling sensitivity of plants. *Nature* **356**, 710–713.

Nordin, K., Heino, P. and Palva, T. (1991). Separate signal pathways regulate the expression of a low-temperature-induced gene in *Arabidopsis thaliana*. *Plant Mol. Biol.* **16**, 1061–1071.

Okuda, T., Matsuda, Y., Yamanaka, A. and Sagisaka, S. (1991). Abrupt increase in the level of hydrogen peroxide in leaves of winter wheat is caused by cold treatment. *Plant Physiol.* **97**, 1265–1267.

Peacock, J.M. (1975). Temperature and leaf growth of *Lolium perenne*. II. The site of temperature perception. *Ann. Appl. Ecol.* **12**, 115–123.

Roberts, D.W.A. (1990). Identification of loci on chromosome 5 of wheat involved in control of cold hardiness, vernalisation, leaf length, rosette growth habit and height of hardened plants. *Genome* **33**, 247–259.

Roupakias, D.G., McMillin, D.E. and Scandalios, J.G. (1980). Chromosomal location of the catalase structural genes in *Zea mays*, using B-A translocation. *Theor. Appl. Genet.* **58**, 211–218.

Schlichting, C.D. (1986). The evolution of phenotypic plasticity in plants. *Annu. Rev. Ecol. Syst.* **17**, 667–93.

Skriver, K., Leah, R., Müller-Uri, F., Olsen, F.-L. and Mundy, J. (1992). Structure and expression of the barley lipid transfer protein gene *Ltp* 1. *Plant Mol. Biol.* **18**, 585–589.

Smith, H. (1990). Signal perception, differential expression within multigene families and the molecular basis of phenotypic plasticity. *Plant Cell Environ.* **13**, 585–594.

Steponkus, P.L., Vemura, M., Balsamo, R.A., Arvinte, T. and Lynch, D.V. (1988). Transformation of the cryobehaviour of rye protoplasts by modification of the plasma membrane lipid composition. *Proc. Natl. Acad. Sci. USA* **85**, 9026–9030.

Sutka, J. and Snape, J.W. (1989). Location of a gene for frost resistance on chromosome 5A of wheat. *Euphytica* **42**, 41–44.

Sutka, J. and Veisz, O. (1988). Reversal of dominance in a gene on chromosome 5A controlling frost resistance in wheat. *Genome* **30**, 313–317.

Sutton, F., Ding, X. and Kenefick, D.G. (1992). Group 3 LEA gene HVA1 regulation by cold acclimation and deacclimation in two barley cultivars with varying freeze resistance. *Plant Physiol.* **99**, 338–340.

Thomas, H. and Stoddart, J.L. (1984). Kinetics of leaf growth in *Lolium temulentum* at optimal and chilling temperatures. *Ann. Bot.* **53**, 341–347.

Thomashow, M.F. (1990). Molecular genetics of cold acclimation in higher plants. *Adv. Genet.* **28**, 99–131.

Whelan, E.D.P. (1990). Differential response to chilling injury of the group 6 chromosomes of cv Chinese spring wheat. *Genome* **34**, 144–150.

Williamson, J.D. and Scandalios, J. (1992). Differential expression of maize catalases and superoxide dismutases to the photoactivated fungal toxin, cercosporin. *Plant J.* **2**, 351–358.

Zhang, L., Dunn, M.A., Pearce, R.S. and Hughes, M.A. (1993). Analysis of organ specificity of a low temperature responsive gene family in rye (*Secale cereale* L.). in press.

CHAPTER 15

Significance of antioxidants in plant adaptation to environmental stress

A. POLLE and H. RENNENBERG

15.1. Introduction

Plants are exposed in their environment to a wide range of different stresses, which can originate from human activities such as air pollutants and acid rain or may have natural causes such as drought, low temperatures, high light intensities and nutritional limitations. The magnitude and extent of individual stress components encountered by plants can show considerable fluctuations. As plants have only limited mechanisms for stress avoidance, they require flexible means for adaptation to changing environmental conditions.

A common feature of different stress factors is their potential to increase the production of reactive oxygen species in plant tissues (Table 15.1). Reactive oxygen species are also generated in plant cells during normal metabolic functions, especially in chloroplasts during photosynthesis (Salin, 1988). To prevent damage, plants possess an antioxidative system composed of low molecular weight antioxidants (glutathione, ascorbate, α-tocopherol) and protective enzymes (superoxide dismutase, ascorbate peroxidase, mono-dehydroascorbate radical reductase, dehydroascorbate reductase, glutathione reductase) (Asada and Takahashi, 1987). This system operates in the cytoplasm and chloroplast in a series of reactions, in which the reduction of activated oxygen ($O_2 \cdot^-$ and H_2O_2) is achieved eventually at the expense of photosynthetically or enzymatically produced reductant NAD(P)H (Fig. 15.1).

For stress compensation, it is important that protection measures are sufficiently available in those cellular compartments in which the increased production of reactive oxygen species takes place. To evaluate the significance of protective systems in different subcellular locations we will discuss the effects of ozone and nutrient deficiency as examples of externally and internally generated oxidative stress.

TABLE 15.1. *Stress factors causing formation of activated oxygen.*

Stressor	Activated oxygen	Reference
(High) light	$O_2 \cdot {}^-$, ${}^1O_2$, H_2O_2	Asada and Takahashi, 1987
Chilling	$O_2 \cdot {}^-$, ${}^1O_2$	Wise and Naylor, 1987
	H_2O_2	MacRae and Ferguson, 1985
Drought	$O_2 \cdot {}^-$	Price and Hendry, 1991
		Quartacci and Navaro-Izzo, 1992
Nutrient deficiency	$O_2 \cdot {}^-$	Cakmak and Marschner, 1988
Wounding	$O_2 \cdot {}^-$	Doke, 1985
Pathogens	H_2O_2	Apostol *et al.*, 1989
Ozone	$O_2 \cdot {}^-$, ${}^1O_2$	Grimes *et al.*, 1983
	$O_2 \cdot {}^-$, ${}^1O_2$, H_2O_2	Heath, 1987
	Organic peroxides	Hewitt *et al.*, 1990
Sulphur dioxide	$O_2 \cdot {}^-$, H_2O_2	Peiser and Yang, 1977
		Tanaka *et al.*, 1982

15.2. Protection from externally and internally generated oxidants

15.2.1. Nutrient deficiency

Leaves of trees and herbaceous plants suffering from nutrient deficiencies such as lack of magnesium, manganese, potassium or nitrogen show typical chlorotic symptoms (Bergmann, 1988). Manganese and magnesium deficiencies impair various metabolic functions because these elements are required as essential cofactors of enzymes and for the functional integrity of photosynthetic membranes (Marschner, 1985). The observation that high light intensities caused the development of chlorosis in bean leaves (Marschner and Cakmak, 1989) and that sun-exposed leaves of trees were more chlorotic than shaded leaves (Osswald *et al.*, 1987; Polle *et al.*, 1992a) implies that chlorosis is not directly caused by lack of magnesium and manganese, but occurs as a secondary event associated with photosynthesis. Investigation of the superoxide dismutase–ascorbate–glutathione pathway revealed increases in antioxidant capacity in chlorotic leaves (Osswald *et al.*, 1987; Cakmak and Marschner, 1992; Polle *et al.*, 1992a). For example, ascorbate concentrations were increased two- to eight-fold in chlorotic compared with green tissue (Cakmak and Marschner, 1992; Polle *et al.*, 1992a), whereas the redox status of ascorbate/dehydroascorbate was not affected (Polle *et al.*, 1992a). Significantly enhanced activities of superoxide dismutase and monodehydroascorbate radical reductase in chlorotic needles indicate an enhanced need for detoxification of $O_2 \cdot {}^-$ and for regeneration of reduced ascorbate, and therefore suggest that manganese- and magnesium-deficient plants suffer from oxidative stress.

Microscopic investigations showed a collapse of sieve elements in the phloem of magnesium-deficient spruce needles (Fink, 1989). This observation

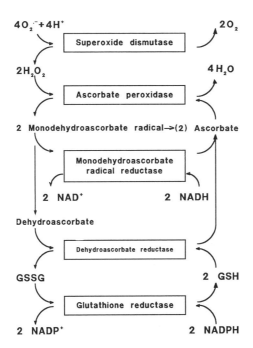

FIGURE 15.1. $O_2 \cdot^-$ and H_2O_2 detoxification pathway. (After the work of Foyer and Halliwell, 1976; Hossain et al., 1984; McCord and Fridovich, 1969.)

implies that the export of photosynthate is hindered. In transgenic plants expressing invertase activity in cell walls, phloem loading was artificially decreased (Stitt et al., 1990). These plants also developed chlorosis, even in the absence of specific stressors. On the metabolic level the Calvin cycle was inhibited in the transgenic plants, and respiration and the content of hexose phosphates were increased (Stitt et al., 1990).

Present evidence suggests the following link between oxidative stress and nutrient deficiency. Magnesium or manganese deficiency causes phloem collapse and thereby prevents effective export of carbohydrates. In consequence, mesophyll cells accumulate photosynthate and slow down the use of ATP and NAD(P)H. When the terminal acceptor for photosynthetic electron transport, $NADP^+$, is missing, electrons are transferred to dioxygen as an alternative electron acceptor, yielding $O_2 \cdot^-$. This in turn requires an increased detoxification capacity, as indicated by increases in superoxide dismutase activity, elevated ascorbate concentrations and enhanced monodehydroascorbate radical reductase activity. Further studies are needed to establish whether the observed chlorosis is a consequence of insufficient increase in antioxidant capacity and subsequent oxidative membrane damage or a result of adaptation to diminished requirement for photosynthetic electron transport.

15.2.2. Ozone

Like most other gaseous air pollutants, ozone enters a plant leaf primarily via the stomata (Lange *et al.*, 1989). Inside the leaves ozone first comes into contact with the aqueous matrix of the apoplastic space in cell walls. Because of its highly oxidizing potential (2.07 V), ozone will either react instantaneously with components present in this compartment or degrade to secondary oxy-products as indicated in Table 15.1.

It has been known for a long time that ascorbate provides protection from ozone injury (Menser, 1964). When ascorbate was found in the apoplastic space of cell walls (Castillo *et al.*, 1987; Castillo and Greppin, 1988), Chameides (1989) calculated that apoplastic ascorbate could provide an efficient ozone protection if present in sufficient concentrations.

Norway spruce is remarkably resistant to ozone-mediated damage (Darrell, 1989). In spruce needles the cell walls probably constitute an important barrier for ozone because they contain ascorbate at concentrations ranging from 1 to 5 mM (Castillo *et al.*, 1987; Polle *et al.*, 1990, 1991a) and also constitute an important compartment by volume (Laisk *et al.*, 1988). Fig. 15.2 presents a model calculation which estimates the oxidation time for apoplastic ascorbate depending on stomatal resistance and external ozone concentration. At an external ozone concentration of 30 ppb, which is a typical mean concentration in the area of Garmisch-Partenkirchen (Calcareous Bavarian Alps; Polle *et al.*, 1992b), the apoplastic reservoir of ascorbate would provide protection for 25 h at a stomatal resistance of $0.0125\,s\,m^{-1}$, which is typical for photosynthetically active needles (cf. Lange *et al.*, 1989). However, it would take only 45 min to oxidize apoplastic ascorbate during peak epi-

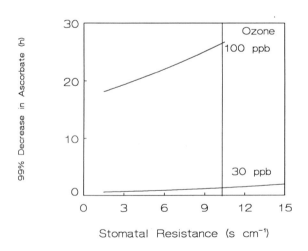

FIGURE 15.2. *Detoxification of ozone by apoplastic ascorbate in dependence on stomatal conductance. Boundary conditions: thickness of cell walls, 1 μm; initial apoplastic ascorbate concentration, 1 μm; rate constant with ozone* $K = 6 \times 10^7$ $m^{-1}s^{-1}$; *no regeneration of oxidized ascorbate. (Courtesy of M. Plöchl.)*

sodes of ozone of 100 ppb (Fig. 15.2) provided that reduced ascorbate is not regenerated and that ascorbate is the only compound reacting with ozone in the apoplastic space.

A regenerating system for reduced ascorbate comparable to that present in chloroplasts or cytosol (cf. Fig. 15.1) was not found in washing fluids obtained from the apoplastic space of spruce needles (Polle *et al.*, 1990). However, if the permeability of the plasma membrane for ascorbate is like that of other small organic molecules such as indole acetic acid (about 10^{-6} to 10^{-8} m s^{-1}; Gimmler *et al.*, 1981), protection mediated by an apoplastic ascorbate pool would increase by one order of magnitude owing to release of symplastic ascorbate.

In addition to ascorbate, the apoplastic compartment contains a variety of other potential reactants for ozone, for example proteins such as peroxidases. Peroxidases can provide protection only if ozone yields H_2O_2 in the aqueous phase. Other ozone products such as OH$\cdot$ (Grimes *et al.*, 1983) or organic peroxides (Hewitt *et al.*, 1990) would presumably destroy peroxidase activity. In Norway spruce needles a reduction in apoplastic peroxidase activity was observed after long-term ozone exposure in open-top chambers (Ogier *et al.*, 1991) and also in needles of mature forest trees exposed to elevated ozone under field conditions (Polle *et al.*, 1991b). These results suggest that peroxidase is more likely to be a target for ozone than a defence against it.

It is still an open question by which mechanism ozone damages leaves. A possibility currently being discussed is that hydrocarbons emitted by plants will react with ozone, yielding highly toxic peroxides, radicals, etc. (Hewitt *et al.*, 1990; Mehlhorn and Wellburn, 1987; see also Chapter 8). If these reactions are prevalent, they would circumvent ascorbate protection.

15.3. Effects of increased antioxidant capacity for stress tolerance

To improve plant protection from oxidative damage it is important to understand both the mode of action of different stress factors and the critical physiological properties that limit ameliorative mechanisms at the subcellular level. The significance of antioxidative compounds and enzymes for removal of oxidants in different compartments is dependent on their concentration in the given compartment and their rate constant for the oxidant. Because of its high rate constant for $O_2\cdot^-$ ($K = 2 \times 10^9$ M^{-1}s^{-1}; Asada and Takahashi, 1987), superoxide dismutase is the most efficient known scavenger for superoxide radicals. The enzyme occurs in different isozymes in different subcellular compartments (Asada and Takahashi, 1987). It was estimated that chloroplasts contain superoxide dismutase, ascorbate and glutathione at concentrations of 10 μM, 10–50 mM and 3 mM, respectively, which will therefore contribute to scavenging of $O_2\cdot^-$ in the order 10:1–5:1 (Rennenberg and Polle, 1992).

When challenged by different stresses, levels of superoxide dismutase isozymes increased specifically in different subcellular compartments (Perl-Treves and Galun, 1991; Tsang *et al.*, 1991). However, positive effects were not always observed when superoxide dismutase levels were increased by

TABLE 15.2. *Effect of an increase in superoxide dismutase or glutathione reductase for stress tolerance.*

Species	Enzyme overproduced	Stress resistance	Reference
Nicotiana tabacum	SOD	=	Pitcher *et al.*, 1991
Nicotiana tabacum	SOD (strongly)	+	Bowler *et al.*, 1991
	SOD (little)	−	Bowler *et al.*, 1991
Lycopersicum esculentum	SOD	=	Tepperman and Dunsmuir, 1990
Nicotiana tabacum	GR	=	Foyer *et al.*, 1991
Nicotiana tabacum	GR	+/=	Aono *et al.*, 1991
Zea mays	GR	+/=	Malan *et al.*, 1990
	SOD	=	Malan *et al.*, 1990
	GR + SOD	+	Malan *et al.*, 1990

GR, glutathione reductase; SOD, superoxide dismutase; +, enhanced stress resistance; =, no effect; −, increased damage.

constructing transgenic organisms (cf. Table 15.2). In bacteria, overproduction of superoxide dismutase activity increased oxidant toxicity (Liochev and Fridovich, 1991; Scott *et al.*, 1987). Deleterious effects also occurred in transgenic plants in which manganese superoxide dismutase was slightly overproduced (Bowler *et al.*, 1991; cf. Table 15.2). It was suggested that over production of superoxide dismutase caused an increased H_2O_2 production which overwhelmed the capacity for H_2O_2 decomposition and, thereby, shifted $O_2 \cdot^-$ and H_2O_2 levels towards a ratio favourable for $OH \cdot$ production. In contrast, strong overproducers of manganese superoxide dismutase were more protected from oxidative stress than controls (Bowler *et al.*, 1991). It was assumed that $O_2 \cdot^-$ concentrations were so far diminished in these plants that $O_2 \cdot^-$ was not available to drive the Haber Weis reaction (Bowler *et al.*, 1991).

In response to naturally occurring stresses such as chilling, drought, stress at high altitude or nutrient deficiencies, plants normally increase several components of the antioxidative system simultaneously (Cakmak and Marschner, 1992; Schöner and Krause, 1990; Jahnke *et al.*, 1991; Smirnoff and Colombe, 1988; Madamachani *et al.*, 1991; Polle *et al.*, 1992a; Polle and Rennenberg, 1992). Conversely cultivars of *Conyza bonariensis*, *Lolium perenne* and *Nicotiana tabacum* in which several components of the antioxidative systems were concurrently increased were more resistant to environmental stresses than cultivars with lower antioxidative capacity (Shaaltiel and Gressel, 1986; Shaaltiel *et al.*, 1988). Crossbreeding of two maize cultivars, which resulted in increased activity of two antioxidant enzymes (superoxide dismutase and glutathione reductase), improved protection from oxidative stress, whereas an increase in glutathione reductase as a single factor had little effect (cf.

Table 15.2). Apparently, a balanced increase in antioxidants is required to obtain increased stress resistance.

15.4. Summary

A common feature of different stress factors is an increased production of reactive oxygen species in plant tissues. Still, their mode of action varies depending on whether oxidants are generated outside (e.g. by oxidizing air pollutants) or inside (e.g. by high radiation, low temperatures or nutrient deficiency) a plant cell. To avoid damage it is necessary that plants contain sufficient concentrations of antioxidants in those subcellular compartments in which reactive oxygen species are generated, and that the regeneration rates for reduced antioxidants cope with the production rate of oxidants. However, current knowledge of the turnover of antioxidants is very limited because static measurements are used for conclusions on dynamic processes.

ACKNOWLEDGEMENTS

We are grateful to M. Plöchl (Institute for Theoretical Chemistry and Physics, University of Frankfurt) for discussion on ozone–ascorbate interaction and for providing the graph shown in Fig. 15.2. The work in the authors' laboratory was financially supported by the Deutsche Forschungsgemeinschaft and by the Bayerisches Staatsministerium für Landesentwicklung und Umweltfragen.

REFERENCES

Aono, M., Kubo, A., Saji, H., Natori, T., Tanaka, K. and Kondo, N. (1991). Resistance to active oxygen toxicity of transgenic *Nicotiana tabacum* that expresses the gene for glutathione reductase from *Escherichia coli*. *Plant Cell Physiol*. **32**, 691–697.

Apostol, I., Heinstein, P. and Low, P. (1989). Rapid stimulation of an oxidative burst during elicitation of cultured plant cells: role in defense and signal transduction. *Plant Physiol*. **90**, 109–116.

Asada, K. and Takahashi, M. (1987). Production and scavenging of active oxygen in photosynthesis. In: *Photoinhibition*, pp. 227–287, eds. D. Kyle, C. Osmond and C. Arntzen. Amsterdam: Elsevier.

Bergmann, W. (1988). Ernährungsstörungen bei Kulturpflanzen. Entstehung und Diagnose. Jena: Gustav Fischer Verlag.

Bowler, C., Slooten, L., Vandenbranden, S., De Rycke, R., Botterman, J., Sybesma, C., Van Montague, M. and Inze, D. (1991). Manganese superoxide dismutase can reduce cellular damage mediated by oxygen radicals in transgenic plants. *EMBO J*. **10**, 1723–1732.

Cakmak, I. and Marschner, H. (1988). Enhanced superoxide radical production in roots of zinc-deficient plants. *J. Exp. Bot*. **39**, 1449–1460.

Cakmak, I. and Marschner, H. (1992). Magnesium deficiency and high light intensity enhance activities of superoxide dismutase, ascorbate peroxidase and glutathione reductase in bean leaves. *Plant Physiol.* **99**, 1222–1227.

Castillo, F. and Greppin, H. (1988). Extracellular ascorbic acid and enzyme activities related to ascorbic acid metabolism in *Sedum album* L. leaves after ozone exposure. *Environ. Exp. Bot.* **28**, 231–238.

Castillo, F., Miller P. and Greppin, H. (1987). Waldsterben: extracellular biochemical markers of photochemical oxidant air pollution damage to Norway spruce. *Experientia* **43**, 111–115.

Chameides, W. (1989). The chemistry of ozone deposition to plant leaves: role of ascorbic acid. *Environ. Sci. Technol.* **23**, 595–600.

Darrell, N. (1989). The effect of air pollutants on physiological processes in plants. *Plant Cell Environ.* **12**, 1–30.

Doke, N. (1985). NADPH-dependent $O_2 \cdot^-$ generation in membrane fractions isolated from wounded tobacco tubers inoculated with *Phytophtera infestans*. *Physiol. Plant Pathol.* **27**, 311–322.

Fink, S. (1989). Pathological anatomy of conifer needles subjected to gaseous air pollutants or mineral deficiencies. *Aquilo Ser. Bot.* **27**, 1–6.

Foyer, C. and Halliwell, B. (1976). The presence of glutathione and glutathione reductase in chloroplasts: a proposed role in ascorbate metabolism. *Planta* **133**, 21–25.

Foyer, C., Lelandais, M., Galap, C. and Kunert, K. (1991). Effects of elevated cytosolic glutathione reductase activity on the cellular glutathione pool and photosynthesis in leaves under normal and stress conditions. *Plant Physiol.* **97**, 863–872.

Gimmler, H., Heilmann, B., Demmig, B. and Hartung, H. (1981). The permeability coefficients of the plasmalemma and the chloroplast envelope of spinach mesophyll cells for phytohormones. *Z. Naturforsch.* **36**, 672–678.

Grimes, H., Perkins, K. and Boss, W. (1983). Ozone degrades into hydroxyl radical under physiological conditions. *Plant Physiol.* **72**, 1016–1020.

Heath, R. (1987). The biochemistry of ozone attack on the plasma membrane of plant cells. *Adv. Phytochem.* **21**, 29–54.

Hewitt, N., Kok, G. and Fall, R. (1990). Hydroperoxide in plants exposed to ozone mediates air pollution damage to alkene emitters. *Nature* **344**, 56–58.

Hossain, M., Nakano, K. and Asada, K. (1984). Monodehydroascorbate reductase in spinach chloroplasts and its participation in the regeneration of ascorbate for scavenging hydrogen peroxide. *Plant Cell Physiol.* **61**, 385–395.

Jahnke, L., Hull, M. and Long, S. (1991). Chilling stress and oxygen metabolizing enzymes in *Zea mays* and *Zea diploperennis*. *Plant Cell Environ.* **14**, 97–104.

Laisk, A., Pfanz, H., Schramm, M. and Heber, U. (1988). Sulfur-dioxide fluxes into different cellular compartments of leaves photosynthesizing in a polluted atmosphere. *Planta* **173**, 230–240.

Lange, O.L., Heber, U., Schulze, E.D. and Ziegler, H. (1989). Atmospheric pollutants and plant metabolism. In: *Forest Decline and Air Pollution*,

Ecological Studies Vol. 77, pp. 238–273, eds. E.D. Schulze, O.L. Lange and R. Oren. Berlin: Springer Verlag.

Liochev, S. and Fridovich, I. (1991). Effects of overproduction of superoxide dismutase on the toxicity of paraquat towards *Escherichia coli*. *J. Biol. Chem.* **266**, 8747–8750.

MacRae, E. and Ferguson, I. (1985). Changes in catalase activity and hydrogen peroxide concentration in plants in response to low temperature. *Physiol. Plant.* **65**, 51–56.

Madamachani, N., Hausladen, A., Alscher, R., Amundson, R. and Fellows, S. (1991). Seasonal changes on antioxidants in red spruce (*Picea rubens*, Sarg.) from three field sites in the northeastern United States. *New Phytol.* **118**, 331–338.

Malan, C., Greyling, M. and Gressel, J. (1990). Correlation between CuZn superoxide dismutase and glutathione reductase and environmental and xenobiotic stress tolerance in maize inbreds. *Plant Sci.* **69**, 157–166.

Marschner, H. (1985). *Mineral Nutrition of Higher Plants*, pp. 195–341. London: Academic Press.

Marschner, H. and Cakmak, I. (1989). High light intensity enhances chlorosis and necrosis in leaves of zinc, potassium and magnesium deficient bean (*Phaseolus vulgaris*) plants. *J. Plant Physiol.* **134**, 308–315.

McCord, J. and Fridovich, I. (1969). Superoxide dismutase: an enzymatic function for erythrocuprein (hemocuprein). *J. Biol. Chem.* **244**, 6049–6055.

Mehlhorn, H. and Wellburn, A. (1987). Stress ethylene formation determines plant sensitivity to ozone. *Nature* **327**, 417–418.

Menser, A. (1964). Response of plants to air pollutants. III. A relation between ascorbic acid levels and ozone susceptibility of light-preconditioned tobacco plants. *Plant Physiol.* **39**, 564–567.

Ogier, G., Greppin, H. and Castillo, F. (1991). Ascorbate and guaiacol peroxidase capacities from apoplastic and cell material extracts of Norway spruce needles after long-term ozone exposure. In: *Molecular, Biochemical and Physiological Aspects of Plant Peroxidases*, pp. 391–400, eds. J. Lobarzewski, H. Greppin, C. Penel and T. Gaspar. Geneva: University Press.

Osswald, W., Senger, H. and Elstner, E. (1987). Ascorbic acid and glutathione contents of spruce needles from different locations of Bavaria. *Z. Naturforsch.* **42**, 879–884.

Peiser, G. and Yang, S. (1977). Chlorophyll destruction by the bisulfite-oxygen system. *Plant Physiol.* **60**, 277–281.

Perl-Treves, R. and Galun, E. (1991). The tomato Cu,Zn superoxide dismutase genes are developmentally regulated and respond to light and stress. *Plant Mol. Biol.* **17**, 745–760.

Pitcher, L., Brennan, E., Hurley, A., Dunsmuir, P., Tepperman, J. and Zilinskas, B. (1991). Overproduction of petunia chloroplastic copper/zinc superoxide dismutase does not confer ozone tolerance in transgenic tobacco. *Plant Physiol.* **97**, 452–455.

Polle, A. and Rennenberg, H. (1992). Field studies on Norway spruce at high altitudes. II. Defence systems against oxidative stress in needles. *New Phytol.* **121**, 635–642.

Polle, A., Chakrabarti, K., Schürmann, W. and Rennenberg, H. (1990). Composition and properties of hydrogen decomposing systems in extracellular and total extracts from needles of Norway spruce (*Picea abies*, L.). *Plant Physiol.* **94**, 312–319.

Polle, A., Chakrabarti, K. and Rennenberg, H. (1991a). Entgiftung von Peroxiden in Fichtennadeln (*Picea abies*, L.) am Schwerpunktstandort Kalkalpen (Wank). *GSF Bericht* 26/91, 151–161.

Polle, A., Chakrabarti, K. and Rennenberg. H. (1991b). Extracellular and intracellular peroxidase activities in needles of Norway spruce (*Picea abies*, L.) under high elevation stress. In: *Molecular, Biochemical and Physiological Aspects of Plant Peroxidases*, pp. 447–453, eds. J. Lobarzewski, H. Greppin, C. Penel and T. Gaspar. Geneva: University Press.

Polle, A., Chakrabarti, K., Chakrabarti, S., Seifert, F., Schramel, P. and Rennenberg, H. (1992a). Antioxidants and manganese deficiency in needles of Norway spruce (*Picea abies* L.) trees. *Plant Physiol.* **99**, 1084–1089.

Polle, A., Mössnang, M., Von Schönborn, A., Sladkovic, R. and Rennenberg, H. (1992b): Field studies on Norway spruce at high altitudes. I. Mineral, pigment and soluble protein content of needles as affected by climate and pollution. *New Phytol.* **121**, 89–99.

Price, A. and Hendry, A. (1991). Iron-catalysed oxygen radical formation and its possible contribution to drought damage in nine native grasses and three cereals. *Plant Cell Environ.* **14**, 477–484.

Quartacci, M. and Navaro-Izzo, F. (1992). Water stress and free radical mediated changes in sunflower seedlings. *J. Plant Physiol.* **139**, 621–625.

Rennenberg, H. and Polle, A. (1992). Metabolic consequences of atmospheric sulfur influx into plants. In: *Air Pollution and Plant Metabolism*, eds. R. Alscher and A. Wellburn. in press.

Salin, M. (1988). Toxic oxygen species and protective systems of the chloroplast. *Physiol. Plant.* **72**, 681–689.

Schöner, S. and Krause H. (1990). Protective systems against active oxygen species in spinach: response to cold acclimation in excess light. *Planta* **180**, 383–389.

Scott, M., Meshnick, S. and Eaton, J. (1987). Superoxide dismutase rich bacteria: paradoxical increase in oxidant toxicity. *J. Biol. Chem.* **262**, 3640–3645.

Shaaltiel, Y. and Gressel, J. (1986). Multienzyme oxygen radical detoxifying system correlated with paraquat resistance in *Conyza bonariensis*. *Pest. Biochem. Physiol.* **26**, 22–28.

Shaaltiel, Y., Glazer, A., Bocion, P. and Gressel, J. (1988). Cross tolerance to herbicidal and environmental oxidants of plant biotypes tolerant to paraquat, sulfur dioxide and ozone. *Pest. Biochem. Physiol.* **31**, 13–23.

Smirnoff, N. and Colombe, S. (1988). Drought influences the activity of enzymes of the chloroplast hydrogen peroxide scavenging system. *J. Exp. Bot.* **39**, 1097–1108.

Stitt, M., Von Schaeven, A. and Willmitzer, L. (1990). 'Sink' regulation of photosynthetic metabolism in transgenic tobacco plants expressing yeast invertase in their cell wall involves a decrease of the Calvin cycle enzymes and an increase of glycolytic enzymes. *Planta* **183**, 40–50.

Tanaka, K., Kondo, N. and Sugahara, K. (1982). Accumulation of hydrogen peroxide in chloroplasts of SO_2-fumigated spinach leaves. *Plant Cell Physiol.* **23**, 999–1007.

Tepperman, J. and Dunsmuir, P. (1990). Transformed plants with elevated levels of chloroplastic SOD are not resistant to superoxide toxicity. *Plant Mol. Biol.* **14**, 501–511.

Tsang, E., Bowler, C., Herouart, D., Van Camp, W., Villaroel, R., Genetello C., Van Montague, M. and Inze, D. (1991). Differential regulation of superoxide dismutases in plants exposed to environmental stress. *Plant Cell* **3**, 783–792.

Wise, R. and Naylor, A. (1987). Chilling enhanced photooxidation: evidence for the role of singlet oxygen and superoxide in the breakdown of pigments and endogenous antioxidants. *Plant Physiol.* **83**, 287–282.

CHAPTER 16

Effects of stress on the genome

K. BACHMANN

16.1. Introduction

Stress from the physical environment plays a major role in evolutionary biology. Survival and reproduction under stress require specific adaptations, which can be treated as design problems to be solved by the organism. The effectiveness of an adaptive response to specific stress conditions can be evaluated in physical terms independently of its contribution to biological fitness. Comparing the two shows how natural selection can achieve results that are hardly distinguishable from conscious design. Adaptation takes place on two levels. One of these is individual phenotypic plasticity; the other selection among genotypes in populations. Individual phenotypic adaptation softens the impact of natural selection among genotypes (Bradshaw, 1965; Schlichting, 1986; Sultan, 1987). Wright has recognized this in his classical paper on evolution in Mendelian populations (1931, p. 147) and has suggested that individual adaptation in itself may be the chief object of natural selection. Adaptive plasticity will reflect the environmental stresses that the species has faced repeatedly in its evolutionary history. Plants can respond adaptively even to severe stress if this stress occurs predictably. The behaviour of deciduous trees in north-temperate regions is an example. Irregular episodes of severe stress, however, are likely to reveal otherwise hidden genetic differences in stress resistance, and they provide some of the best documented examples for selection in nature. Recently, Hoffmann and Parsons (1991; Parsons, 1992) have emphasized the role of severe environmental stress in evolution.

Genetic variation in stress resistance is primarily allelic variation at single gene loci acting individually or as multigenic systems. These genes and their alleles arise by random mutation. They cannot be specifically induced. Therefore, a great deal of biotechnological research is aimed at identifying, mapping, isolating and transferring resistance genes. Besides heritable stress resistance due to existing genes and alleles, there are indications that some stress factors directly affect genomes and induce heritable changes, which may even be adaptive under the stress conditions that caused them.

Recently, there have been several reports that in bacteria, under conditions involving nutrient stress in stationary phase, mutations can occur more often when they are advantageous than when they are neutral (Shapiro, 1984; Cairns et al., 1988; Hall, 1988, 1989, 1990; Drake, 1991). The effect seems to be independent of the nature of the mutations, because it has been found for transposon-mediated mutations (Shapiro, 1984; Hall, 1988) and for point mutations (Hall, 1990). It looks as though an individual adaptation to the stress situation has been incorporated into the genome, and the suggestive name 'directed mutations' has been applied to these cases (Cairns et al., 1988). Several explanations for the effect based on known molecular mechanisms have been proposed (Hall, 1988, 1990; Foster and Cairns, 1992). No evidence is yet available for the existence of such 'Cairnsian mutations' in even the simplest eukaryote (Hall, 1990).

Direct adaptive responses of eukaryote genomes to stress are known. The classical case is the amplification of the gene for dihydrofolate reductase in tissue-cultured murine S180 cells resistant to methotrexate (Schimke et al., 1978). Similar reactions seem to be possible in plant cells. Shah et al. (1986) have found that Petunia cells selected in tissue culture for resistance against the herbicide glyphosate showed a 20-fold amplification of the gene for 5-enolpyruvylshikimate-3-phosphate synthase (EPSP synthase).

Gene amplification is one of various mechanisms thought to contribute to the evolutionary flexibility or 'fluidity' of the eukaryotic genome (Hohn and Dennis, 1985; Walbot and Cullis, 1985). We know that genome size in eukaryotes is not related to genetic complexity and that most of the nuclear DNA does not code for gene products. In this chapter, I am going to examine the claims that genome size, aside from the coding content of the DNA, has phenotypic effects, which play a role in organismic adaption, especially under stress conditions; that there is heritable variation in genome size within species that is visible as phenotypic variation; that dramatic changes in genome size can occur quickly, sometimes within a generation and sometimes in response to environmental stress; and that this may constitute an important factor in the stress resistance of plants.

Although the idea of a genomic response to stress is not new (summaries in McClintock, 1984; Cullis, 1985, 1986, 1991; Grime, 1989; Price, 1991), the evidence for it is still largely anecdotal. I shall illustrate the various component claims with relevant examples from the literature and summarize some of the better documented case histories. I shall conclude that the data justify a more thorough examination of the problem, and I shall outline some possible approaches to the question of adaptive genome size variation.

16.2. The nucleotype

The variation in haploid nuclear genome sizes (C values) among the angiosperms is enormous (summaries in Sparrow et al., 1972; Bennett and Smith,

1976, 1991; Bennett *et al.*, 1982a). The smallest genomes are found in the Brassicaceae, where a haploid genome size of 0.05 pg has been reported for *Cardamine amara* (S.R. Band, cited after Bennett, 1987) and 0.15 pg for *Arabidopsis thaliana* (Arumuganathan and Earle, 1991a), and in the Rosaceae (Dickson *et al.*, 1992). The largest genomes are found in some monocots such as *Fritillaria assyriaca* (Liliaceae) with 127.4 pg. Species within one genus can have appreciably different nuclear DNA contents (Rothfels *et al.*, 1966; Chooi, 1971; Price and Bachmann, 1975; Narayan, 1982, 1983; Sims and Price, 1985; Kenton *et al.*, 1986) up to a nine-fold range in *Crepis* (Jones and Brown, 1976). These differences are mainly due to the amounts of non-coding repetitive DNA (Hutchinson *et al.*, 1980; Raina and Narayan, 1984; Flavell, 1986), which can vary widely while the number and even the chromosomal locations of the coding genes remain more or less constant.

Qualitative similarities or differences among plants seem to be remarkably unaffected by differences in genome size (Hutchinson *et al.*, 1979; Price *et al.*, 1986). However, interspecific variation in DNA amounts is correlated with various quantitative properties of cells, and these may secondarily affect quantitative characters of the whole plant (Bennett, 1973, 1987; Cavalier-Smith, 1985a,b; Bachmann *et al.*, 1985). Positive correlations have been documented between nuclear DNA amounts and the total length or volume of mitotic or meiotic metaphase chromosomes (Rees *et al.*, 1966; Price and Bachmann, 1976; Anderson *et al.*, 1985), volume and protein content of interphase nuclei (Baetcke *et al.*, 1967; Edwards and Endrizzi, 1975; Sunderland and McLeish, 1961; Pegington and Rees, 1970), cell volume and weight (Martin, 1966; Lawrence, 1985) and chloroplast number per guard cell (Butterfass, 1983, 1988). Moreover, genome size is positively correlated with mitotic cycle time (Van't Hof and Sparrow, 1963; Van't Hof, 1965, 1975; Evans *et al.*, 1972; Price and Bachmann, 1976) and the duration of meiosis (Bennett, 1971, 1987). Finally, there is an apparently very close correlation between nuclear genome size and radiation sensitivity (mutations per locus per rad; Sparrow and Miksche, 1961; Baetcke *et al.*, 1967; Abrahamson *et al.*, 1973).

Bennett (1971, 1972) has coined the term 'nucleotype' for 'those conditions of the nuclear DNA which affect the phenotype independently of its encoded informational content' (Bennett, 1987). The nucleotypic effects seem to be primarily consequences of the mass of the DNA and the associated proteins. As Bennett (1987) has illustrated, it is impossible to increase the nuclear DNA amount by a factor of 40 without a visible effect on minimal chromosome and nuclear volume. The term 'minimal' is crucial, because nuclear and cell volume normally vary within an individual, depending mainly on transcriptional activity. This variation can exceed by far the differences in nuclear and cell size between species due to genome size differences. These are revealed when homologous cell types in equivalent functional stages are compared. Nucleotypic effects on cell size can be recognized even in cell types where the volume of the nuclear DNA is only a fraction of the total volume. Effects besides those of differences in bulk of DNA, e.g. proportional increases in the non-chromosomal cellular protein content, must contribute to some nucleotypic correlations. The determination of characters 'under nucleotypic control' is nearly always an interaction between quantitative (nucleotypic) and qualitative (genotypic) influences. Selection pressures can effect genotypic or nucleotypic responses in any combination. Murray *et al.* (1992) have shown highly significant differences in pollen diameter (81.4–94.8µm) among cultivars of sweet pea (*Lathyrus odoratus*) that do not differ in nuclear genome size. Development of the sweet pea has essentially taken place since 1868. During this time there has been genotypic selection on pollen size and an increase in the number of chloroplasts per guard cell, i.e. of characters that show nucleotypic correlations with genome size in interspecific comparisons (Bennett, 1972; Butterfass, 1983, 1988). In all interspecific correlations between genome size and a cellular parameter, some species deviate strongly from expectations based on genome size alone. The closest correlations, of course, are those that are most directly dependent on DNA bulk (mitotic chromosome volume and, apparently, minimum mitotic cycle time).

16.3. Organismic correlates with genome size

The enormous range of nuclear DNA values across all angiosperms is accompanied by nucleotypic effects that far exceed the regulatory potential of the genotype. Differences in cell size and cell division rate among the angiosperms have a profound influence on the anatomy and physiology of the entire plant. This influence is more likely to set upper or lower limits for physiological or anatomical parameters than to determine one-to-one relationships. In a comparison of seed weights and diploid nuclear genome sizes among 24 British legume species, Mowforth (1985, cited after Bennett, 1987) found that species with small or large genomes (approx. 1–20 pg) can produce heavy seeds (weights between 20 and 30 mg), but only species with small genomes (below 4 pg) produce light seeds (less than 1 mg). Other organismic limits are set by the minimum mitotic cycle time and the duration of meiosis. Bennett (1973, 1987; Bennett *et al.*, 1982b) has discussed the relationship between nuclear DNA amounts and the minimum generation time. There is a maximum nuclear DNA content (roughly 10 pg for a 4C nucleus) compatible with a 7 week ephemeral life cycle, and all plants with more than about 85 pg per 4C nucleus are obligate perennials. Species with intermediate genome sizes can determine their annual or perennial habit genotypically. The estimates for the limiting genome sizes differ somewhat among the results by various authors (Lawrence, 1985, for *Senecio*; Grime and Mowforth, 1982, for 161 British angiosperms), but the reality and general validity of the effect is clearly illustrated by the small genome sizes of the plants (and animals) that have become model systems in breeding genetics because of their short life cycles.

The involvement of genome size in the phenology of plants has been examined by Grime and Mowforth (1982) for British angiosperms. They found that large genomes are particularly associated with Mediterranean geophytes and grasses in which growth is confined to the cool conditions of winter and early spring. Such growth is achieved by the rapid inflation of large cells formed during a preceding warm and dry season. Where sufficient moisture allows growth to occur in the summer, temporal separation of mitosis and cell expansion confers no advantage, the long mitotic cycle correlated with large genome size is likely to restrict the rate of development, and smaller genome sizes will be favoured. The differential effect of low temperature on cell division and cell expansion and its consequences for genome size and seasonal timing has been confirmed within a species-rich limestone grassland community in northern England by Grime *et al.* (1985).

The organismic correlates of differences in genome size are likely to have consequences for the ecological and geographic distribution of plants (Bennett, 1987). There are several reports correlating genome sizes and distribution patterns within angiosperm groups (e.g. *Microseris*; Price and Bachmann, 1975) or among geographic regions (tropics versus temperate: Levin and Funderberg, 1979; South Georgia and the Antarctic peninsula: Bennett *et al.*, 1982b; British flora: Grime and Mowforth, 1982). Bennett (1976) examined the distribution of cereal grains, pasture grasses and pulses under cultivation. He showed that the cultivation of species with high $2x$ DNA amounts tends to be localized in temperate latitudes or to seasons and regions at lower latitudes, with temperate conditions. At successively lower latitudes, humans have tended to choose species with increasingly lower DNA amounts for cultivation. A comparison with the sites of domestication suggests that this is a natural cline modified and exaggerated in agriculture.

16.4. Levels of genome size variation

For genome sizes to respond to natural selection, there has to be intraspecific genome size variation of a magnitude to have nucleotypic effects on the phenotype. Very little is known about the mutational mechanisms creating such variation, and it is virtually impossible to predict the genomic responses to artificial selection, especially since simultaneous genotypic responses can never be excluded.

An examination of the small genome of *Arabidopsis thaliana* (Meyerowitz, 1987) shows that selection, probably for rapid reproduction, seems to have opportunistically fixed any non-lethal random loss of DNA. The haploid genome of *A. thaliana* contains about 145 000 kb of DNA (Arumuganathan and Earle, 1991a) with a relatively small proportion of repetitive sequences (Leutwiler *et al.*, 1984). There are only about 570 copies of rDNA with 9.9–10.7 kb per copy. There is only one gene for alcohol dehydrogenase, not two as in maize, with only six introns totalling 576 bp per gene (there are nine introns totalling 1837 bp in maize), and there are only three genes for chlorophyll *a/b* binding proteins. However, not all gene families are smaller in *Arabidopsis* than in other angiosperms. Snustad *et al.* (1992), for instance, report the presence of nine β-tubulin genes.

Unfortunately, we do not know how long it took to accumulate these DNA losses. Williams (1986) has obtained *Brassica* strains with very short generation times by strong artificial selection. These may be entirely due to genotypic responses, but there seem to be no detailed comparisons of the genome structure of the 'rapid-cycling' *Brassica* populations and the strains from which they have been derived. There are indications that, at least in some cases, genome sizes can change quickly.

Intraspecific variability in the diploid DNA content is probably not as rare as are well documented and confirmed examples of it. Accurate and comparable measurements of genome sizes are difficult to obtain. A useful protocol for DNA determination in single nuclei via the microspectrophotometric determination of Feulgen stain is given by Price (1988b). For a long time this method, if handled properly, has been the only accurate one for the determination of genome sizes in plants. The method is subject to various sources of error (Bennett and Smith,1976; Greilhuber, 1988) and is very laborious and slow. Recently, laser flow cytophotometry has been adapted for plant cell nuclei (Galbraith et al., 1983; Michaelson et al., 1991a; Arumuganathan and Earle, 1991b).

Laser flow cytophotometry of sunflower (Helianthus annuus) nuclei has shown that the 2C nuclear DNA amount in second leaves of 13 diploid ($2n = 34$) cultivars and inbred lines varied from 6.01 to 7.95 pg (i.e. by 32%). Mean DNA content varied up to 19% within lines. Mean DNA content varied among leaves from different nodes by up to 27% in the open pollinated variety, Californicus, and up to 48% in the inbred line RHA299 (Michaelson et al., 1991b). Intraspecific DNA amount variation in Helianthus annuus had been demonstrated before with Feulgen densitometry (Cavallini et al., 1986). Although the results from the two laboratories differ in some details, it is noteworthy that both find variations in genome sizes even within individual plants.

Other well documented examples of intraspecific genome size variation are cited by Price (1988a) and a complete list of the reported cases of intraspecific variation in DNA content has been compiled by Cavallini and Natali (1991). A variation in C values from 1.84 to 7.14 pg among populations of Collinsia verna (Scrophulariaceae) seems to be the widest intraspecific range on record (Greenlee et al., 1984). Variations in nuclear DNA amounts during the normal development of individual plants have been repeatedly reported. The relevant literature has recently been reviewed by Bassi (1990). The compilation contains many suggestive results, but there is no pattern indicative of general mechanisms.

The activity of transposable elements, unequal crossing over and saltatory amplifications of DNA are generally considered likely mechanisms for fast changes in genome size, but direct evidence for the molecular events involved in sudden genome size changes in plants is still minimal. I have mentioned gene amplification in tissue culture above (Shah et al., 1986). Massive gene amplification seems to occur in hybrids between Nicotiana tabacum and N. otophora or N. plumbaginifolia, where one or more chromosomes enlarge up to 20 times normal size (Gerstel and Burns, 1966; Moav et al., 1968). Apparently spontaneous genome size changes have been recorded in plants grown from achenes of a single head of a highly inbred sunflower, Helianthus annuus, strain (Cavallini et al., 1989). The differences in DNA amounts observed in this case involve more DNA than is contained in the entire genome of Arabidopsis thaliana.

The involvement of transposable elements in sudden genome size change is very likely, but at present a more detailed discussion would be mostly speculative. We may note the following observations. Transposable elements are activated under circumstances likely to lead to genome restructuring (McClintock, 1984; Peschke *et al.*, 1987; Peschke and Philips, 1991; Brettell and Dennis, 1991). Transposition frequency is subject to all kinds of environmental influences, which seem to act mainly indirectly via signal transmission pathways. As a result, the same environmental signal (e.g. elevated temperature) reduces transposition frequency in some cases and increases it in others (Fincham and Sastry, 1974; Doodeman *et al.*, 1985). Finally, at least in *Drosophila*, transposition has been linked to the generation of genetic variability for quantitative characters (Torkamanzehi *et al.*, 1992).

16.5. Adaptive value of intraspecific variation in grasses

The preceding section has shown that intraspecific variation in DNA amount is anything but rare. There is much less evidence to show that nucleotypic effects of this variation on the phenotype are adaptive and that intraspecific genome size variation is shaped by natural selection (Cavallini and Natali, 1991). I shall summarize the observations on flax genotrophs and on the annual species of *Microseris* in the following sections. Additional examples come mainly from various cultivated and wild grasses.

As high as 38.8% variation in genome size has been reported for cultivated corn (*Zea mays* ssp. *mays*; Laurie and Bennett, 1985; Rayburn *et al.*, 1985). Corn lines from the higher latitudes of North America have significantly lower nuclear DNA amounts than those of lower latitudes (Rayburn *et al.*, 1985). Rayburn and Auger (1990) have determined the nuclear DNA content of 12 south-western US Indian maize populations collected at various altitudes and observed a significant positive correlation between genome size and altitude. Higher DNA amounts at higher elevation have also been found in teosinte, the closest relative of cultivated corn (Laurie and Bennett, 1985) and among species in the genus *Secale* (Bennett, 1976). However, there are several reports of an inverse, rather than a positive, correlation between the number of heterochromatic knobs and altitude (Longley, 1938; Mangelsdorf and Cameron, 1942), whereas the south–north decrease of knob number in the USA (Brown, 1949) parallels the decrease in nuclear DNA amount.

Mowforth and Grime (1989) have documented an 80% variation in nuclear DNA amount (2.9–5.2 pg) at constant 2C chromosome number between families of seed progeny derived from established plants of *Poa annua* taken from a single pasture population in North Wales. The variation was positively associated with variation in cell size and negatively correlated with variation in seedling growth rate, but was not associated with variation in life history.

Highly significant differences of up to 32% in nuclear DNA amount were found by Ceccarelli *et al.* (1992) in meristems of seedlings grown from seeds collected from 35 natural populations of hexaploid *Festuca arundinacea* in Italy. The genome sizes are positively correlated with the mean temperature during the year and with that of the coldest month at the various stations and correlate negatively with their latitudes.

16.6. Environmental induction of genome size changes in flax

Heritable phenotypic changes are induced by various combinations of nitrogen, phosphorus and potassium in the fertilizer in selfing lines of *Nicotiana rustica* (Hill, 1965) and in some, but not all, strains of flax (*Linum usitatissimum;* Durrant, 1962, 1971). The effect has been studied in detail in flax where the phenotypically altered lines, called 'genotrophs', derived from the variety Stormont Cirrus, differ in nuclear DNA content (Evans, 1968; Cullis, 1981). The molecular changes occur during the vegetative growth of the plants under inducing conditions. By the time the plants flower, the changes that will be observed in the next generation are already apparent in the apical cells. All seeds from inbred individuals grown under one set of inducing conditions are similar to each other but different from the parental line. The majority of the variation occurred in the highly repetitive fraction of the genome. Representative members of highly repetitive sequence families have been cloned and have been used to determine the quantitative changes in the various fractions. The variation in any one of these families was independent of that in the others and one of the families did not change (Cullis, 1985). Variation was also detected in the copy number of the genes for the 18S and 25S ribosomal RNA and the genes for the 5S ribosomal RNA (Timmis and Ingle, 1973). There are 117 000 copies of the 5S RNA in diploid nuclei of variety Stormont Cirrus. With a basic repeating unit of 350 basepairs, they constitute about 3% of the total nuclear DNA. These genes are arranged in tandem arrays, which are dispersed throughout the genome at many chromosomal sites. In this, flax differs from other plants in which the 5S genes tend to be clustered at one locus. Subsets of the 5S genes are differentially variable during genotroph induction. One subset, characterized by the absence of a site for the restriction enzyme TaqI in the repeat unit, is preferentially (completely?) lost in genotrophs that reduce the number of 5S genes (Goldsborough *et al.*, 1981). A cloned probe representative of a certain fraction (group 4) of 5S genes identifies a well defined set of restriction fragment length polymorphisms (RFLPs) between Stormont Cirrus and four independently induced genotrophs of the 'small phenotype'. Apparently specific DNA rearrangments are predictably induced under defined conditions together with specific phenotypes (Schneeberger and Cullis, 1991).

16.7. Microseris

The asteraceaen genus *Microseris* of western North America contains two monophyletic groups of diploid ($2n = 18$) plants, the perennials with bigger cells and haploid genome sizes around 3.5 pg and the annuals with smaller cells and about 1.5 pg in the haploid genome (Price and Bachmann, 1975; Wallace and Jansen, 1990). The perennials occur in cooler, more mesic habitats, whereas the annuals are adapted to drier ephemeral habitats (Chambers, 1955). Castro-Jimenez *et al.* (1989) have compared anatomical and physiological responses to drought stress in one of the perennial and one of the annual species. Both species respond to low water availability by maintaining turgor with small cell volumes, higher tissue elasticity and osmotic adjustment. Enhanced tissue elasticity and small cell volumes appear to be inherent characteristics of the annual *M. biqelovii*, and drought-induced responses of the perennial *M. laciniata*.

DNA amounts vary over 20% within the annual species *Microseris bigelovii* (Price *et al.*, 1981a) and *M. douglasii* (Price *et al.*, 1980, 1981b). *M. bigelovii* has a nearly linear distribution along the Pacific coast of North America from Southern California to mid-Oregon, with a disjunct occurrence on Vancouver Island, BC (Chambers, 1955). The lower DNA values were detected in plants at the low-rainfall southern end of the distribution range at Point Sal, and in the plants from Vancouver Island, which grow on thin soil over barren rock outcrops with high rainfall. In the Californian inland species, *M. douglasii*, 24 populations have been sampled. Here, high DNA values are restricted to plants growing in more mesic habitats, generally in well developed soil (Price *et al.*, 1981b). In addition, at some population sites, temporal changes in DNA content over several generations have been found, which correlate with amount of precipitation and the length of time to complete the life-cycle (Price *et al.*, 1986).

High and low DNA content biotypes have been maintained as inbred lines for many generations under greenhouse or growth chamber conditions. They have bred true for the DNA amounts that are present in the nuclei of plants grown directly from field-collected achenes (Price *et al.*, 1981b). To determine whether the natural variation in genome sizes is due to instant induction analogous to the flax genotrophs, Price *et al.* (1986) have grown plants from six biotypes of *M. douglasii* with relatively high DNA amounts under severe water and heat stress. Progeny from self-pollination of the stressed plants and from controls grown under favourable conditions were cultured under normal conditions and their DNA amounts were compared. In each case, offspring of stressed plants had slightly lower Feulgen absorption values, on average 0.8% less than the controls. While this corresponds to about 10 000 kb, it is below the resolution of the Feulgen procedure and not anywhere near the level of intraspecific variation found in nature.

16.8. Inheritance of variable genome sizes in hybrids

Genome size differences within a species or in closely related species can be subjected to genetic analysis.

Hutchinson *et al.* (1979) crossed *Lolium* species differing by 40% in nuclear DNA amounts. The F1 plants had intermediate DNA amounts, and the distribution of DNA amounts in F2 and backcross progeny were compatible with the segregation of differences at many unlinked loci.

Strikingly different results were obtained in crosses between plants with different DNA amounts in flax, annual species of *Microseris*, and *Helianthus*.

Reciprocal crosses between high-DNA and low-DNA genotrophs of the flax variety Stormont Cirrus to a variety with intermediate DNA content, Liral Monarch, gave F1 hybrids with DNA contents similar to the Liral Monarch parent (Durrant, 1981). Price *et al.* (1983) have analysed offspring from several crosses between biotypes of *Microseris douglasii* differing by about 10% in nuclear DNA amounts. In one case, 25 F2 plants obtained from selfing an F1 hybrid all had essentially the same nuclear DNA amount, which was slightly higher than the mid-parent value. In another case, F2 offspring families from two sister F1 hybrids grown from achenes harvested from the same maternal capitulum differed markedly. One F2 consisted of plants with DNA amounts near that of the low-DNA parent; in the other F2 the DNA values trailed from the mid-parent value to that of the low-value parent, with the four plants from one head of the F1 plant having the smallest genome sizes. In an interspecific cross between strains of *M. douglasii* and *M. bigelovii* differing by about 10% in nuclear DNA content, the DNA amounts in 12 F1 plants encompassed nearly the full range between the parental values, and the DNA amounts in F2 families derived from five of these F1 plants corresponded to the F1 parental values. In a similar study, Price *et al.* (1985) tested F3 offspring from an F2 plant segregating for values between the mid-parent value and that of the high-DNA parent and found that the DNA values were apparently stable by the F3 generation.

Crossing plants with different genome sizes can obviously induce readjustments in genome size. These changes seem to take place during vegetative growth of the F1 individual, so that different F1 plants and different meristems of the same F1 plant can have different genome sizes whereas genome sizes are usually stabilized in the F2 offspring. In *Microseris*, the outcome of the interaction of different sized genomes in the F1 hybrid appears to be unpredictable; in the flax crosses, the parent with the non-inducible stable genome determined the F1 genome size. In sunflowers, the inheritance of genome size seems to be under developmental control: plants growing from achenes with a peripheral position in the capitulum have a higher nuclear DNA amount than those grown from central achenes (Cavallini and Natali, 1991). A radial morphogenetic gradient across the composite capitulum is well documented (Bachmann, 1983; Vlot and Bachmann, 1991).

16.9. Summary and outlook

This survey of a literature that spans the last 25 years shows that considerable changes in genome size seem to occur regularly within many species. Such changes may be part of the normal developmental programme in some species; they seem to occur in the vegetative development of F1 hybrids from 'wide crosses', and in some cases they may occur in response to environmental stress. These changes often have quantitative effects on the phenotype whereas, with the exception of the flax genotrophs (Durrant and Nicholas, 1970; Cullis and Kolodynska, 1975), there is no indication of their influence on the inheritance and segregation of qualitative characters. Such effects would be difficult to detect in the wide genetic crosses likely to involve genome size restructuring, and it is not at all impossible that they play a role in deviations from Mendelian segregation ratios, differential lethality and hybrid breakdown.

Although evidence for the frequency of genome size rearrangements and for their possible adaptiveness is steadily accumulating, it is very difficult to make a general assessment of the importance of genome size changes in nature and of ways in which they could be used in plant improvement. Much of the evidence is anecdotal, and there are more differences among the details of the various cases than there are common effects. More details, especially experimental studies of selected cases, and information about the specific molecular mechanisms involved in these cases are urgently needed.

It has been a severe hindrance of such investigations that there are no methods for fast and reliable routine determinations of small DNA amount differences. Feulgen DNA determinations are slow, their error limit is of the order of a few per cent, and they are very prone to considerably larger systematic errors. The literature on genome sizes contains several cases of non-reproducible or contradictory results. It is to be hoped that flow cytometric techniques will increase the speed and reliability of DNA amount comparisons.

In some cases of intraspecific genome size changes a stabilization of the nuclear genome size has been recorded after one or two generations of restructuring. The wide crosses in *Microseris* are an example. Since the original observations were made, molecular techniques have been introduced that provide new analytical approaches. We have recently repeated some of these wide crosses and are now (1992/93) raising several large F2 families. Primarily, these will be used for a thorough analysis of the segregation of qualitative and quantitative phenotypical characters combined with thorough mapping of molecular markers. Special attention will be paid to deviations from expected Mendelian segregation ratios and to sources of hybrid lethality (postzygotic isolating mechanisms). One result of this work will be the availability of well characterized recombinant inbred lines. We expect these lines to have stabilized genome size changes, so that they can be assessed with the necessary precision and compared with those of the parental inbred lines. Virtually unlimited material should then be available for a detailed molecular and cytological investigation of their genome rearrangements on one hand, and for comparative studies of stress resistance on the other.

REFERENCES

Abrahamson, S., Bender, M.A., Conger, A.D. and Wolff, S. (1973). Uniformity of radiation-induced mutation rates among different species. *Nature* **245**, 460–462.

Anderson, L.K., Stack, S.M., Fox, M.H. and Chuanshan, Z. (1985). The relationship between genome size and synaptonemal complex length in higher plants. *Exp. Cell Res.* **156**, 367–377.

Arumuganathan, K. and Earle, E.D. (1991a). Nuclear DNA contents of some important plant species. *Plant Mol. Biol. Reporter* **9**, 208–218.

Arumuganathan, K. and Earle, E.D. (1991b). Estimation of nuclear DNA content of plants using flow cytophotometry. *Plant Mol. Biol. Reporter* **9**, 229–241.

Bachmann, K. (1983). Evolutionary genetics and the genetic control of morphogenesis in flowering plants. *Evol. Biol.* **16**, 157–208.

Bachmann, K., Chambers, K.L. and Price, H.J. (1985). Genome size and natural selection: observations and experiments in plants. In: *The Evolution of Genome Size*, pp. 267–276, ed. T. Cavalier-Smith. Chichester: Wiley.

Baetcke, K.P., Sparrow, A.H., Naumann, C.H. and Schwemmer, S.S. (1967). The relationship of DNA content to nuclear and chromosome volumes and radiosensitivity (LD_{50}). *Proc. Natl. Acad. Sci. USA* **58**, 533–540.

Bassi, P. (1990). Quantitative variations of nuclear DNA during plant development: a critical analysis. *Biol. Rev.* **65**, 185–225.

Bennett, M.D. (1971). The duration of meiosis. *Proc. R. Soc. Lond. B* **178**, 259–275.

Bennett, M.D. (1972). Nuclear DNA content and minimum generation time in herbaceous plants. *Proc. R. Soc. Lond. B.* **181**, 109–135.

Bennett, M.D. (1973). Nuclear characters in plants. *Brookhaven Symposia in Biology* **25**, 344–366.

Bennett, M.D. (1976). DNA amount, latitude, and crop plant distribution. *Env. Exp. Bot.* **16**, 93–108.

Bennett, M.D. (1987). Variation in genomic form in plants and its ecological implications. *New Phytol.* **106** (Suppl.), 177–200.

Bennett, M.D. and Smith, J.B. (1976). Nuclear DNA amounts in angiosperms. *Phil. Trans. R. Soc. Lond. B* **274**, 227–274.

Bennett, M.D. and Smith, J.B. (1991). Nuclear DNA amounts in angiosperms. *Phil. Trans. R. Soc. Lond. B* **334**, 309–345.

Bennett, M.D., Smith, J.B. and Heslop-Harrison, J.S. (1982a). Nuclear DNA amounts in angiosperms. *Proc. R. Soc. Lond. B* **216**, 179–199.

Bennett, M.D., Smith, J.B. and Lewis Smith, R.I. (1982b). DNA amounts of angiosperms from the Antarctic and South Georgia. *Env. Exp. Bot.* **22**, 307–318.

Bradshaw, A.D. (1965). Evolutionary significance of phenotypic plasticity in plants. *Adv. Genet.* **13**, 115–155.

Brettell, I.S. and Dennis, E.S. (1991). Reactivation of a silent *Ac* following tissue culture is associated with heritable alterations in its methylation pattern. *Mol. Gen. Genet.* **229**, 365–372.

Brown, W.L. (1949). Numbers and distribution of chromosome knobs in United States maize. *Genetics* **34**, 524–536.

Butterfass, T. (1983). A nucleotypic control of chloroplast reproduction. *Protoplasma* **118**, 71–74.

Butterfass, T. (1988). Nuclear control of chloroplast division. In: *The Division and Segregation of Organelles*, pp. 21–37, eds. S.A. Boffey and D. Lloyd. Cambridge: Cambridge University Press.

Cairns, J., Overbaugh, J. and Miller, S. (1988). The origin of mutants. *Nature* **335**, 142–145.

Castro-Jimenez, Y., Newton, R.J., Price, H.J. and Halliwell, R.S. (1989). Drought stress responses of *Microseris* species differing in nuclear DNA content. *Am. J. Bot.* **76**, 789–795.

Cavalier-Smith, T. (1985a). Eucaryote gene numbers, non-coding DNA and genome size. In: *The Evolution of Genome Size*, pp. 69–103, ed. T. Cavalier-Smith. Chichester: Wiley.

Cavalier-Smith, T. (1985b). Cell volume and the evolution of eukaryotic genome size. In: *The Evolution of Genome Size*, pp. 105–184, ed. T. Cavalier-Smith. Chichester: Wiley.

Cavallini, A. and Natali, L. (1991). Intraspecific variation of nuclear DNA content in plant species. *Caryologia* **44**, 93–107.

Cavallini, A., Zolfino, C., Cionini, G., Cremonini, R., Natali, L., Sassoli, O. and Cionini, P.G. (1986). DNA changes within *Helianthus annuus* L.: cytophotometric, karyological and biochemical analyses. *Theor. Appl. Genet.* **73**, 20–26.

Cavallini, A., Zolfino, C., Natali, L., Cionini, G. and Cionini, P.G. (1989). Nuclear DNA changes within *Helianthus annuus* L.: origin and control mechanism. *Theor. Appl. Genet.* **77**, 12–16.

Ceccarelli, M., Falistocco, E. and Cionini, P.G. (1992). Variation of genome size and organization within hexaploid *Festuca arundinacea*. *Theor. Appl. Genet.* **83**, 273–278.

Chambers, K.L. (1955). A biosystematic study of the annual species of *Microseris*. *Contrib. Dudley Herb.* **4**, 207–312.

Chooi, W.Y. (1971). Variation in nuclear DNA content in the genus *Vicia*. *Genetics* **68**, 195–211.

Cullis, C.A. (1981). Environmental induction of heritable changes in flax: defined environments inducing changes in rDNA and peroxidase isozyme band pattern. *Heredity* **47**, 87–94.

Cullis, C.A. (1985). Sequence variation and stress. In: *Genetic Flux in Plants*, pp. 157–168, eds. B. Hohn and E.S. Dennis. Vienna: Springer Verlag.

Cullis, C.A. (1986). Unstable genes in plants. In: *Plasticity in Plants*, pp. 77–84, eds. D.H. Jennings and A.J. Trewavas. Cambridge: Society for Experimental Biology.

Cullis, C.A. (1991). Molecular characterization of plant responses to stress. In: *Ecological Genetics and Air Pollution*, pp. 245–263, eds. G.E. Taylor, L.F. Pitelka and M.T. Clegg. New York: Springer Verlag.

Cullis, C.A. and Kolodynska, K. (1975). Variations in the isozymes of flax (*Linum usitatissimum*) genotrophs. *Biochem. Genet.* **13**, 687–697.

Dickson, E.E., Arumuganathan, K., Kresovich, S. and Doyle, J.J. (1992). Nuclear DNA content variation within the Rosaceae. *Am. J. Bot.* **79**, 1081–1086.

Doodeman, M., Bino, R.J., Uytewaal, B. and Bianchi, F. (1985). Genetic analysis of instability in *Petunia hybrida*. 4. The effect of environmental factors on the reversion rate of unstable alleles. *Theor. Appl. Genet.* **69**, 489–495.

Drake, J.W. (1991). Spontaneous mutation. *Annu. Rev. Genet.* **25**, 125–146.

Durrant, A. (1962). The environmental induction of heritable changes in *Linum*. *Heredity* **17**, 27–61.

Durrant, A. (1971). Induction and growth of flax genotrophs. *Heredity* **27**, 277–289.

Durrant, A. (1981). Unstable genotypes. *Phil. Trans. R. Soc. Lond.* B **292**, 467–474.

Durrant, A. and Nicholas, D.B. (1970). An unstable gene in flax. *Heredity* **25**, 513–527.

Edwards, G.A. and Endrizzi, J.E. (1975). Cell size, nuclear size and DNA content relationships in *Gossypium*. *Can. J. Genet. Cytol.* **17**, 181–186.

Evans, G.M. (1968). Nuclear changes in flax. *Heredity* **23**, 25–38.

Evans, G.M., Rees, H., Snell, C.L. and Sun, S. (1972). The relationship between nuclear DNA amount and the duration of the mitotic cycle. *Chromosomes Today* **3**, 24–31.

Fincham, J.R.S. and Sastry, G.R.K. (1974). Controlling elements in maize. *Annu. Rev. Genet.* **8**, 15–50.

Flavell, R.B. (1986). Repetitive DNA and chromosome evolution in plants. *Phil. Trans. R. Soc. Lond. B* **312**, 227–242.

Foster, P.L. and Cairns, J. (1992). Mechanisms of directed mutation. *Genetics* **131**, 783–789.

Galbraith, D.W., Harkins, K.R., Maddox, J.M., Ayres, N.M., Sharma, D. P. and Firoozabady, E. (1983). Rapid flow cytophotometric analysis of the cell cycle in intact plant tissues. *Science* **220**, 1049–1051.

Gerstel, D.U. and Burns, J.A. (1966). Chromosomes of unusual length in hybrids between two species of *Nicotiana*. In: *Chromosomes Today*, Vol. 1, pp. 41–56, eds. C.D. Darlington and K.R. Lewis. New York: Plenum Press.

Goldsborough, P.B., Ellis, T.H.N. and Cullis, C.A. (1981). Organization of the 5S genes in flax. *Nucl. Acids Res.* **9**, 5895–5904.

Greenlee, J.K., Rai, K.S. and Floyd, A.D. (1984). Intraspecific variation in nuclear DNA content in *Collinsia verna* Nutt. (Scrophulariaceae). *Heredity* **52**, 235–242.

Greilhuber, J. (1988). 'Self-tanning': a new and important source of stoichiometric error in cytophotometric determination of nuclear DNA content in plants. *Plant Syst. Evol.* **158**, 87–96.

Grime, J.P. (1989). Whole plant responses to stress in natural and agricultural systems. In: *Plants under Stress: Biochemistry, Physiology and Ecology and their Application to Plant Improvement*, SEB Lancaster Meeting, March 1988, eds. H.G. Jones, T.J. Flowers and M.B. Jones.

Grime, J.P. and Mowforth, M.A. (1982). Variation in genome size: an ecological interpretation. *Nature* **299**, 151–153.

Grime, J.P., Shacklock, J.M.L. and Band, S.R. (1985). Nuclear DNA contents, shoot phenology and species coexistence in a limestone grassland community. *New Phytol.* **100**, 435–445.

Hall, B.G. (1988). Adaptive evolution that requires multiple spontaneous mutations. I. Mutations involving an insertion sequence. *Genetics* **120**, 887–897.

Hall, B.G. (1989). Selection, adaptation, and bacterial operons. *Genome* **31**, 265–271.

Hall, B.G. (1990). Spontaneous point mutations that occur more often when advantageous than when neutral. *Genetics* **126**, 5–16.

Hill, J. (1965). Environmental induction of heritable changes in *Nicotiana rustica*. *Nature* **207**, 732–734.

Hoffmann, A.A. and Parsons, P.A. (1991). *Evolutionary Genetics and Environmental Stress*. New York: Oxford University Press.

Hohn, B. and Dennis, E.S., eds. (1985). *Genetic Flux in Plants*. Vienna: Springer Verlag.

Hutchinson, J., Rees, H. and Seal, A.G. (1979). An assay of the activity of supplementary DNA in *Lolium*. *Heredity* **43**, 411–421.

Hutchinson, J., Narayan, R.K.J. and Rees, H. (1980). Constraints upon the composition of supplementary DNA. *Chromosoma* **78**, 137–145.

Jones, R.N. and Brown, L.M. (1976). Chromosome evolution and DNA variation in *Crepis*. *Heredity* **36**, 91–104.

Kenton, A.Y., Rudall, P.J. and Jounson, A.R. (1986). Genome size variation in *Sisyrinchium* L. (Iridaceae) and its relationship to phenotype and habitat. *Bot. Gaz.* **147**, 342–354.

Laurie, D.A. and Bennett, M.D. (1985). Nuclear DNA content in the genus *Zea* and *Sorghum*: intergeneric, interspecific and intraspecific variation. *Heredity* **55**, 307–313.

Lawrence, M.E. (1985). *Senecio* L. (Asteraceae) in Australia: nuclear DNA amounts. *Aust. J. Bot.* **33**, 221–232.

Leutwiler, L.S., Hough-Evans, B.R. and Meyerowitz, E.M. (1984). The DNA of *Arabidopsis thaliana*. *Mol. Gen. Genet.* **194**, 15–23.

Levin, D.A. and Funderberg, S.W. (1979). Genome size in angiosperms: temperate *versus* tropical species. *Am. Naturalist* **114**, 784–795.

Longley, A.E. (1938). Chromosomes of maize from North American Indians. *J. Agric. Res.* **56**, 177–196.

Mangelsdorf, P.C. and Cameron, J.W. (1942). Western Guatemala: a secondary center of origin of cultivated maize varieties. *Bot. Mus. Leafl. Harv. Univ.* **10**, 217–252.

Martin, P.G. (1966). Variation in the amounts of nucleic acids in the cells of different species of higher plants. *Exp. Cell Res.* **44**, 84–98.

McClintock, B. (1984). The significance of responses of the genome to challenge. *Science* **26**, 792–801.

Meyerowitz, E.M. (1987). *Arabidopsis thaliana*. *Annu. Rev. Genet.* **21**, 93–111.

Michaelson, M.J., Price, H.J., Ellison, J.R. and Johnston, J.S. (1991a). Comparison of plant DNA contents determined by Feulgen microspectrophotometry and laser flow cytometry. *Am. J. Bot.* **78**, 183–188.

Michaelson, M.J., Price, H.J., Johnston, J.S. and Ellison, J.R. (1991b). Variation of nuclear DNA content in *Helianthus annuus* (Asteraceae). *Am. J. Bot.* **78**, 1238–1243.

Moav, J., Moav, R. and Zohary, D. (1968). Spontaneous morphological alterations of chromosomes in *Nicotiana* hybrids. *Genetics* **59**, 57–63.

Mowforth, M.A. and Grime, J.P. (1989). Intra-population variation in nuclear DNA amount, cell size and growth rate in *Poa annua* L. *Functional Ecol.* **3**, 289–295.

Murray, B.G., Hammett, K.R.W. and Standring, L.S. (1992). Genomic constancy during development of *Lathyrus odoratus* cultivars. *Heredity* **68**, 321–327.

Narayan, R.K.J. (1982). Discontinuous DNA variation in the evolution of plant species: the genus *Lathyrus*. *Evolution* **36**, 877–891.

Narayan, R.K.J. (1983). Chromosome changes in the evolution of *Lathyrus* species. In: *Kew Chromosome Conference*, pp. 243–250, eds. P.E. Brandham and M.D. Bennett. Boston, Sydney.

Parsons, P.A. (1992). Evolutionary adaptation and stress. *Evol. Biol.* **26**, 191–223.

Pegington, C. and Rees, H. (1970). Chromosome weights and measures in the Triticinae. *Heredity* **25**, 195–205.

Peschke, V.M. and Philips, R.L. (1991). Activation of the maize transposable element *Suppressor–mutator* (*Spm*) in tissue culture. *Theor. Appl. Genet.* **81** 90–97.

Peschke, V.M., Philips, R.L. and Gengenbach, B.G. (1987). Discovery of transposable element activity among progeny of tissue culture-derived maize plants. *Science* **238**, 804–807.

Price, H.J. (1988a). Nuclear DNA content variation within angiosperm species. *Evol. Trends Plants* **2**, 53–60.

Price, H.J. (1988b). DNA content variation among higher plants. *Ann. Missouri Bot. Garden* **75**, 1248–1257.

Price, H.J. (1991). Genomic stress, genome size, and plant adaptation. In: *Ecological Genetics and Air Pollution*, pp. 277–287, eds. G.E. Taylor, L.F. Pitelka and M.T. Clegg. New York: Springer Verlag.

Price, H.J. and Bachmann, K. (1975). DNA content and evolution in the Microseridinae. *Am. J. Bot.* **62**, 262–267.

Price, H.J. and Bachmann, K. (1976). Mitotic cycle time and DNA content in annual and perennial Microseridinae (Compositae, Cichoriaceae). *Plant Syst. Evol.* **126**, 323–330.

Price, H.J., Bachmann, K., Chambers, K.L. and Riggs, J. (1980). Detection of intraspecific variation in nuclear DNA content in *Microseris douglasii* (Asteraceae). *Bot. Gaz.* **141**, 195–198.

Price, H.J., Chambers, K.L. and Bachmann, K. (1981a). Genome size variation in diploid *Microseris bigelovii* (Asteraceae). *Bot. Gaz.* **142**, 156–159.

Price, H.J., Chambers, K.L. and Bachmann, K. (1981b). Geographic and ecological distribution of genomic DNA content variation in *Microseris douglasii* (Asteraceae). *Bot. Gaz.* **142**, 415–426.

Price, H.J., Chambers, K.L., Bachmann, K. and Riggs, J. (1983). Inheritance of nuclear 2C DNA content variation in intraspecific and interspecific hybrids of *Microseris* (Asteraceae). *Am. J. Bot.* **70**, 1133–1138.

Price, H.J., Chambers, K.L., Bachmann, K. and Riggs, J. (1985). Inheritance of nuclear 2C DNA content in a cross between *Microseris douglasii* and *M. bigelovii. Biol. Zbl.* **104**, 269–276.

Price, H.J., Chambers, K.L., Bachmann, K. and Riggs, J. (1986). Patterns of mean nuclear DNA content in *Microseris douglasii* (Asteraceae) populations. *Bot. Gaz.* **147**, 496–507.

Raina, S. and Narayan, R.K. (1984). Changes in DNA composition in the evolution of *Vicia* species. *Theor. Appl. Genet.* **68**, 187–192.

Rayburn, A.L. and Auger, J.A. (1990). Genome size variation in *Zea mays* ssp. *mays* adapted to different altitudes. *Theor. Appl. Genet.* **79**, 470–474.

Rayburn, A.L., Price, H.J., Smith, J.D. and Gold, J.R. (1985). C-banded heterochromatin and DNA content in *Zea mays. Am. J. Bot.* **72**, 1610–1617.

Rees, H., Cameron, F.M., Hazarika, M.H. and Jones, G.H. (1966). Nuclear variation between diploid angiosperms. *Nature* **211**, 828–830.

Rothfels, K., Sexsmith, E., Heimburger, M. and Krause, M. (1966). Chromosome size and DNA content of species of *Anemone* L. and related genera (Ranunculaceae). *Chromosoma* **20**, 54–74.

Schimke, R.T., Kaufman, R.J., Alt, F.W. and Kellems, R.F. (1978). Gene amplification and drug resistance in cultured murine cells. *Science* **202**, 1051–1055.

Schlichting, C.D. (1986). The evolution of phenotypic plasticity in plants. *Annu. Rev. Ecol. Syst.* **17**, 667–693.

Schneeberger, R.G. and Cullis, C.A. (1991). Specific DNA alterations associated with the environmental induction of heritable changes in flax. *Genetics* **128**, 619–630.

Shah, D.M., Horsch, R.B., Klee, H.J., Kishore, G.M., Winter, J.A., Tumer, N.E., Hironaka, C.M., Sanders, P.R., Gassere, C.S., Aykent, S., Siegel, N.R., Rogers, S.G. and Fraley, R.T. (1986). Engineering herbicide tolerance in transgenic plants. *Science* **233**, 478–481.

Shapiro, I.A. (1984). Observations on the formation of clones containing *araB-lacZ* cistron fusions. *Mol. Gen. Genet.* **194**, 79–90.

Sims, L. and Price, H.J. (1985). Nuclear DNA content variation in *Helianthus* (Asteraceae). *Am. J. Bot.* **72**, 1213–1219.

Snustad, D.P., Haas, N.A., Kopczak, S.D. and Silflow, C.D. (1992). The small genome of *Arabidopsis* contains at least nine expressed β-tubulin genes. *Plant Cell* **4**, 549–556.

Sparrow, A.H. and Miksche, J.P.M. (1961). Correlation of nuclear volume and DNA content with higher plant tolerance to chronic radiation. *Science* **134**, 282–283.

Sparrow, A.H., Price, H.J. and Underbrink, A.G. (1972). A survey of DNA content per cell and per chromosome of prokaryotic and eukaryotic organisms: some evolutionary considerations. *Brookhaven Symp. Biol.* **23**, 451–494.

Sultan, S.E. (1987). Evolutionary implications of phenotypic plasticity in plants. *Evol. Biol.* **21**, 127–178.

Sunderland, N. and McLeish, J. (1961). Nucleic acid content and concentration in root cells of higher plants. *Exp. Cell Res.* **24**, 541–554.

Timmis, J.N. and Ingle, J. (1973). Environmentally induced changes in rRNA gene redundancy. *Nature New Biol.* **244**, 235–236.

Torkamanzehi, A., Moran, C. and Nicholas, F.W. (1992). *P* element transposition contributes substantial new variation for a quantitative trait in *Drosophila melanogaster*. *Genetics* **131**, 73–78.

Van't Hof, J. (1965). Relationship between mitotic cycle duration, S period duration and the average rate of DNA synthesis in the root meristem cells of several plants. *Exp. Cell Res.* **39**, 48–58.

Van't Hof, J. (1975). The duration of chromosomal DNA synthesis of the mitotic cycle, and of meiosis of higher plants. In: *Handbook of Genetics*, pp. 363–377, ed. R.C. King. New York: Plenum Press.

Van't Hof, J. and Sparrow, A.H. (1963). A relationship between DNA content, nuclear volume, and minimum cell cycle time. *Proc. Natl. Acad. Sci. USA* **49**, 897–902.

Vlot, E.C. and Bachmann, K. (1991). Genetic and non-genetic variation in the number of pappus parts in *Microseris douglasii* strain D37 (Asteraceae, Lactuceae). *Ann. Bot.* **68**, 235–241.

Walbot, V. and Cullis, C.A. (1985). Rapid genomic change in higher plants. *Annu. Rev. Plant Physiol.* **36**, 367–396.

Wallace, R.S. and Jansen, R.K. (1990). Systematic implications of chloroplast DNA variation in the genus *Microseris* (Asteraceae, Lactuceae). *Syst. Bot.* **15**, 606–616.

Williams, P.H. (1986). Rapid-cycling populations of *Brassica. Science* **232**, 1385–1389.

Wright, S. (1931). Evolution in mendelian populations. *Genetics* **16**, 97–159.

Interactive stresses

CHAPTER 17

Stress responses in plants infected by pathogenic and mutualistic fungi

P. G. AYRES and H. M. WEST

17.1. Introduction

Autotrophic green plants support the growth of heterotrophic microorganisms in a variety of prolonged and physically intimate associations. The microbe typically exploits the plant's carbon metabolism and, in addition, pathogens may damage the plant in other ways that also limit its growth, phenotypic plasticity and capacity to cope with abiotic stress. However, in some associations, such as mycorrhizas, the plant gains mutual benefit because aspects of microbial activity more than compensate for the loss of carbon.

Among the many and diverse groups of pathogens, and of mutualists, some fungi in each group are obligately biotrophic. These are the powdery mildew, rust and smut fungi, which are pathogens of leaves (and, in the case of smuts, reproductive organs), and the vesicular–arbuscular mycorrhizal (VAM) fungi, which are root-infecting mutualists promoting the uptake of nutrients, especially phosphorus, and probably water. Although the former group is composed of many species which are typically host-specific, whereas the latter comprises a few species of worldwide distribution, which inhabit the soil and are host non-specific, and although the pathogenic and mutualistic fungi are taxonomically unrelated, there are fundamental functional similarities between their members. None causes significant direct physical injury to the host. In both groups, production of polysaccharidase enzymes is limited and toxin production is absent. Moreover, individuals of both groups withdraw similar amounts of carbon from their host, and there is no extensive aggregation of hyphae during growth or reproduction. Attention is focused on these groups.

17.2. Carbon budget

Infection is a biotic stress upon the plant, particularly on its carbon budget, the extent of which depends on plant age and the severity and age of infection. Pathogenic infection damages the budget in three ways.

First, the total amount of carbon available for translocation is reduced because fixation within infected areas is progressively reduced as time after infection increases. Infection is rarely uniform and differences occur between diseases. Even among rusts, the rate of net photosynthesis may be reduced within lesions but be normal between lesions, as in the cases of bluebell infected by *Uromyces muscari* (Scholes and Farrar, 1985) and leek infected by *Puccinia allii* (Roberts and Walters, 1988), or depressed between lesions but enhanced within lesions, as in the case of barley infected by *P. hordei* (Scholes and Farrar, 1986). Some compensation may occur because photosynthesis may be stimulated in young uninfected leaves of infected plants, e.g. barley infected by *Erysiphe graminis* f.sp. *hordei* (Walters and Ayres, 1984), and leaf area ratio may increase, e.g. in groundsel (*Senecio vulgaris*) infected by *P. lagenophorae* (Paul and Ayres, 1986a).

Secondly, carbon is withdrawn by the fungus for its growth and respiration. There are few measurements of fungal biomass *in planta*, probably because it is difficult to distinguish chemically between plant and fungus. However, in some powdery mildews, e.g. *Erysiphe pisi* on pea, mycelium can be peeled from the leaf surface (haustoria remain within the epidermis). In 7 days after infection, the leaf dry weight of pea declined from 1.47 to 1.38 mg cm^{-2}, while there developed 0.23 mg mycelium cm^{-2}. Thus, the mildew accounted for at least 14% of the leaf + fungal mass; at this time, net photosynthesis was inhibited by approximately 25% (Ayres, 1976). A detailed analysis by microscopy (Kneale and Farrar, 1985) indicated that 7 days after infection of barley by *P. hordei* there was 0.26 mg of fungal material per cm^2 leaf area. Assuming a total dry weight of 4 mg cm^{-2} leaf area, there was a fungus biomass: plant ratio of 1:15. Later in infection, after the peak of sporulation, the ratio may have been closer to unity because, from studies of *P. recondita* on wheat over an equivalent period, Mehta and Zadoks (1970) found that spore dry weight may be approximately equal to half the dry weight of the leaf.

Thirdly, extra carbon is consumed because dark respiration in the host increases two- to three-fold, clearly exceeding normal capacity and requiring the synthesis of extra respiratory machinery (Farrar, 1992). Both cytochrome and alternative pathways of electron transport contribute to this increase, which is associated with an increased demand for extra carbon skeletons and energy (from the cytochrome pathway) made by induced defence mechanisms, whether these are successful in stemming infection or not (see below). These include the synthesis of callose and lignin destined for localized thickenings of cell walls, phytoalexins and other pathotoxic compounds, and proteins some of which have a function that is clearly understood (e.g. the hydroxyproline-rich glycoproteins, which also strengthen host cell walls, or the chitinases and b1–3 glucanases that attack fungal cell walls) and others whose function is yet to be discovered (e.g. pathogenesis-related proteins).

A budget for barley seedlings infected by brown rust at the onset of sporulation suggests that approximately 10% of net daily photosynthate of an infected leaf is withdrawn by the fungus and, after other drains are accounted for, the infected plant has less than 60% of the sucrose available to the healthy plant for translocation (Farrar and Lewis, 1987; Farrar, 1992) (Fig. 17.1). As noted above, however, some compensation may occur through enhanced photosynthetic activity in uninfected leaves.

The problem of distinguishing between plant and fungal biomass in infected tissues, and the source of respired carbon dioxide originating from such tissues, applies to VAM as well as to pathogenic infections. On the basis that intra- and extra-radical mycelium may be 10% (Harley and Smith, 1983) and 1% (Bevege *et al.*, 1975) of root weight, respectively, about 10% of the leaf's daily production of photosynthate may be withdrawn by the fungus (Fig. 17.1), a value similar to that given above for pathogens. Estimates for VAM fungi range from 7% (Pang and Paul, 1980) to 20% (Jakobsen and Rosendahl, 1990), and *inter alia* will depend on the intensity and stage of infection. The drain may be most important at the seedling stage, when it may cause a lag in development (Duce, 1987; cited by Allen, 1991).

Kucey and Paul (1982) found that the rate of carbon fixation by mycorrhizal *Vicia faba* was greater than that by non-mycorrhizal plants, an observation subsequently made for other species, at least where rates are calculated per unit leaf weight. The increase may in part be explained by an increase in specific leaf area resulting from VAM infection. In addition, where the rate

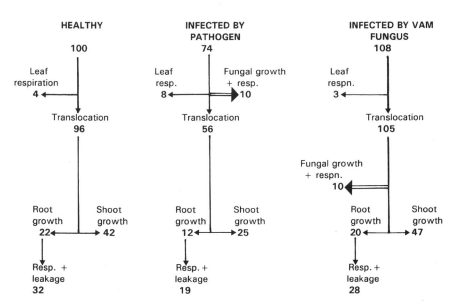

FIGURE 17.1 *Generalized patterns of carbon allocation. A leaf on a healthy plant is compared with a leaf infected by a biotrophic pathogen and a leaf on a plant infected by a vesicular–arbuscular mycorrhizal (VAM) fungus. Comparisons assume 100 units of photosynthate produced in the healthy leaf and are based on Kucey and Paul (1982), Farrar and Lewis (1987) and Whipps (1990).*

of photosynthesis is limited by low concentrations of cytosolic phosphate, infection probably enhances phosphorus uptake, so that photosynthesis becomes potentially sink-limited; the fungus supplies an extra sink, and thus lifts the limit on the photosynthetic rate.

The extent to which additional carbon is consumed by the host is unknown, but is probably less than that in pathogenic infection because defence and repair processes will be less important. A portion of the carbon translocated to roots will be respired by the fungus. Pang and Paul (1980) found that, although mycorrhizal *V. faba* had lower root : shoot ratios than non-mycorrhizal plants, the infected roots respired 12% more of incorporated $^{14}CO_2$ than did controls. Snellgrove *et al.* (1982) calculated that, on the basis of fungal biomass being 10% of root weight in leek and the additional carbon dioxide production in mycorrhizal roots being the result of fungal respiration, the fungal metabolic rate would be 11.0 mg $CO_2 g^{-1} h^{-1}$. In contrast, Silsbury *et al.* (1983) concluded that VAM infection did not alter the carbon economy of 5-week-old swards of subterranean clover; neither the specific rate of dark respiration nor growth efficiency was affected by infection. The greater age of the clover plants may explain the contrast with earlier results, but it is also possible that the ^{14}C pulse-chase method underestimated long-term maintenance respiration and daily fluctuations in overall dark respiration.

Consonant with the suggestion that a critical drain occurs during the early stages of infection, Cooper (cited by Silsbury *et al.*, 1983) found the fraction of ^{14}C-labelled assimilates entering young infected roots ($< 10\%$ infection) was 2.5 times that in controls, but in older roots ($> 20\%$ infection) was similar to that in controls.

17.2.1. Allometry

Although studies of the effect of VAM infection on root : shoot ratios are equally divided between those reporting an increase and those reporting a decrease (Allen, 1991), the situation is very different in pathogenic infections. The reduction in photosynthate available for translocation restricts the growth of all organs but for some, particularly roots, the situation is made relatively worse because of changes in allometry. This is exemplified in groundsel, where infection by *P. lagenophorae* increased partitioning to leaves at the expense of roots (see Section 17.3.3); reproductive organs were buffered from changes in partitioning (Paul, 1992). (Where some smut infections cause massively increased growth of host reproductive organs, e.g. *Ustilago maydis* on maize (Billett and Burnett, 1978), this may be largely at the expense of root growth.) In groundsel, reduction in the growth of root length was only partly ameliorated by an increase in specific root length (length per unit dry weight), and by changes in specific absorption rates (see Section 17.3.2).

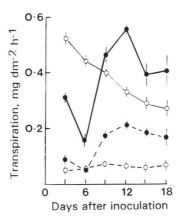

FIGURE 17.2 *Transpiration per unit area of healthy (○) and rust-infected (●) leaves of* Vicia faba *in light (——) and dark (----). Bars show s.e.m. of four replicates. (From Tissera and Ayres, 1986.)*

17.3. Interaction with other stresses

17.3.1. Drought

Diseased plants can explore a smaller volume or depth of soil than healthy plants because of the absolute reduction in root growth that infection causes. This alone is sufficient to make plants prone to the effects of drought stress or (see below) nutrient deficiency. Such susceptibility to drought may increase in the later stages of infection, when the pathogen sporulates, because stomatal closure is inhibited, e.g. in *Poa pratensis* infected by *Ustilago striiformis* (Nus and Hodges, 1986), or the epidermis is torn around the erupting fungus, e.g. in *V. faba* infected by *Uromyces viciae-fabae* (Tissera and Ayres, 1986) leading to increased transpiration, particularly at night, and progressive loss of turgor (Fig. 17.2).

The situation relating to infection by VAM fungi is less clear. Infection enhanced water stress in some studies (e.g. Levy and Syvertsen, 1983), whereas in others it improved drought tolerance (Allen and Allen, 1986). Most importantly, closely related fungi may have different effects on the same host. Thus, *Glomus fasciculatum* improved drought tolerance in wheat, whereas *G. mosseae* reduced it (Allen and Boosalis, 1983).

Most studies show that mycorrhizal infection results in increased stomatal conductance and transpiration rates in well watered plants (Fitter, 1988). There are two popular explanations for this:

1. *Direct flow via hyphae.* Extraradical hyphae may be responsible for the direct uptake of water under drought stress conditions. Hyphae with a diameter of 2–5 μm can penetrate soil pores inaccessible to root hairs

(10–20 μm diameter). Total hyphal length can reach $50\,m\,cm^{-3}$ soil (Allen, 1991). In *Bouteloua gracilis* infected by *G. fasciculatum*, leaf resistance was reduced by 50% and transpiration increased by 100% without there being any change in leaf or root water potentials (Allen, 1982). Mycorrhizal and non-mycorrhizal plants had similar leaf areas and root lengths, although roots of infected hosts had fewer and shorter root hairs. Allen estimated that the rate of transport from extraradical hyphae to the root was $2.8 \times 10^{-5}\,mg\,s^{-1}$ per entry point, this being sufficient to maintain normal water relations. It has been suggested by others, however, that direct water uptake to such a significant level is not hydraulically possible (see Fitter, 1985).

2. *Nutritional effects*. Altered water relations resulting from mycorrhizal infection may be a function of enhanced plant phosphorus status and, thus, a secondary consequence of colonization. Fitter (1988) showed that uninfected non-fertilized *Trifolium pratense* had low stomatal conductances and leaf phosphorus concentrations, whereas plants subjected to high phosphorus additions, and hence those with greatest leaf phosphorus concentrations, exhibited high stomatal conductance. In contrast, Augé *et al.* (1986) observed higher conductance in mycorrhizal rose plants fertilized with low levels of phosphorus than in plants subjected to high phosphorus treatment. They concluded that leaf phosphorus concentrations did not affect conductance but, rather, infection was the important factor. However, as Fitter (1988) observed, the leaf phosphorus concentrations were generally low, despite the level of phosphorus nutrition, and treatments were not sufficiently different from each other (the non-mycorrhizal high-phosphorus treatment produced a leaf phosphorus concentration of $0.82\,mg\,g^{-1}$ (dry wt), compared with $0.61\,mg\,g^{-1}$ for the low-phosphorus treatment) to effect a nutritional response. Endophyte infection levels were lower in the high-phosphorus than in the low-phosphorus treatment group, however, justifying the conclusion by Augé *et al.* that increased conductance was correlated with infection levels.

Complicating any explanation of increased water fluxes in VAM-infected plants, there is the possibility that infection might directly affect the signals that communicate changes in soil water status from root to shoot (see Chapter 11). Thus, it has been suggested that mycorrhizal fungi may be associated with altered levels of abscisic acid (ABA; Allen *et al.*, 1982), although firm evidence is lacking. It is intriguing therefore that, when roses were grown in a split-root system that allowed one side of the root system to be colonized with *Glomus intraradices* while the corresponding side remained uninfected, Augé and Duan (1991) recently found that the stomatal conductance of plants with the mycorrhizal half of the system water stressed (but the uninfected half watered) was reduced to 80% of that of the control plants (both root halves watered) within 11 days of the stress being initiated. In contrast, 17 days were required before stomatal conductance was reduced below that of the controls in plants in which the uninfected root portion was stressed. Most significantly, no treatment-related differences in leaf water potential or relative water content were observed during the period of water stress.

Although infection may have led to localized differences in soil drying and root water relations, it is possible that infection may have interfered directly with ABA concentrations in root or shoot.

Each of the above mechanisms is probably important at some time and under some circumstances, the balance between them reflecting the dynamic nature of the mycorrhizal association. Indeed, further mechanisms may be operative. Thus, Kothari *et al.* (1990) suggested that higher water uptake rates in mycorrhizal plants could be a result of anatomical alterations in infected roots; for example, more differentiated metaxylem vessels and changes in root exodermis formation. Interestingly, it has been suggested that increases in the hydraulic conductance of roots from rusted barley and *V. faba* are associated with a reduction in root cortical thickness (Tissera and Ayres, 1988).

17.3.2. Nutrient deficiencies

Tissues infected by biotrophic pathogens become sites where phosphorus, often nitrogen, and other nutrients accumulate (Paul, 1989). Recycling from shoot to root is decreased for phloem-mobile solutes containing phosphorus and nitrogen but, as with carbon, amounts in plant and fungal tissues at specific sites cannot be separately determined. Different pathogens may sequester different proportions of available elements and it may be for this reason that changes in nutrient concentrations in uninfected tissues are not always consistent, even for similar diseases such as rusts. What does appear to be consistent, however, is that there is an increase in specific absorption rate (SARw), i.e. nutrient absorbed per unit dry weight in roots (but not per unit length). The increase may result from increased shoot demand because SARw is linearly related to the root : shoot ratio in both healthy and infected plants (Paul and Ayres, 1986a).

There is no evidence that pathogenic infection causes or exacerbates nutrient deficiency in tissues when plants are grown in a medium supplying uniformly distributed nutrients and in the absence of competition; in neither case is the absolute reduction in the size of the root system a disadvantage, but in other situations disease can have damaging consequences. Recent studies of *V. faba* infected by rust show that, whereas growth of roots of healthy plants increased in zones where nutrients were concentrated within 1 m deep soil columns, roots of rusted plants were unable to respond similarly (Table 17.1). As noted below, the smaller root system of rusted plants may disadvantage them as they compete in mixtures with healthy plants for nutrients in lower regions of the soil profile.

The benefits of VAM infection have been unequivocally demonstrated in pot cultures where the improved growth of mycorrhizal plants is often clearly associated with enhanced uptake of phosphate relative to non-mycorrhizal controls. This is particularly evident under stress conditions, where the costs of infection are more than outweighed by the benefit of increased nutrient uptake. Phosphate ions are relatively immobile in soil and hence diffusion rates are slow. This, coupled with the high phosphate demand of plants, causes the production of depletion zones around roots. The fine extraradical hyphae of a mycorrhizal fungus forage for nutrients (as for water, see Section 17.3.1)

TABLE 17.1. *Effect of nutrient distribution on the distribution of length (m) of secondary roots in Vicia faba grown in soil columns 1 m deep, 0.1 m radius, and healthy or infected by rust.*

Depth of roots (cm)	Nutrients throughout 0–100cm		Nutrients confined to 33–100cm		Nutrients confined to 60–100 cm	
	Healthy	Rust-infected	Healthy	Rust-infected	Healthy	Rust-infected
0–8	7.59	6.37‡	7.30	6.52‡	6.69	4.74*
8–16	5.49	4.47‡	4.26*	5.29§	5.48	3.93*
16–25	4.97	4.60	4.18	4.71	5.08	3.99*
25–33	5.98	5.71	6.10	5.74	4.46*	4.28
33–41	4.83	5.21	5.34	4.98	4.18	3.54
41–50	2.06	2.91	5.22†	3.91‡	3.46†	3.09
50–58	1.63	2.48§	6.38†	2.09‡	3.39†	1.38‡
58–66	1.19	1.68§	2.33†	0.51‡	2.26†	0.44‡
66–75	0	1.05§	0	0.04	2.51†	0.02‡
75–83	0	0.11	0	0	1.20†	0‡
83–91	0	0	0	0	0	0
91–100	0	0	0	0	0	0

Soil was a 50:50 mix of sterilized field soil : sand, with 3.2 g nitrogen, 1.9 g phosphorus and 5.2 g potassium added per column. Each value is the mean of four replicates. At $P < 0.05$, *significantly less than healthy nutrients throughout; † significantly greater than healthy nutrients throughout; ‡ significantly less than paired healthy; § significantly greater than paired healthy; analysis by ANOVAR.

beyond those depletion zones much more effectively than do root hairs. Cooper and Tinker (1981) calculated that, even at modest rates of transpiration, the phosphorus flux in hyphae of *G. mosseae* infecting white clover was 2×10^{-9} mol cm^{-2} s^{-1}. Phosphorus translocation increased at higher rates of transpiration.

It is important to note that it is often difficult to find any benefits of VAM infection under natural conditions (Fitter, 1986; McGonigle and Fitter, 1988). Although Koide (1991) has proposed that responsiveness (defined as a function of increased phosphorus uptake and the utilization efficiency of the plant) will be least in plants of slow growth rate, or having a low ratio of root to shoot, and may differ within, as well as between species, Fitter (1990) has emphasized the importance of phenology. Thus, infection may be beneficial to plants only when their phosphorus deficit is greatest, for example during flowering. The effectiveness of the fungus may also determine whether a beneficial response is measured after VAM infection. Jakobsen *et al.* (1992) found that of three fungal species examined on subterranean clover, hyphae of *Acaulospora laevis* spread the most rapidly and phosphorus inflow per unit hyphal length was almost three times greater than that observed in *Glomus* sp. and *Scutellospora calospora*.

A factor complicating simple interpretation of the benefits of VAM infection is that phosphorus availability also affects infection levels. Low concentrations may increase infection (Schubert and Hayman, 1986), whereas high phosphorus levels are often inhibitory because they may limit entry point formation and infection density (Amijee *et al.*, 1989). Using a split-root experimental design, Thomson *et al.* (1991) found that phosphorus addition to one half of the root system of *Trifolium subterraneum* decreased VAM infection in both halves. They concluded that deficient and moderately deficient phosphorus treatments limited infection indirectly, through plant factors, by regulating levels of soluble carbohydrates in both portions of the root, thereby controlling a substrate utilized by the fungus. In contrast, their luxurious phosphorus level directly inhibited fungal infection processes.

17.3.3. Winter stresses

Winter at high latitudes presents plants with a cocktail of abiotic stresses. Days are short and temperatures low (restricting photosynthesis), tissues may freeze (rupturing their structure), and soil may be waterlogged or frozen (both damaging normal plant water relations). Disease commonly either reduces plant survival over winter (Frank, 1973; and see below) or checks the rate of growth in the following spring by reducing the area of green leaf that survives winter, e.g. as in barley infected by powdery mildew (McAinsh *et al.*, 1990a) or by brown rust (McAinsh *et al.*, 1990b). The specific ways in which disease interacts with individual stresses have rarely been examined. One proven interaction, however, is through the temperature at which freezing occurs.

Buds of apple infected by powdery mildew (*Podosphaera leucotricha*) suffer freeze injury at higher temperatures than healthy buds (Spotts *et al.*, 1981), e.g. whereas 100% of terminal buds broke after a freezing treatment at $-18\,°C$, only 55% of infected buds broke. In laboratory studies of both groundsel (Paul and Ayres, 1991) and barley (McAinsh *et al.*, 1990b), rust infection, particularly after sporulation, raised the temperature at which leaves suffered freezing injury. In groundsel, healthy leaves supercooled by 3.1–5.8 °C, whereas rusted leaves supercooled by only 1.8–4.9 °C. Supercooling of control leaves was reduced by dusting with aeciospores, suggesting that fungal structures, or bacteria that they carry, may act as ice-nucleating factors.

Another way in which pathogens may interact with winter stresses is through the depletion of the carbohydrate reserves needed for respiration during winter and for supporting production of the first green shoots in spring. Although the causes of mortality were not defined, such depletion may at least in part explain why perennial flax (*Linum marginale*) suffered a high mortality rate after rust infection the previous summer (mortality rate was proportional to the severity of infection), as did *Phlox* spp. infected by *Erysiphe cichoracearum* (Burdon and Jarosz, 1988). Rust (*Uromyces rumicis*) infection of *Rumex crispus* in summer similarly increased host mortality rate (Frank, 1973).

Much more is known about the carbohydrate relations of groundsel, an overwintering annual, than of these perennials. In rust-infected groundsel, non-structural carbohydrates (NSC) accumulate in leaves at the expense of stems and roots. After 10 weeks' growth at normal temperatures, roots of healthy plants contained twice the amount of NSC (ethanol-soluble, starch and fructan fractions) found in roots of infected plants; in the next 3 weeks, amounts in healthy plants doubled, even though flowering was beginning, while amounts in rusted plants changed little (Fig. 17.3).

In groundsel grown under cold-acclimating conditions (12 h day; day/night temperature 8/4 °C), over 60% of NSC in rust-infected plants was present in leaves, compared with less than 50% in healthy plants. During 4 weeks when night temperature was reduced to − 5 °C, the increased loss of leaves induced by rust infection deprived the plant of over half its NSC reserves (Paul and Ayres, unpublished data). Although groundsel suffers significant mortality

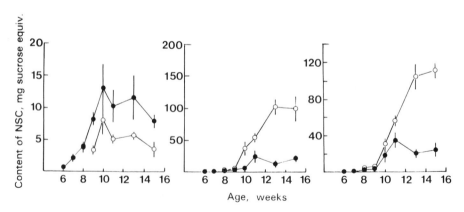

FIGURE 17.3 *Contents of non-structural carbohydrates (NSC) in (a) leaves, (b) roots and (c) stems of* Senecio vulgaris. *Plants were healthy (○) or infected by* Puccinia lagenophorae *(●) when 4 weeks old. Bars show s.e.m. of five replicates. (From Paul, 1992.)*

over winter, it is largely caused by the secondary infection of rust lesions by more damaging necrotrophic fungi, and the main effect of carbohydrate depletion is to cause the growth of rusted plants to lag behind that of healthy plants when warmer temperatures prevail (Paul and Ayres, 1986b).

Little is known of how VAM infection may affect plant responses to overwintering stresses, which is especially surprising in view of the drain that VAM infection places on the plant's carbon budget. Mycorrhiza formation is favoured by higher temperatures, but Chilvers and Daft (1982) reported that, once formed, the symbiosis could function at 5 °C despite a reduction in host plant growth that temperature caused.

Some clue as to the responses of VAMs to winter may be gleaned from their responses to low irradiance. In *Allium cepa*, the potential benefits of enhanced phosphate uptake that plants gained from VAM infection were not realized when plants were grown at low irradiance (Son and Smith, 1988).

17.4. Costs and fitness

The various drains that pathogenic infection puts on a plant's carbon balance were enumerated above. In the balance sheet there is some reciprocity between investment in induced defence, its success or failure, and the extent of pathogen development. Even in a fully susceptible interaction there is some investment in defence. Where defence is wholly successful (and otherwise outside the scope of this chapter), it has a cost, clearly identified in one instance. Grain yield of barley repeatedly inoculated with an avirulent race of powdery mildew was reduced by 7%, and the yield of grain protein was reduced by 30% (Smedegaard-Petersen, 1984).

There is another drain, not yet mentioned, which is into constitutive defence. This may be virtually impossible to quantify, because the component mechanisms often have additional unrelated functions in the plant, e.g. a thick waxy and reflective cuticle, like flavonoids in epidermal cells, may confer resistance to ultraviolet radiation as well as to pathogens. Moreover, apparent cost may depend on the environment in which components are assessed. Thus, Burdon and Muller (1987) found that lines of *Avena fatua* susceptible to *Puccinia coronata* were more fecund than resistant lines when grown in mixtures at 25 °C, because they germinated more rapidly, but the difference in fecundity was reversed at 15 °C, where germination rates were similar in both types. These and other difficulties inherent in any attempt to cost constitutive resistance are discussed by Parker (1992).

Gulmon and Mooney (1986) regarded heavy investment in constitutive defence (against herbivory) as having 'opportunity' costs, i.e. space was surrendered to more rapidly growing competitors early in the season, never to be recovered. Opportunity costs must apply equally to the cumulative effects of the other drains on the carbon budget described previously because their effect is to reduce the size of roots and shoots, leading to reduced competitive fitness and increased susceptibility to abiotic stresses.

Rust infection caused groundsel to lose the competitive advantage that it held over Shepherd's purse (*Capsella bursa-pastoris*) when the two were grown in mixture in high-nutrient conditions. In low-nutrient conditions, however, *C. bursa-pastoris* was the stronger competitor, but that advantage was not affected by infection of groundsel (Paul and Ayres, 1990). The explanation is that in high-nutrient conditions, healthy groundsel develops a low root : shoot ratio and competes strongly for light; this advantage is lost after infection. In low-nutrient conditions, *C. bursa-pastoris* was the stronger because it had 2.5–3.5 more length of root per volume of soil than did groundsel. However, because infection did not decrease the root : shoot ratio, had no deleterious effects on specific root length and increased specific absorption rates of nitrogen, phosphorus and potassium, the competitive disadvantage of groundsel was not exacerbated by infection.

In mixtures of healthy and rusted groundsel of the same susceptible seed line, growing in well fertilized soil, infection caused a loss of competitive fitness only when water was withheld from plants (Paul and Ayres, 1987). Although root growth was not measured, reduced root growth in rusted plants was probably a major factor contributing to loss of fitness.

As the emphasis of this chapter implies, biotrophic pathogens alter their host typically by affecting its budget for carbon and, to a related extent, for other nutrients. In a few smut diseases, there is good evidence that they may directly, or indirectly via host metabolism, disturb the balance of plant growth regulators, but generally evidence is unreliable by modern criteria. Structural damage is rare, but a notable and important exception is the rupture of the plant's epidermis caused by rust fungi immediately before sporulation.

VAM infection has costs to the plant and, as noted above, may reduce the growth of seedlings or of mature plants at low irradiances but, generally, where effects have been demonstrated on plant growth they have been beneficial. Offspring vigour may also be affected by the symbiosis. Shoot weight, root weight, height and leaf area of *Abutilon theophrasti* seedlings were increased as a result of maternal mycorrhizal infection but not affected when non-mycorrhizal plants were supplied with phosphorus (Lewis and Koide, 1990). A similar result was observed in *Avena fatua* but, additionally, offspring reproductive success was enhanced by maternal infection (Koide and Lu, 1992). Individual seed weight of offspring was reduced, whereas seed numbers and phosphorus content were increased. Interestingly, seeds from infected plants possessed a greater allocation of phytate than seeds from non-mycorrhizal plants.

It is a paradox that the benefits of mycorrhizal infection are often difficult to demonstrate in the field because approximately 70% of plant species form symbioses with VAM fungi and no attempt has been made by the hosts to select against the association, suggesting that benefits are derived by the plant. Fitter (1985) proposes various reasons for the lack of mycorrhizal efficacy under field conditions, including the presence of ineffective endophyte strains and grazing of extraradical hyphae by soil invertebrates. Koide et al. (1988) showed that *A. fatua* responded less to infection and phosphorus treatment than did cultivated oat, suggesting that wild oat was inherently adapted to nutrient deficiency, hence limiting the effectiveness of VAM infection. Colonization levels may not necessarily be correlated with mycorrhizal dependency, expressed as improved phosphorus nutrition.

Recent work by West et al. (1993a,b) illustrates another benefit that may derive from VAM infection. Fitness in a natural population of *Vulpia ciliata* ssp. *ambigua* was measured while mycorrhizal infection was manipulated by using fungicides. Benomyl reduced endophyte infection levels but, despite this, fecundity was not affected. Phosphorus inflow rates were not increased by infection and the authors concluded that VAM infection could be parasitic, neutral or mutualistic depending on site conditions. It is possible that, because *V. ciliata* did not benefit from enhanced phosphorus nutrition, mycorrhizal infection protected roots from attack by fungal pathogens. For instance, *Fusarium oxysporum* was abundant at a site from which neutral VAM effects were observed. *F. oxysporum* was also controlled by benomyl treatment, hence any potential benefit of VAM at that site was counterbalanced by the reduction in abundance of *F. oxysporum*.

REFERENCES

Allen, M.F. (1982). Influence of vesicular–arbuscular mycorrhizae on water movement through *Bouteloua gracilis* (H.B.K.) Lag Ex Steud. *New Phytol.* **91**, 191–196.

Allen, M.F. (1991). *The Ecology of Mycorrhizae*. Cambridge: Cambridge University Press.

Allen, E.B. and Allen, M.F. (1986). Water relations of xeric grasses in the field: interactions of mycorrhizas and competition. *New Phytol.* **104**, 559–571.

Allen, M.F. and Boosalis, M.G. (1983). Effects of two species of VA mycorrhizal fungi on drought tolerance of winter wheat. *New Phytol.* **93**, 67–76.

Allen, M.F., Moore, W.K. and Christensen, M. (1982). Phytohormone changes in *Bouteloua gracilis* infected by vesicular–arbuscular mycorrhizae. II. Altered levels of gibberellin-like substances and abscisic acid in the host plant. *Can. J. Bot.* **60**, 468–471.

Amijee, F. Tinker, P.B. and Stribley, D.P. (1989). The development of endomycorrhizal root systems. VII. A detailed study of effects of soil phosphorus on colonization. *New Phytol.* **111**, 435–446.

Augé, R.M. and Duan, X. (1991). Mycorrhizal fungi and nonhydraulic root signals of soil drying. *Plant Physiol.* **97**, 821–824.

Augé, R.M., Schekel, K.A. and Wample, R.L. (1986). Greater leaf conductance of well-watered VA mycorrhizal rose plants is not related to phosphorus nutrition. *New Phytol.* **103**, 107–116.

Ayres, P.G. (1976). Patterns of stomatal behaviour, transpiration, and CO_2 exchange in pea following infection by powdery mildew (*Erysiphe pisi*). *J. Exp. Bot.* **101**, 354–363.

Bevege, D.I., Bowen, G.D. and Skinner, M.F. (1975). Comparative carbohydrate physiology of ecto- and endomycorrhizas. In: *Endomycorrhizas*, pp. 149–174, eds. F.E. Sanders, B. Mosse and P.B. Tinker. London: Academic Press.

Billett, E.E. and Burnett, J.H. (1978). The host–parasite physiology of the maize smut fungus, *Ustilago maydis*. II. Translocation of ^{14}C-labelled assimilates in smutted maize plants. *Physiol. Plant Pathol.* **12**, 103–112.

Burdon, J.J. and Jarosz, A.M. (1988). The ecological genetics of plant–pathogen interactions in natural communities. *Phil. Trans. R. Soc. B* **321**, 349–363.

Burdon, J.J. and Muller, W.J. (1987). Measuring the cost of resistance to *Puccinia coronata* in *Avena fatua. J. Appl. Ecol.* **24**, 191–200.

Chilvers, M.T. and Daft, M.J. (1982). Effects of low temperature on development of the vesicular–arbuscular mycorrhizal association between *Glomus caledonium* and *Allium cepa. Trans. Br. Mycol. Soc.* **79**, 153–157.

Cooper, K.M. and Tinker, P.B. (1981). Translocation and transfer of nutrients in vesicular–arbuscular mycorrhizas. IV. Effect of environmental variables on movement of phosphorus. *New Phytol.* **88**, 327–339.

Duce, D.H. (1987). Effects of vesicular–arbuscular mycorrhizae on *Agropyron smithii* grown under drought stress and their influence on organic phosphorus mineralization. MS thesis, Utah State University, Logan, Utah.

Farrar, J.F. (1992). Beyond photosynthesis: the translocation and respiration of diseased leaves. In: *Pests and Pathogens: Plant Responses to Foliar Attack*, pp. 107–128, ed. P.G. Ayres. Oxford: Bios.

Farrar, J.F. and Lewis, D.H. (1987). Nutrient relations in biotrophic infections. In: *Fungal Infection of Plants*, pp. 92–132, eds. G.F. Pegg and P.G. Ayres. Cambridge: Cambridge University Press.

Fitter, A.H. (1985). Functioning of vesicular–arbuscular mycorrhizas under field conditions. *New Phytol.* **99**, 257–265.

Fitter, A.H. (1986). Effect of benomyl on leaf phosphorus concentration in Alpine grasslands: a test of mycorrhizal benefit. *New Phytol.* **103**, 767–776.

Fitter, A.H. (1988). Water relations of red clover *Trifolium pratense* L. as affected by VA mycorrhizal infection and phosphorus supply before and during drought. *J. Exp. Bot.* **39**, 595–603.

Fitter, A.H. (1990). The role and ecological significance of vesicular–arbuscular mycorrhizas in temperate ecosystems. *Agric. Ecosyst. Environ.* **29**, 137–151.

Frank, P.A. (1973). A biological control agent for *Rumex crispus* L. *Common. Instit. Biol. Control, Misc. Publ.*, **6**, 121–126.

Gulmon, S.L. and Mooney, H.A. (1986). Economics of biotic interactions. In: *On the Economy of Plant Form and Function*, pp. 667–698, ed. T. Givnish. Cambridge: Cambridge University Press.

Harley, J.L. and Smith, S.E. (1983). *Mycorrhizal Symbiosis*. London: Academic Press.

Jakobsen, I. and Rosendahl, L. (1990). Carbon flow into soil and external hyphae from roots of mycorrhizal cucumber plants. *New Phytol.* **115**, 77–83.

Jakobsen, I., Abbott, L.K. and Robson, A.D. (1992). External hyphae of vesicular–arbuscular mycorrhizal fungi associated with *Trifolium subterraneum* L. I. Spread of hyphae and phosphorus inflow into roots. *New Phytol.* **120**, 371–380.

Kneale, J. and Farrar, J.F. (1985). The localization and frequency of haustoria in colonies of brown rust on barley leaves. *New Phytol.* **101**, 495–505.

Koide, R.T. (1991). Nutrient supply, nutrient demand and plant response to mycorrhizal infection. *New Phytol.* **117**, 365–386.

Koide, R.T. and Lu, X. (1992). Mycorrhizal infection of wild oats: maternal effects on offspring growth and reproduction. *Oecologia* **90**, 218–226.

Koide, R., Li, M., Lewis, J. and Irby, C. (1988). Role of mycorrhizal infection in the growth and reproduction of wild *vs.* cultivated oats. *Oecologia* **77**, 537–543.

Kothari, S.K., Marschner, H. and George, E. (1990). Effect of VA mycorrhizal fungi and rhizosphere microorganisms on root and shoot morphology, growth and water relations in maize. *New Phytol.* **116**, 303–311.

Kucey, R.M.N. and Paul, E.A. (1982). Carbon flow, photosynthesis and N_2 fixation in mycorrhizal and nodulated faba beans (*Vicia faba* L.). *Soil Biol. Biochem.* **14**, 407–412.

Levy, Y. and Syvertsen, J.P. (1983). Effect of drought stress and vesicular–arbuscular mycorrhiza on citrus transpiration and hydraulic conductivity of roots. *New Phytol.* **93**, 61–66.

Lewis, J.D. and Koide, R.T. (1990). Phosphorus supply, mycorrhizal infection and plant offspring vigour. *Functional Ecol.* **4**, 695–702.

McAinsh, M.R., Ayres, P.G. and Hetherington, A.M. (1990a). The effects of soil phosphate on injury to winter barley caused by mildew infection (*Erysiphe graminis* f. sp. *hordei*). *Ann. Bot.* **65**, 417–423.

McAinsh, M.R., Ayres, P.G. and Hetherington, A.M. (1990b). The effects of brown rust and phosphate on cold acclimation in winter barley. *Ann. Bot.* **66**, 101–110.

McGonigle, T.P. and Fitter, A.H. (1988). Growth and phosphorus inflows of *Trifolium repens* L. with a range of indigenous vesicular mycorrhizal infection levels under field conditions. *New Phytol.* **108**, 59–65.

Mehta, Y.R and Zadoks, J.C. (1970). Uredospore production and sporulation period of *Puccinia recondita* f. sp. *triticini* on primary leaves of wheat. *Neth. J. Plant Pathol.* **76**, 267–276.

Nus, J. and Hodges, C.F. (1986). Comparative water-use rates and efficiencies, leaf diffusive resistances, and stomatal action of healthy and stripe-smutted Kentucky bluegrass. *Crop Sci.* **26**, 321–324.

Pang, P.C. and Paul, E.A. (1980). Effects of vesicular arbuscular mycorrhiza on ^{14}C and ^{15}N distribution in nodulated fababeans. *J. Soil Sci.* **60**, 241–250.

Parker, M.A. (1992). Constraints on the evolution of resistance to pests and pathogens. In: *Pests and Pathogens: Plant Responses to Foliar Attack*, pp. 181–194, ed. P.G. Ayres. Oxford: Bios.

Paul, N.D. (1989). Effects of fungal pathogens on nitrogen, phosphorus and sulphur relations of individual plants and populations. In: *Nitrogen, Phosphorus and Sulphur Utilization by Fungi*, pp. 155–180, eds. L. Boddy, R. Marchant and D.J. Read. Cambridge: Cambridge University Press.

Paul, N.D. (1992). Partitioning to storage, regrowth and reproduction. In: *Pests and Pathogens: Plant Responses to Foliar Attack*, pp. 129–142, ed. P.G. Ayres. Oxford: Bios.

Paul, N.D. and Ayres, P.G. (1986a). The effects of infection by rust (*Puccinia lagenophorae*) on the growth of groundsel (*Senecio vulgaris*) cultivated under a range of nutrient conditions. *Ann. Bot.* **58**, 321–331.

Paul, N.D. and Ayres, P.G. (1986b). The impact of a pathogen (*Puccinia lagenophorae*) on populations of groundsel (*Senecio vulgaris*) overwintering in the field. I. Mortality, vegetative growth and the development of size hierarchies. *J. Ecol.* **74**, 1069–1084.

Paul, N.D. and Ayres, P.G. (1987). Water stress modifies intraspecific interference between rust (*Puccinia lagenophorae*)-infected and healthy groundsel (*Senecio vulgaris*). *New Phytol.* **106**, 555–566.

Paul, N.D. and Ayres, P.G. (1990). Effects of interactions between nutrient supply and rust infection of *Senecio vulgaris* L. in competition with *Capsella bursa-pastoris* (L.) Medic. *New Phytol.* **114**, 667–674.

Paul, N.D. and Ayres, P.G. (1991). Changes in tissue freezing in *Senecio vulgaris* infected by rust (*Puccinia lagenophorae*). *Ann. Bot.* **68**, 129–133.

Roberts, A.M. and Walters, D.R. (1988). Photosynthesis in discrete regions of leek infected by rust, *Puccinia allii* Rud. *New Phytol.* **110**, 371–376.

Scholes, J.D. and Farrar, J.F. (1985). Photosynthesis and chloroplast functioning within individual pustules of *Uromyces muscari* on bluebell leaves. *Physiol. Plant Pathol.* **27**, 387–400.

Scholes, J.D. and Farrar, J.F. (1986). Increased rates of photosynthesis in localized regions of a barley leaf infected with brown rust. *New Phytol.* **104**, 601–612.

Schubert, A. and Hayman, D.S. (1986). Plant growth responses to vesicular-arbuscular mycorrhizae. XVI. Effectiveness of different endophytes at different levels of soil phosphate. *New Phytol.* **103**, 79–90.

Silsbury, J.H., Smith, S.E. and Oliver, A.J. (1983). A comparison of growth efficiency and specific rate of dark respiration of uninfected and vesicular-arbuscular mycorrhizal plants of *Trifolium subterraneum* L. *New Phytol.* **93**, 555–566.

Smedegaard-Petersen, V. (1984). The role of respiration and energy generation in diseased and disease-resistant plants. In: *Plant Diseases: Infection, Damage and Loss*, pp. 73–85, eds. R.K.S. Wood and G.J. Jellis. Oxford: Blackwell Scientific Publications.

Snellgrove, R.C., Splittstoesser, W.E., Stribley, D.P. and Tinker, P.B. (1982). The distribution of carbon and the demand of the fungal symbiont in leek plants with vesicular–arbuscular mycorrhizas. *New Phytol.* **92**, 75–87.

Son, C.L. and Smith, S.E. (1988). Mycorrhizal growth responses: interactions between photon irradiance and phosphorus nutrition. *New Phytol.* **108**, 305–314.

Spotts, R.A., Covey, R.P. and Chen, P.M. (1981). Effect of low temperature on survival of apple buds infected with the powdery mildew fungus. *Hort. Sci.* **16**, 781–783.

Thomson, B.D., Robson, A.D. and Abbott, L.K. (1991). Soil mediated effects of phosphorus supply on the formation of mycorrhizas by *Scutellospora calospora* (Nicol. & Gerd.) Walker & Sanders on subterraneum clover. *New Phytol.* **118**, 463–469.

Tissera, P. and Ayres, P.G. (1986). Transpiration and the water relations of faba bean (*Vicia faba*) plants infected by rust (*Uromyces viciae-fabae*). *New Phytol.* **102**, 385–395.

Tissera, P. and Ayres, P.G. (1988). Hydraulic conductance and anatomy of roots of *Vicia faba* L. plants infected by *Uromyces viciae-fabae* (Pers.) Shroet.

Physiol. Mol. Plant Pathol. **32**, 192–207.

Walters, D.R. and Ayres, P.G. (1984). Ribulose bisphosphate carboxylase and enzymes of CO_2 assimilation in healthy leaves of barley infected by powdery mildew. *Phytopathol. Z.* **109**, 208–218.

West, H.M., Fitter, A.H. and Watkinson, A.R. (1993a). Response of *Vulpia ciliata* ssp. *ambigua* to removal of mycorrhizal infection and to phosphate application under field conditions. *J. Ecol.* in press.

West, H.M., Fitter, A.H. and Watkinson, A.R. (1993b). The influence of four biocides on the fungal associates of the roots of *Vulpia ciliata* ssp. *ambigua* under natural conditions. *J. Ecol.* in press.

Whipps, J.M. (1990). Carbon economy. In: *The Rhizosphere*, pp. 59–98, ed. J.M. Lynch. Chichester: Wiley.

CHAPTER 18

Plant strategies and environmental stress: a dialectic approach

L. OKSANEN

18.1. Introduction

In spite of the failures of some of its political applications, Hegel's idea of dialectics as the model for emergence and fall of ideas captures essential aspects of the scientific process (see Haila, 1982; Haila and Järvinen, 1982). Ideas are normally conceived in response to existing ideas and tend to be formulated as their polar opposites. Thereafter, a period of intense debate and, hopefully, critical empirical testing ensues. Sometimes, either the original idea (thesis) or the new one (antithesis) emerges as victorious and the other becomes rejected. Often, however, both ideas turn out to contain both realistic elements and less realistic ones and are replaced by a synthesis.

The dialectic view of science is consistent with the hypotheticodeductive one, where science starts from observations, followed by speculations, which are consolidated into a hypothesis (Popper, 1963; Fretwell, 1975). The dialectic view only adds that the reason for some observations leading to speculations is that they are unexpected on the basis of the prevailing theory. The concept of antithesis also implies that in the process of formulating a new hypothesis it is legitimate initially to take a maximalistic standpoint. Otherwise, it would be difficult to break the human tendency of conformism and to explore the strengths and limitations of new ideas. It is thus no coincidence that modern science has grown out of the confrontationist Greek–West European culture, rather than from cultures emphasizing consensus and shared ideas and values.

On the issue of how plant strategies and vegetation processes are related to environmental stresses, we are in the middle of an intense and potentially fruitful conflict between contesting theories. Unfortunately, the interaction between the theories has been hampered by the different way in which they have been presented. The scope of the present chapter is to formalize all three alternatives as variants of the theory of succession and to explore their predictions and implications for land management.

18.2. Development of the three alternatives

According to the traditional view of European continental ecologists, there is no general relation between the intensities of stress and competition. All undisturbed communities are structured by competition and dominated by competitively superior species. However, what is competitive depends on environmental conditions (Cajander, 1906, 1909). For instance, evergreenness is supposed to be competitive in boreal areas and in winter rain–summer drought climates, whereas temperate climates with warm and moist summer favour the deciduous habit (Cajander, 1916). Allocation to a tall woody stem is competitive in humid areas, where light is a limiting resource, whereas small graminoids with a copious root system are supposed to be competitive on arid steppes (Walter, 1964) and in nutrient-poor habitats (Chapin, 1980). This view is currently championed by Tilman (1982, 1988), who has translated the verbal arguments of European classics to rigorous mathematical models and computer algorithms. In essence, these authors emphasize exploitation competition, where the species that tolerates the lowest resource levels is the most competitive one (MacArthur, 1972). This view implies that competitiveness and stress-tolerance are two aspects of the same property.

In 1977, the European tradition was challenged from two different directions. Fretwell (1977) proposed that decreasing primary productivity leads to shortening of terrestrial food chains, which altogether changes the nature of vegetation processes. In productive habitats, herbivores are controlled by carnivores. Consequently, natural grazing pressure is light and plant communities are structured by competition, as proposed by Hairston et al. (1960). In barren areas, however, the third trophic level ceases to have an impact on grazer–plant dynamics. Plants live under intense grazing pressure and the vegetation is structured by apparent competition (grazer-mediated indirect interactions; see Holt, 1977) rather than by classic resource competition. In extremely barren areas, where grazers cannot persist, classic resource competition becomes again the main structuring force for the scanty vegetation, presumably mainly in the form of pre-emptive competition for the few microsites that can support plant growth (Oksanen, 1980).

Although initially connected to the food chain views of Odum (1971), Fretwell's proposition can be derived from mechanistic models of carnivore–herbivore–plant interaction (Oksanen et al., 1981). In relatively barren areas, the grazer density that carnivores would need to break even is more than enough to consume the scanty growth of the vegetation. Thus, carnivores have no way to prevent grazers from having a strong impact on the vegetation (see Oksanen, 1988; Oksanen and Oksanen, 1989).

The other challenge was presented by Grime (1977, 1979), who argued that competitiveness can be treated as an absolute quantity, which is inversely related to stress tolerance. According to Grime, shifts in the identity and characteristics of dominating plants along gradients of increasing stress reflect tolerance limits of more competitive species, which gradually become replaced by more stress-tolerant ones. This part of Grime's theory corresponds to the theory of competitive hierarchies and centrifugal (or radial) community organization (Rosenzweig and Abramsky, 1986; Keddy, 1989). The theory is thus based on the assumption that the main mechanism of competition is interference, where the outcome depends on attributes analogous to physical strength in animals, which acts in the same way in all environments and can be assumed to be inversely related to tolerance of low resource levels.

Grime (1977, 1979) proposed that stress not only excludes competitive species but also reduces the importance of competition as a force structuring plant communities. Along gradients of intensifying environmental stress, interactions between plants thus change from competitive to mutualistic (Callaghan and Emmanuelsson, 1985). Such changes in the intensity of competition are not a necessary condition for the existence of competitive hierarchies. However, the two ideas are so tightly intertwined in Grime's writing that no attempt is made here to dissociate them.

Grime's (1977, 1979) classic works also include a third concept, which is just as radically different from the traditional European view as the two discussed above. He states that stress-tolerant plants have experienced intense natural selection for resistance to herbivory. The proposed cause lies in the low productivity of stressful habitats: plants grow slowly and can thus be severely damaged even by relatively infrequent grazing. This part of Grime's theory is so closely related to the arguments of Fretwell (1977) that it might be warranted to consider them as independent origins of the same theory. Unfortunately, Grime never specifies what part of the stress-tolerant strategy is selected for by stress *per se* and which features represent adaptations to grazing. To clarify the situation, I will split the theory into two components. One, referred to as Grime I, is derived from the assumptions of competitive hierarchies and decreasing intensity of competition along gradients of intensifying stress. The other, Grime II, states that stress is accompanied by intense grazing pressure.

Grime is not alone in straddling different camps. Tilman (1988) discovered that the outcome of simulated interactions between plants along gradients of increasing loss-rate closely mimicked the outcome of interactions along gradients of decreasing nutrient availability. Tilman also accepted the idea that loss-rates may vary along gradients of primary productivity in the way proposed by Fretwell (1977). I will thus distinguish between Tilman I, which is a formalization of the traditional European view, and Tilman II, which is a variant of Fretwell's (1977) theory.

18.3. Three theories in relation to the theory of succession

To formulate all three theories in the same language, let us look at what all of them agree will happen in a productive habitat after a devastating disturbance (e.g. crown fire, landslide). When fertile land is laid bare, there will be a period of transient dynamics, where quick invaders are gradually replaced by more competitive plants. Depending on their somewhat different operational definition of the word 'competitive', Grime and Tilman would describe the events in a somewhat different way (see Grace, 1990), but all plant ecologists are likely to agree that biomass expansion during this process can be described by following equation:

$$dB/dt = r_p Bg\,(B, K_p, B_{cp}) \tag{18.1}$$

where B is the plant biomass, r_p is its maximum rate of expansion per biomass unit in a productive habitat, K_p is the maximum amount of biomass that can be supported in the productive habitat, and B_{cp} is the critical level at which plant–plant interactions change from mutualistic to competitive (i.e. $dg/dB > 0$ when $B < B_{cp}$; $dg/dB < 0$ when $B > B_{cp}$).

The process is often simulated by the logistic equation, where the parameter B_{cp} is zero and where $g(B, K_p) = 1 - B/K$. Thus, $dg/dB < 0$ holds for all B. The assumption $B_{cp} > 0$ implies that, when the density of plants is sufficiently low, plants do not compete but help each other (e.g. by giving protection against desiccation and snow abrasion). All three theories seem to imply that, in productive habitats, B_{cp} is very small (possibly zero) and the time required for plant biomass to reach B_{cp} after disturbance (T_{cp}) is much shorter than the mean time interval between successive events of major disturbance (T_{dp}). With these assumptions, equation 18.1 generates a growth curve, which is similar to the logistic one (Fig. 18.1). For most of the time, the plant community will be structured by intense competitive interactions.

Also, in stressful habitats, the vegetation gradually recovers from devastating disturbance. Moreover, there is some maximum biomass, which the habitat can support (K_s), some intrinsic rate of biomass expansion (r_s), some mean time interval between successive disturbance events (T_{ds}) and some threshold biomass value at which competitive interactions start to prevail (B_{cs}).

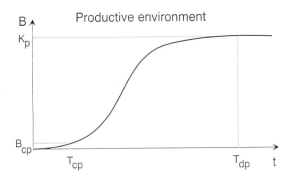

FIGURE 18.1 *Development of plant biomass (B) as a function of time (t) elapsed after devastating disturbance in a productive habitat, according to Equation 18.1, assuming $0 < B_{cp} \ll K_p$. The mean time interval between successive events of devastating disturbance is denoted by T_{dp}. T_{cp} refers to the time interval between disturbance and the build-up of plant biomass to the critical level (B_{cp}), where competitive interactions start to prevail over mutualistic and commensalistic ones.*

Development of plant biomass in stressful habitats will thus proceed in accordance to the equation

$$dB/dt = r_s Bg(B, K_s, B_{cs})$$ (18.2)

which is structurally similar to Equation 18.1 but with different parameter values. Following Grime's definition of stress as any factor, which limits the primary productivity of the habitat, I assume that $r_s < r_p$ and $K_s < K_p$. This will not influence the shape of the curve for biomass accumulation, provided that the following equations hold:

$$r_s T_{ds}/K_s = r_p T_{dp}/K_p$$ (18.3)

and

$$B_{cs}/K_s = B_{cp}/K_p$$ (18.4)

Equations 18.3 and 18.4 thus create a situation corresponding to Tilman I, where the intensity of competition remains unchanged along gradients of intensifying stress (Fig. 18.2, dashed curve). Notice that this curve only captures that part of Tilman's model which deals with transient dynamics. Over a longer time perspective, Tilman argues that stressful systems normally change towards the direction of higher nutrient availability (lower stress), which will increase the values of K_s and r_s. Literally interpreted, Equations 18.3 and 18.4 represent an unnecessarily strong condition for Tilman I. A

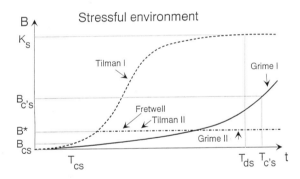

FIGURE 18.2 *Development of plant biomass (B) as a function of time (t) elapsed after devastating disturbance in a barren habitat, according to Equation 18.1. Tilman I (----) refers to a situation where Equations 18.3 and 18.4 hold, i.e. the only difference between barren habitats and productive ones is the absolute scale on the B-axis. Grime I (——) refers to a situation where Inequalities 18.5 and 18.6 hold, and where the time ($T_{c's}$) required for plant biomass to reach the critical level ($B_{c's}$) where competitive interactions would start to prevail exceeds the mean time interval (T_{ds}) between two successive disturbance events. Fretwell + Tilman II (−·−·−·−·) refers to a situation, where the plant biomass first develops according to Tilman I but, instead of approaching K_s, becomes locked at the hunger threshold of grazers (B^*). Grime II refers to a similar situation, except that the build-up of plant biomass initially follows Grime I.*

L. OKSANEN

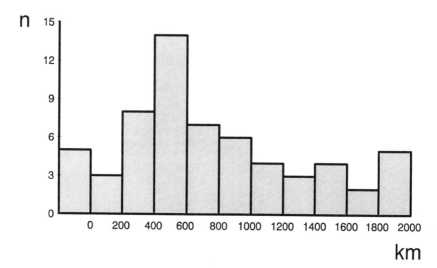

FIGURE 18.3 *Distribution of distances between the northernmost point of problem-free cultivation in Finland and the northern limit of natural occurrence for all 61 reasonably widespread European temperate woody plants with cultivation records in Finland. (Compiled from Hämet-Ahti et al., 1989.)*

sufficient condition is that neither rT_d/K nor K/B_c has a systematic tendency to decrease with decreasing primary productivity.

Grime I, in turn, implies that the following inequalities hold:

$$r_s T_{ds}/K_s \ll r_p T_{dp}/K_p \tag{18.5}$$

and

$$B_{cs}/K_s \gg B_{cp}/K_p \tag{18.6}$$

In extreme cases, $T_{ds} < T_{cs}$ and, consequently, competitive interactions practically never occur. This situation is illustrated by the curve labeled Grime I in Fig. 18.2. Recall that r is a shorthand for the difference between maximum rate of expansion and minimum rate biomass loss. Thus, one way to achieve a situation corresponding to Grime I is to assure that shoot mortality is heavy and inversely density dependent over a wide range of B. However, inequalities 18.5 and 18.6 are only necessary conditions for Grime I. It is also essential that there are competitive hierarchies and that adaptations to stress, competition and disturbance form a zero-sum game. If inequalities 18.5 and 18.6 hold, but the assumption of a zero-sum game does not, stressful habitats will be occupied by some variants of the classic r strategy (Kautsky, 1988; Taylor *et al.*, 1990).

According to Fretwell and Tilman II, relatively stressful habitats are characterized by an interactive laissez-faire predator–prey system between grazers and plants. The dynamics of such a system can be modelled by the

318

following equations (Oksanen *et al.*, 1981):

$$dB/dt = r_sBg(B,K_s,B_{cs}) - a\,f(B)BH \qquad (18.7)$$

$$dH/dt = -mH + ka\,f(B)BH \qquad (18.8)$$

where H is the grazer density, a the maximum attack rate of grazers, $f(B)$ the relation between the grazers' attack rate and plant biomass, m the mortality rate of starving grazers, and k the efficiency with which grazers convert ingested plant biomass to energy for maintenance and reproduction. The equilibrium can be found by setting the derivatives on the left sides of Equations 18.7 and 18.8 equal to zero, which yields equations

$$H^* = (r_s/a)\,[g(B,K_s,B_{cs})/f(B)] \qquad (18.9)$$

and

$$B^* = m/(kaf(B^*)) \qquad (18.10)$$

where $f(B^*)$ refers to the value of $f(B)$, evaluated at $B = B^*$. Equation 18.9 describes the plant isocline, which indicates how many grazers it will take to hold consumption and production in balance. Equation 18.10 describes the grazer isocline and tells how much plant biomass grazers need in order to break even. Note that the parameter H does not appear in Equation 18.10. This implies that plant biomass will be locked at a constant level, representing the hunger threshold of grazers (B^*, horizontal line in Fig. 18.2, labelled Fretwell + Tilman II). According to Oksanen *et al.* (1981), this level is well below K_s in moderately stressful habitats (e.g. tundras, steppes).

This result can be directly applied in the context of plant strategies. Direct plant–plant interactions are predicted to be weak, which implies that the value of $g(B,K_s,B_{cs})$ will be close to unity in the neighbourhood of B^*. On the other hand, grazers cannot be saturated when they are barely able to break even. Thus, the value of $f(B)$ must also be close to unity in this neighbourhood. The net expansion rate of plants in the neighbourhood of $B = B^*$ is approximately r_s/a. For an average plant this rate will be zero, but there will be variation between and within species in features influencing r_s and a. Oksanen (1990a) and Oksanen and Ranta (1992) thus argued that plants of moderately stressful habitats have lived under intense selection pressure for maximizing r_s/a. They also propose that those features, which Grime regards as stress-tolerant, actually represent adaptations to intense natural grazing pressure in nutrient-poor habitats, where low a has priority over high r_s (see Bryant *et al.*, 1983). In relatively barren but nutrient-rich habitats (e.g. alpine tundras and arctic and arid lands with lime-rich bedrocks), intense grazing favours plants with high r_s (e.g. graminoids, relatively palatable prostrate dicotyledons).

With respect to the relation between B^* and B_{cs}, Grime II can be expected to diverge from Fretwell and from Tilman II. Grime emphasizes mutualism and lack of competition in stressful habitats, which translates to the assumption that $B^* < B_{cs}$. Conversely, the idea of mutualism as a major community structuring force is totally against the spirit of the writings of Tilman, Fretwell and Oksanen. Thus, although Grime II, Tilman II and

Fretwell agree about the importance of natural grazing in relatively barren areas, the theories are by no means identical.

In extremely barren habitats, $K < B^*$. In other words, even the maximum plant biomass that can be supported in the habitat is below the hunger threshold of grazers. With regard to such extremely barren habitats (e.g. polar deserts and high-alpine boulderfields), Fretwell and Tilman II converge with Tilman I and predict that competition is again the main structuring force for the scanty vegetation. Indeed, competition for scattered microsites favours different kinds of plant (e.g. robust cushion or rosette plants capable of covering an entire microsite) than competition for light in closed vegetation. Nevertheless, Fretwell and Tilman imply that plants of extreme habitats are just as competitive as broad-leaved temperate herbs, shrubs and trees, but for different resources and under different environmental constraints. Conversely, Grime II should converge with Grime I, predicting strongly mutualistic plant–plant interactions and prevalence of extreme S strategists.

18.4. How to resolve the debate?

In the contest between different theories, the standard scientific practice is to look for situations where the predictions of the theories are different and to test these experimentally. Relevant experiments have been conducted in the past by, for example, de Vries (1934) and Ellenberg (1953). The plant sociological implications of such early experiments were summarized by Ellenberg (1954). Some results support the existence of absolute competitive hierarchies. The tall grass *Arrhenatherum elatius* performed just as well in pure cultures as in mixed ones under all experimental conditions, whereas species with lower stature often thrived best in benign conditions when grown alone, but in more stressful conditions when grown in mixtures. However, other species seemed to be maximally competitive in the vicinity of their physiological optima, whereas others still had a bimodal performance profile in mixed cultures. The message of the studies is thus somewhat ambiguous, maybe because their scope was usually to show the difference between physiological and ecological optima rather than resolve the issue of absolute competitive hierarchies.

During the past decade, Grime and Tilman have tried to resolve the issue by conducting independent experiments. Tilman (1982) convincingly demonstrated that the outcome of competitive interactions between algae can depend on external conditions. As some of the trade-offs involved (e.g. ability versus inability to fix nitrogen) are directly transferrable to terrestrial vegetation, the relevance of these results is not restricted to pelagic systems. In his experiments with vascular plants at Cedar Creek, Tilman (1988) showed that species with elevated foliage responded positively when nitrogen was added to natural vegetation, whereas the response of prostrate and semi-prostrate plants to this treatment was negative. Unfortunately, these results are compatible with both Tilman's and Grime's viewpoints. Campbell and Grime (1992), in turn, conducted a series of experiments, where different grasses were grown in pure and mixed cultures under different soil conditions. As in Ellenberg's (1953) study,

Arrhenatherum elatius turned out to be the best competitor both in benign conditions and in stressful ones, whereas the narrow-leaved bunchgrass *Festuca ovina* was consistently in the bottom of the competitive hierarchy.

Grime (1979 and personal communication) suggests that the explanation for the existence of competitive hierarchies can be found in plant physiology. Plants that have rapid shoot growth capture most light energy and, thus, become superior in exploiting pockets with high nutrient availability. Slow-growing plants become trapped in depletion zones with respect to both light and soil resources. However, they can survive by means of their conservative energy and nutrient physiology (i.e. stress tolerance). Later, when conditions become uniformly stressful with respect to both light and nutrients, the quick but 'leaky' competitors will be replaced by stress tolerators, which have been biding their time as small and stunted individuals. This scenario implies that interference competition prevails during early phases of vegetation development in both shoot and root environments, whereas exploitation competition (included in Grime's concept of stress) takes over later on. The chief trade-off would then be between speed and efficiency, as proposed by Smith (1976), but allocation patterns might become critical for the slow plants, because relative shortages of light and soil resources in mature vegetation should vary under different environmental conditions.

However, the results of Campbell and Grime (1992) do not fit easily with this interpretation. The percentage reduction of plant biomass due to competition was similar on nutrient-rich and nutrient-poor soils, suggesting that the intensity of competition did not vary along gradients of environmental stress. Moreover, the supposedly quick but leaky *A. elatius* was not excluded by stress from even the most nutrient-poor soils but turned out to be the superior competitor, even under conditions where it does not thrive in nature. This suggests that either the time-scale of the experiment was too short or that some important prevailing factor was excluded from the experimental setup.

The time-scale argument is difficult to deal with. The evidence from the long-term plots at Rothamsted suggests that it may take 50–100 years before the final impact of a treatment can be detected in the composition of herbaceous vegetation (Tilman, 1982), and responses of woody plants must be even slower. An experimentalist does not normally have this much time available. Fortunately, forestry, park management and horticulture constitute a rich source of relevant experimental evidence. Both for economic and aesthetic reasons, trees and shrubs have frequently been planted in conditions that are more stressful than those prevailing within the natural range of the species. This is especially true for northern Europe, where the number of native woody plants is low and where noblemen and industrial aristocrats have, for centuries, tried to enrich their environment by introducing species from Central Europe.

Hämet-Ahti *et al.* (1989) provide information of successes and problems with the cultivation of woody plants in Finland, along with information on the natural ranges of the species. Excluding local species (mean range diameter less than 300 km) and subalpine ones, the material includes 61 species from temperate Europe with abundant cultivation records. Fig. 18.3 summarizes the distances between the northernmost natural occurrences of the species in

Europe and their northern limit of problem-free cultivation in Finland. We see that five species have tolerance problems at their northern limit of natural occurrence. However, the vast majority of species can be grown easily in more stressful areas than are encountered within the natural range. Instead of becoming excluded by the stressfulness of the boreal climate, these species become poor competitors in the taiga. They grow slowly and often only reproduce during especially favourable years. Thus, they are outperformed by conifers both physiologically and in terms of reproduction. The victorious conifers have several traits, which Grime (1979) regards as stress-tolerant (e.g. evergreenness, needle-shaped and long-lived leaves, and ability to opportunistically acquire resources whenever conditions are favourable). However, these traits do not just allow conifers to survive in the taiga, they also make conifers competitively superior to broad-leaved deciduous trees. As several deciduous trees (e.g. birches, aspens, alders and larches) also thrive in boreal areas, the key to success seems to be appropriate phenology, not evergreenness *per se*.

Although the transition between temperate deciduous forests and the boreal taiga supports Tilman's (1982, 1988) idea of shifting terms of competition, there are some crucial differences between this case and the system studied by Campbell and Grime (1992). The mechanism behind the competitiveness of conifers and quickly leafing deciduous trees in the taiga is obvious: their foliage is deployed ready to photosynthetize in the winter, or almost immediately after the snowmelt, whereas broad-leaved deciduous trees lose the benefit of several weeks in spring and early summer, when days are long and skies are normally clear, because they have not yet produced a functional canopy. Conversely, the advantages of narrow-leaved bunchgrasses like *Festuca ovina* over tallgrasses like *Arrhenatherum elatius* appear obscure. Even if nutrient-poor habitats favoured plants with copious investment in below-ground organs, it is not obvious why such an allocation pattern should be accompanied by the shoot morphology of a narrow-leaved bunchgrass. According to Tilman's (1988) own simulations (see also Givnish, 1982; Oksanen, 1990a), shoot competition is almost always strong enough to make it optimal for a plant to invest some resources in support structures. It should then also be advantageous to take maximal use of this investment in elevating leaves above those of neighbours. The design of a tallgrass, with leaves radiating from a central culm, seems much more efficient than the design of a bunchgrass or rhizomatous shortgrass, where the youngest parts of the leaves lie close to the ground in the shadow of both neighbouring plants and the plant's own senescing leaf tips.

Different types of shoot design thus appear to constitute absolute competitive hierarchies in all habitats with closed and continuous vegetation. In the bottom region of the competitive rank order, there are entirely prostrate plants, which only use the stem to elevate the inflorescence (plant 5 in Fig. 18.4). Basal-leaved graminoids and narrow-leaved rosette herbs form the next step in the rank order (plant 4). The next rank consists of tallgrasses and herbs with leaves attached to a central stem or stem-like petiole (plant 3) because this design reduces self-shading and makes more efficient use of support structures. The maximally competitive shoot design for a herb is to attach all leaves apically to the tip of a stem or stems (plant 2). However, it is still more competitive to accumulate growth increments from year to year in the form of a woody stem (plant 1).

Leaf morphologies also constitute competitive hierarchies. To retain proper orientation, each leaf needs internal support structures. As bending momentum increases rapidly with increasing distance from the point of support (Givnish, 1982), plants can minimize the amount of resources allocated to internal support structures by constructing disc-shaped leaves, attached to the stem or petiole at their midpoints (plant 2 in Fig. 18.4). Small deviations from this design (e.g. lobed leaves, clusters of broad leaves in shoot tips, clusters of basal leaves with erect stem-like petioles and blades radiating in different directions) are not especially costly, as long as the principle of design is to entirely fill a circular area around the point of support. However, narrow or finely lobed leaves constituting a foliage with big gaps close to support structures (plants 3–5 in Fig. 18.4) represent clearly suboptimal use of internal support tissues.

Many plant communities are dominated by plants with competitively inferior shoot and leaf design. Thus, competitiveness must carry a cost, which is not always worth paying. The arguments of Walter (1964) and Tilman I imply that efficiency in below-ground competition somehow constrains shoot and leaf design. Grime I implies that competitive shoot and leaf design would

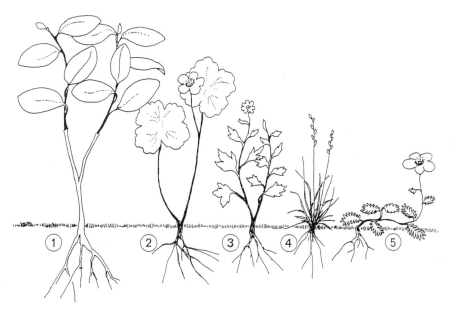

FIGURE 18.4 *Suggested competitive hierarchy in shoot design. Competitiveness increases leftwards and grazing tolerance increases rightwards, due to lack of perennating above-ground parts, exposed to browsing and girdling in the winter (2–5), narrow or finely lobed leaves with low risk of secondary injury after grazing (3–5), basal leaf-producing meristems (4–5) and prostrateness (5).*

323

be incompatible with stress tolerance. There is clear empirical evidence against both conjectures. *Ranunculus glacialis*, which is a dominating high-alpine plant in Scandinavia (Du Rietz, 1925; Nordhagen, 1928; Gjaerevoll, 1956; Oksanen and Virtanen, unpublished data) and in the Alps (Reisgl and Pitschmann, 1958; Ellenberg, 1978), is a mesomorphic broad-leaved herb with leaves attached to a central stem or supported by stem-like petioles. So is *Oxyria digyna*, which has achieved the status of model plant in ecophysiological arctic–alpine studies (Mooney and Billings, 1961) and *Papaver radicatum* s.l., which predominates in true polar deserts (Aleksandrova, 1983; Bliss *et al.*, 1984). The high-alpine elevations of tropical mountains are characterized by giant rosette plants with broad and seemingly mesomorphic leaves (Troll, 1959; Hedberg, 1964; Smith, 1980), and malacophyllous shrubs prevail in many arid deserts (Knapp, 1965; Walter, 1968).

A more plausible cost carried by competitive shoot and leaf designs is sensitivity to grazing. An erect woody plant exposes resources accumulated over several years to browsers and bark-gnawers during the unfavourable season, when herbaceous plants are hiding below the soil surface. A single grazing or trampling incident can destroy the entire foliage of a herb with apically attached leaves, whereas herbs with more basally attached and less differentiated leaves have at least parts of their foliage left intact. Graminoids with basal leaf-producing meristems can gain from the removal of senescent leaf tips (McNaughton, 1979) and, furthermore, are quick to restore the lost leaf surfaces. Prostrate plants can escape grazing altogether, especially if grazers are large and the ground surface is uneven. Also, narrow or finely lobed leaves are advantageous under heavy grazing pressure, because partial consumption of broad and mesomorphic leaves produces long wounds, which expose remaining tissues to desiccation and fungal attack.

Empirical evidence for the importance of natural grazing for the vegetation of relatively barren areas has been discussed by Oksanen (1988, 1990a; Oksanen and Oksanen, 1989). Descriptive ecological studies performed in steppe and tundra habitats yield direct evidence for major impacts of grazing vertebrates on the vegetation (Lavrenko, 1952: Tihomirov, 1959; Kalela, 1961, 1971; Thorn, 1982; McNaughton, 1985; Oksanen and Oksanen, 1981; Oksanen and Ericson, 1987; Oksanen *et al.*, 1987; Moen *et al.*, 1993). Experimental exclusion of grazers leads to dramatic expansion of tallgrasses, broad-leaved herbs or robust cryptogams, whereas prostrate dicotyledons and basal-leaved graminoids suffer (McNaughton, 1979; Batzli *et al.*, 1980; Huntly, 1987; Oksanen, 1988, 1990a; Moen and Oksanen, unpublished data). On isolated islands without native vertebrate grazers, increasing aridity leads to replacement of tall trees with successively smaller and more shrubby ones; and with increasing altitude, woodlands are replaced by herbfields or by stands of giant tussock grasses or rosette trees (Cockayne, 1958; Burrell, 1965; Mueller-Dombois, 1967, 1987). On grazer-free arctic and subantarctic islands, in turn, scrublands and herbfields grade directly to stands of giant cushion plants or moss banks (Werth, 1928; Wilson, 1952; Wace, 1961; Lid, 1964; Engelskjön and Schweitzer, 1970; Virtanen *et al.*, unpublished data). In all these gradients, prostrate dicotyledons and basal-leaved graminoids are rare and restricted to physically disturbed habitats. Introduction of grazers has quickly created communities, where giant rosette and cushion plants and robust, mesomorphic herbs are replaced by basal-leaved graminoids and pro-strate or semi-prostrate dicotyledons (Werth, 1928; Leader-Williams, 1988).

The available evidence thus suggests that there might not be any simple solution to the competition between the contesting theories. In some cases, the seemingly stress-tolerant plants are actually better competitors under stressful conditions, as proposed by Tilman (1982, 1988). In other cases, genuine com-petitive hierarchies appear to exist, due to trade-offs between competitiveness and grazing tolerance. The importance of competition in moderately stressful habitats appears to be modest, because of the overriding impacts of grazing and apparent competition (Holt, 1977), but the direct plant–plant interactions seem to be competitive rather than mutualistic (Fowler, 1986; Moen and Oksanen, unpublished data).

What happens in extremely stressful habitats, where grazers cannot persist, is still unclear. There is evidence suggesting that competition for below-ground resources structures the scanty shrub cover of arid deserts (Phillips and Mac-Mahon, 1981; Cody, 1986). The study of Carlsson and Callaghan (1991) on interactions in high-altitude vegetation of North Fennoscandian mountains suggests that the sedge *Carex bigelowii* is favoured by being associated with the evergreen dwarf shrub *Cassiope tetragona*. This result is consistent with the mutualism hypothesis. However, the authors note that the result could also be due to patchy occurrence of nutrients. Moreover, the study was conducted in middle alpine vegetation, where reindeer can be expected to graze at least during the peak of the mosquito season. Thus, *C. bigelowii* might also be favoured by association with the highly unpalatable *C. tetragona* (Aleksandrova *et al.*, 1964). Experiments of Moen (1993) in a still more extreme boulderfield habitat suggest that the relationships between two high-alpine herbs (*Ranun-culus glacialis* and *Oxyria digyna*) vary between neutral and competitive. The idea that plant–plant interactions in extremely stressful habitats are mutualistic rather than competitive, owing to high and density-independent or inversely density-dependent shoot mortality, is logical and interesting, but so far lacking solid empirical support.

The situation calls for further empirical tests. It should be highly advan-tageous if such studies could be integrated with descriptive studies and carried out in areas where at least a substantial part of the vertebrate grazer com-munity consists of native species and is not under direct human influence. Useful experiments on competition and grazing can indeed be carried out in laboratory-like conditions, but without a natural reference system, the results of such experiments are difficult to interpret. The reference system should cover fairly large areas, because dynamics in the grazing chain are scale-dependent (Oksanen, 1990b). Moreover, the descriptive studies should be carried out on a long-term basis, because relatively rare events may play a major role in structuring communities. Horticultural experience suggests that plants are rarely killed by average climatic conditions. If they are eliminated by stress, this normally happens during extreme years (Hämet-Ahti, personal communication). Also grazing intensity can vary dramatically. Norwegian lemmings, which play a key role in preventing the build-up of moss banks in the tundra and mountain areas of northern Fennoscandia (Oksanen, 1983), were practically absent from the Kilpisjärvi area in the heart of the mountains during 1946–59 (Lahti *et al.*, 1976; Laine and Henttonen, 1983). A short- or medium-term study conducted during this period could have led to grossly incorrect conclusions on the influence of vertebrate grazers on the mountain vegetation of northern Fennoscandia.

Present conditions tend to favour short-term studies, carried out in the vicinity of the base institutions of researchers, i.e. normally in temperate or boreal parts of Europe and North America, where key native grazers have either been eliminated or are directly managed. The main exceptions are studies conducted in African national parks (e.g. McNaughton, 1979, 1985), but the implications of these studies on systems with lower grazer diversity are unclear. Besides the arid parts of eastern and southern Africa, the other region where relatively undisturbed stressful habitats abound is the Arctic.

In defining the key issues to be studied, the following questions appear especially appropriate:

1. Are the direct plant–plant interactions in stressful habitats competitive or mutualistic? This could be studied by means of reciprocal manipulations of plant populations within grazer-proof exclosures or on grazer-free islands.
2. Are the dominating plant–plant interactions in stressful habitats direct or indirect? This could parallel studies under 1.
3. Does the plant cover of stressful habitats recover from disturbance more slowly than the plant cover of productive habitats, as predicted by Grime? It will be essential that the rate of recovery is measured in terms of percentage of biomass prior to disturbance; all theories agree that absolute rates of recovery are faster in productive habitats than in stressful ones.
4. Does addition of resources to plants of stressful habitats lead to an expansion of plant biomass as predicted by Grime I and Tilman I, or does it only lead to intensified grazing, as predicted by Fretwell?
5. Do absolute competitive hierarchies exist between plants with different shoot designs, as predicted by Grime and by Fretwell? This issue could be studied by means of long-term field or laboratory manipulations similar to those in point 1, but spanning sufficiently wide environmental gradients to ensure that some contesting plants become eliminated by the rigors of the environmental stress.
6. Does exclusion of grazers allow plants with competitive shoot and leaf morphology (in accordance to Fig. 18.4) to thrive in stressful habitats, as predicted by Fretwell? It will be necessary to combine grazer exclosures with seeding and transplantation, because of the potentially long time lag before propagules of plants, normally restricted to entirely different habitats, find their way to an exclosure located in the middle of a steppe or a tundra (Moen and Oksanen, unpublished data).

It should be noted that the results predicted by Fretwell's approach depend upon the level of stress. Natural grazers are predicted to be important in moderately stressful habitats (e.g. steppes, arctic–alpine tundras) but not in extremely stressful ones (e.g. extreme arid deserts, polar deserts, high-alpine boulderfields). Conversely, both Grime I and Tilman I imply that the response of the vegetation to intensifying stress is monotonic (the higher the intensity of stress, the stronger the response). Conducting parallel experiments in productive, moderately stressful and extremely stressful habitats should thus be an especially useful way to find out the strengths and limitations of the three theories.

18.5. Conclusions and practical implications

The three prevailing theories create altogether different perspectives on management problems of stressful habitats. If Grime I is correct, these systems should be extremely sensitive to disturbance. Caution should thus be applied to all potentially disturbing forms of land use, including grazing. According to Tilman I, there are no systematic differences in the resilience of the plant cover between stressful and benign habitats. Fretwell, in turn, predicts that plant communities of relatively stressful habitats are especially resilient and pre-adapted to such types of disturbance, which are similar to the impact of native grazers.

According to Fretwell's view, the very concept of overgrazing is not easily applicable to stressful habitats. The actual productivity of the vegetation may be somewhat below the potential one, because of grazing (Oksanen et al., 1981) and the vegetation may be dominated by relatively unpalatable species, especially in habitats with shortage of mineral nutrients (Bryant et al., 1983; Oksanen, 1990a). By these criteria, many stressful habitats should be naturally overgrazed. If, however, overgrazing is defined as a state where forage production is reduced below the level prevailing in corresponding natural vegetation, then sustained overgrazing of stressful habitats should be impossible. Problems labelled as overgrazing should be due to misgrazing: that is, the use of introduced grazers much heavier than native ones or which move and graze in a way not corresponding to the behaviour of native species. Misgrazing may also be caused by replacement of a native grazer fauna by a single species (usually cattle) that cannot utilize some kinds of native plants. Consequently, populations of these plants explode and render the range useless for the introduced species. It is essential to distinguish misgrazing from overgrazing, because the remedy for misgrazing is to return to native grazers or to a more diverse grazer stock that mimics the native grazers. Mere reduction of the grazer stock might only make the problem worse by allowing still more selective grazing, as pointed out by M. Westoby (personal communication).

Fretwell's approach also implies that undergrazing could be a major problem for preserving natural plant communities and biological diversity in stressful habitats, if their dominating native grazers have been replaced by managed lifestock. In the present time, when grazing of marginal lands is often too unprofitable to pay for labour costs in countries with a high standard of living, grazing pressure in barren areas is rapidly declining in many European countries. If native grazers are not reintroduced and allowed to reach carrying capacity, the likely result is that competitive plants will take over, eliminating the competition-shy native species and reducing species diversity (Grime, 1973). The Baltic island of Gotland provides a sorry example of this: its arid limestone heaths (alvars) are disappearing, along with many characteristic species, including large numbers of spectacular orchids and even some endemic species (Pettersson, 1959).

Management problems of various kinds abound in arctic, alpine and arid areas and, if anything, are becoming more acute. Thus, testing the three theories is not only an academic exercise but an essential task, which should be tackled without delay. The current reindeer controversy in Sweden (Nyman and Jennersten, 1992) is an excellent case in point. Nature protection agencies are concerned about what they perceive as overgrazing of Swedish mountains and would like to reduce reindeer numbers or exclude reindeer from botanically valuable areas (as has been done in the north-westernmost corner of Finnish Lapland). The Fretwell approach implies that such measures create an unnatural situation and are counterproductive for nature protection. The natural situation should be a reindeer population that fluctuates about the carrying capacity (Caughley, 1976). Such fluctuations do indeed create short-term aesthetic problems and declines in the populations of arctic–alpine plants. However, the long-term result should be improved nutrient circulation and seedling recruitment and thus higher abundances of nutrient-requiring and competition-shy arctic–alpine plants than in areas with lower reindeer numbers.

At present, ecologists cannot give reliable advice on how stressful habitats should be managed. Different schools have different ideas; some may be realistic, but others can be potentially disastrous. We do, however, have a relatively clear picture of how the problem could be resolved. We should thus proceed with critical tests in appropriate systems. As pointed out by Grime (1989), the unifying concept of stress is highly advantageous in this context, because it allows us to operate on a general level and learn much more from each experiment than could be possible by a system-specific approach. A generalistic and theoretical approach is also a maximally practical one, provided that the theories do not remain in their own spheres, but are subjected to rigorous and critical tests.

ACKNOWLEDGEMENTS

Sincerest thanks to Leena Hämet-Ahti for instructive discussions on planted woody plants and their relevance to the theory of competitive hierarchies and to Philip Grime for useful clarifying statements concerning his theory.

The figures were drawn by Katarina Stemman.

REFERENCES

Aleksandrova, V.D. (1983). *Rastitel'nost Poljarnyh Pustyn SSSR*. Moscow: Nauka.
Aleksandrova, V.D., Andreev, V.N., Vahtina, V.T., Dydina, R.A., Karev, G.L, Petrovskij, V.V. and Samarin, V.T. (1964). Kormovaja harakteriska

rostennij Krajnego Severa. *Rastitel'nost Krajnego Severa i ee Osvoenie* **5**, 1–484.

Batzli, G.O., White, R.G., McLean, S.F. Jr, Pitelka, F.A. and Collier, B.D. (1980). The herbivore-based trophic system. In: *An Arctic Ecosystem: The Coastal Tundra at Barrow, Alaska*, pp. 335–440, eds. J. Brown *et al.* Stroudsburg, Penn.: Dowden, Hutchinson & Ross.

Bliss, L.C., Svoboda, J. and Bliss, D.I. (1984). Polar deserts, their plant cover and plant production in the Canadian High Arctic. *Holarctic Ecol.* **7**, 305–324.

Bryant, J.P., Chapin, S.F. III and Klein, D.J. (1983). Carbon/nutrient balance of boreal plants in relation to vertebrate herbivory. *Oikos* **40**, 357–368.

Burrell, J. (1965). Ecology of Leptospermum in Otago. *N Z J. Bot.* **3**, 3–16.

Cajander, A.K. (1906). The struggle between plants in nature. *Luonnon Ystävä* **9**, 296–300 (in Finnish, English translation: *Trends Ecol. Evol.* **6**, 295).

Cajander, A.K. (1909). Über die Waldtypen. *Acta For. Fenn.* **1**, 1–175.

Cajander, A.K. (1916). *Foundations of Forestry*. Porvoo, Finland: Söderstöm (in Finnish).

Callaghan, T.V. and Emmanuelsson, U. (1985). Population structure processes of tundra plants and vegetation. In: *The Population Structure of Vegetation*, pp. 399–439, ed. J. White. Dordrecht: Junk.

Campbell, B.D. and Grime, J.P (1992). An experimental test of the plant strategy theory. *Ecology* **73**, 15–29.

Carlsson, B.A. and Callaghan, T.V. (1991). Positive plant interactions in the tundra vegetation and the importance of shelter. *J. Ecol.* **79**, 973–983.

Caughley, G. (1976). Wildlife management and the dynamics of ungulate populations. *Appl. Biol.* **1**, 183–246.

Chapin, S.F. III (1980). The mineral nutrition of wild plants. *Annu. Rev. Ecol. Syst.* **11**, 233–260.

Cockayne, L. (1958). *The Vegetation of New Zealand*, 3rd edn. Leipzig: Engelmann.

Cody, M.L. (1986). Structural niches in plant communities. In: *Community Ecology*, pp. 381–405, eds. J. Diamond and T.J. Case. New York: Harper & Row.

de Vries, O. (1934). Unkräuter und Säuregrad. *Z. Pflanzennährung, Düngung und Bodenkunde* Beitr. 13.

Du Rietz, E. (1925). Studien über die Höhengrenzen der hochalpinen Gefässpflanzen im nördlichen Lappland. *Veröff. Geobotanischer Inst. Rübel* **3**, 67–86.

Ellenberg, H. (1953). Physiologisches und ökologisches Verhalten derselben Pflanzenarten. *Ber. Dtsch. Bot. Ges.* **65**, 350.

Ellenberg, H. (1954). Über einige Fortschritte der kausalen Vegetationskunde. *Vegetatio* **5/6**, 199–211.

Ellenberg, H. (1978). *Vegetation Mitteleuropas mit den Alpen*, 2nd edn. Stuttgart: Ulmer.

Engelskjön, T. and Schweitzer, H.J. (1970). Studies on the flora of Bear Island (Björnöya). I. Vascular plants. *Astarte* **3**, 1–36.

Fowler, N. (1986). The role of competition in plant communities in arid and semiarid regions. *Annu. Rev. Ecol. Syst.* **17**, 89–110.

Fretwell, S.D (1975). The impact of Robert MacArthur on ecology. *Annu. Rev. Ecol. Syst.* **6**, 1–13.

Fretwell, S.D. (1977). Regulation of plant communities by food chains exploiting them. *Perspect. Biol. Med.* **20**, 169–185.

Givnish, T.J. (1982). On adaptive significance of leaf height in forest herbs. *Am. Naturalist* **120**, 353–381.

Gjaerevoll, O. (1956). The plant communities of Scandinavian alpine snow-beds. *Kongelige Norske Videnskaps Selskabets Skrifter* **1956**, 1–405.

Grace, J.B. (1990). On the relationship between plant traits and competitive ability. In: *Perspectives on Plant Competition*, pp. 51–65, eds. J.B. Grace and D. Tilman. San Diego, Calif.: Academic Press.

Grime, J.P. (1973). Competitive exclusion in herbaceous vegetation. *Nature* **250**, 26–31.

Grime, J.P. (1977). Evidence for the existence of three primary strategies in plants and its relevance to ecological and evolutionary theory. *Am. Naturalist* **111**, 1169–1194.

Grime, J.P. (1979). Plant Strategies and Vegetation Processes. Chichester: Wiley.

Grime, J.P. (1989). The stress debate: symptom of impending synthesis. *Biol. J. Linn. Soc.* **37**, 3–17.

Haila, Y. (1982). Hypothetico deductivism and the competition controversy in ecology. *Ann. Zool. Fenn.* **19**, 255–263.

Haila, Y. and Järvinen, O. (1982). The role of theoretical concepts in understanding the ecological theatre: a case study on island biogeography. In: *Conceptual Issues in Ecology*, pp. 261–278, ed. E. Saarinen. Reidel.

Hairston, N.G, Smith, F.E. and Slobodkin, L.B. (1960). Community structure, population control and competition. *Am. Naturalist* **94**, 421–425.

Hämet-Ahti, L., Palmén, A., Alanko, P. and Tigerstedt, P.M.A. (1989). *The Tree and Shrub Flora of Finland*, Finnish Dendrological Society Bulletin 5. Helsinki: Finnish Dendrological Society (in Finnish).

Hedberg, O. (1964). Features of afroalpine plant ecology. *Acta Phytogeogr. Suec.* **49**, 1–144.

Holt, R.D. (1977). Predation, apparent competition and structure of prey communities. *Theor. Popul. Biol.* **12**, 197–299.

Huntly, N. (1987). Influence of a refuging consumer on resources: pikas (*Ochotona princeps*) and subalpine meadow vegetation. *Ecology* **68**, 274–283.

Kalela, O. (1961). Seasonal change of habitat in the Norwegian lemming, *Lemmus lemmus* L. *Ann. Acad. Sci. Fenn. AIV* **55**, 1–72.

Kalela, O. (1971). Seasonal difference in habitats of the Norwegian lemming, *Lemmus lemmus* (L.) in 1959 and 1960 at Kilpisjärvi, Finnish Lapland. *Ann. Acad. Sci. Fenn. AIV* **178**, 1–22.

Kautsky, L. (1988). Life strategies of aquatic soft-bottom macrophytes. *Oikos* **53**, 126–135.

Keddy, P.A. (1989). *Competition*. London: Chapman & Hall.

Knapp, R. (1965). *Die Vegetation vom Nord- und Mittelamerika und der Hawaii-Inseln*. Stuttgart: Gustav Fischer.

Lahti, S., Tast, J. and Uotila, H. (1976). Fluctuations in small rodent populations in the Kilpisjärvi fjeld area, Finnish Lapland. *Luonnon Tutkija* **84**, 19–23 (in Finnish, English summary).

Laine, K. and Henttonen, H. (1983). The role of plant production in microtine cycles in northern Fennoscandia. *Oikos* **40**, 407–418.

Lavrenko, E.M. (1952). *Mikrokompleksnost' i mozaicnost rastitel'nogo pokrova stepej kak rezul'tat žiznedejatel'nosti životnyh i rastenij*, Geobotanical Bulletin 8, Moscow: Botan. Inst. Akad. Nauk SSSR.

Leader-Williams, N. (1988). *Reindeer on South Georgia*. Cambridge: Cambridge University Press.

Lid, J. (1964). *The Flora of Jan Mayen*. Norwegian Polar Institute Bulletin 130. Oslo: Norwegian Polar Institute.

MacArthur, R.H. (1972). *Geographical Ecology*. New York: Harper & Row.

McNaughton, S.J. (1979). Grazing as an optimization process: grass–ungulate relationships in the Serengeti. *Am. Naturalist* **113**, 691–703.

McNaughton, S.J. (1985). Ecology of a grazing system: the Serengeti. *Ecological Monographs* **55**, 259–294.

Moen, J. (1993). Positive versus negative interactions in a high alpine Fennoscandian boulder-field. *Arctic Alpine Res.* in press.

Moen, J., Lundberg, P.A. and Oksanen, L. (1993). Lemming grazing on snowbed vegetation during a population peak. *Arctic Alpine Res.* in press.

Mooney, H.A. and Billings, W.D. (1961). Comparative physiological ecology of arctic and alpine populations of *Oxyria digyna*. *Ecol. Monogr.* **31**, 1–29.

Mueller-Dombois, D. (1967). Ecological relationships in the alpine and subalpine vegetation of Mauna Loa, Hawaii. *J. Indian Bot. Soc.* **46**, 403–411.

Mueller-Dombois, D. (1987). Forest dynamics in Hawaii. *Trends Ecol. Evol.* **2**, 216–220.

Nordhagen, R. (1928). Die Vegetation und Flora des Sylenegebietes. I. Die Vegetation. *Norske Videsnk. Akad. Skr. I. Mat.-Naturv. Kl.* **1927**, 1–612.

Nyman, L. and Jennersten, O., eds. (1992). *WWFs renbeteskonferens 1992*. Stockholm: Världs-naturfomden WWF (in Swedish, English summary).

Odum, E.P. (1971). *Fundamentals of Ecology*, 3rd edn. Philadelphia, Penn.: W.B. Saunders.

Oksanen, L. (1980). Abundance relationships between competitive and grazing-tolerant plants in productivity gradients on Fennoscandian mountains. *Ann. Bot. Fenn.* **17**, 410–429.

Oksanen, L. (1983). Trophic exploitation and arctic phytomass patterns. *Am. Naturalist* **122**, 45–52.

Oksanen, L. (1988). Ecosystem organization: mutualism and cybernetics or plain Darwinian struggle for existence? *Am. Naturalist* **131**, 417–422.

Oksanen, L. (1990a). Predation, herbivory and plant strategies along gradients of primary productivity. In: *Perspectives on Plant Competition*, pp. 445–474, eds. J.B. Grace and D. Tilman. San Diego, Calif.: Academic Press.

Oksanen, T. (1990b). Exploitation ecosystems in heterogeneous habitat complexes. *Evol. Ecol.* **4**, 220–234.

Oksanen, L. and Ericson, L. (1987). Dynamics of tundra and taiga populations of herbaceous plants in relation to the Tihomirov–Fretwell and Kalela–Tast hypotheses. *Oikos* **50**, 381–388.

Oksanen, L. and Oksanen, T. (1981). Lemmings (*Lemmus lemmus*) and grey-sided voles (*Clethrionomys rufocanus*) in interaction with their resources and predators on Finnmarksvidda, northern Norway. *Rep. Kevo Subarctic Res. Stat.* **17**, 7–31.

Oksanen, L. and Oksanen, T. (1989). Natural grazing as a factor shaping out barren landscapes. *J. Arid. Environ.* **17**, 219–233.

Oksanen, L. and Ranta, E. (1992). Plant strategies along mountain gradients: a test of two theories *J. Vegetation Sci.* **3**, 175–186.

Oksanen, L., Fretwell, S.D., Arruda, J. and Niemelä, P. (1981). Exploitation ecosystems in gradients of primary productivity. *Am. Naturalist* **118**, 240–261.

Oksanen, L., Oksanen, T., Lukkari, A. and Sirén, S. (1987). The role of phenol-based inducible defense in the interaction between tundra populations of the vole *Clethrionomys rufocanus* and the dwarf shrub *Vaccinium myrtillus*. *Oikos* **50**, 371–380.

Oksanen, T., Oksanen, L. and Gyllenberg, M. (1992). Exploitation ecosystems in heterogeneous habitat complexes. II. Impact of small-scale spatial heterogeneity on predator–prey dynamics. *Evol. Ecol.* **6**, 383–398.

Pettersson, B. (1959). Dynamik und Konstanz in der Flora und Vegetation von Gottland. *Acta Phytogeogr. Suec.* **40**, 1–288 (in Swedish, with German summary).

Phillips, D.L. and MacMahon, J.A. (1981). Competition and spacing patterns in desert shrubs. *J. Ecol.* **69**, 97–115.

Popper, K.R. (1963). *Conjectures and Refutations*. London: Routledge.

Reisgl, H. and Pitschmann, H. (1958). Oberen Grenzen von Flora und Vegetation in der Nivalstufe der zentralen Ötztaler Alpen (Tirol). *Vegetatio* **8**, 93–129.

Rosenzweig, M.L. and Abramsky, Z. (1986). Centrifugal community organization. *Oikos* **46**, 339–348.

Smith, C.C. (1976). When and how much to reproduce: the trade-off between power and efficiency. *Am. Zool.* **16**, 763–774.

Smith, A.P. (1980). The paradox of plant heights in Andean giant rosette species. *J. Ecol.* **68**, 63–73.

Taylor, D.R., Aarssen, L.W. and Loehle, C. (1990). On the relationship between r/K selection and carrying capacity: a new habitat templet for plant life history strategies. *Oikos* **58**, 239–250.

Thorn, C.E. (1982). Gopher disturbance: its variability by Braun-Blanquet vegetation units in the Niwot Ridge alpine tundra zone, Colorado Front Range, USA. *Arctic Alpine Res.* **14**, 45–51.

Tihomirov, B.A. (1959). *Vzajmosvjazy Životnogo Mira i Rastitel'nogo Pokrova Tundry*. Moscow: Trudy Biol. Inst. 'Komarova' Akad. Nauk SSSR.

Tilman, D. (1982). *Resource Competition and Community Structure*. Princeton, NJ: Princeton University Press.

Tilman, D. (1988). *Dynamics and Structure of Plant Communities*. Princeton, NJ: Princeton University Press.

Troll, C. (1959). *Die tropischen Gebirge, ihre dreidimensionale klimatische und pflanzensoziologische Zonierung*. Bonn: University of Bonn Press.

Wace, N.M. (1961). The vegetation of Gough Island. *Ecol. Monogr.* **31**, 337–367.

Walter, H. (1964). *Vegetation der Erde in öko-physiologischer Betractung. I. Die tropischen und subtropischen Zonen*. Jena: Gustav Fischer.

Walter, H. (1968). *Vegetation der Erde in öko-physiologischer Betractung. II. Die gemässigten und arktischen Zonen*. Jena: Gustav Fischer.

Werth, E. (1928). Überblick über die Vegetetationsgliederung von Kerguelen sowie von Posession-Eiland (Crozet-Gruppe) und Heard-Eiland. In: *Deutsche Südpolar-Expedition 1901–1903*, Vol. 8, pp. 127–176, ed. E. von Drygalski. Berlin: De Gruyter.

Wilson, J.W. (1952). Vegetation patterns associated with soil movement on Jan Mayen Island. *J. Ecol.* **40**, 17–264.

Index

INDEX

INDEX